Divert!

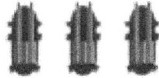

NUMEC, Zalman Shapiro and the diversion of US weapons grade uranium into the Israeli nuclear weapons program

Published by the Institute for Research: Middle Eastern Policy, Inc.
Calvert Station
PO Box 32041
Washington, DC 20007

First published in 2012 by the Institute for Research: Middle Eastern Policy

7 9 10 8 6
Copyright Institute for Research: Middle Eastern Policy, Inc.
All Rights Reserved

Paperback ISBN-13 978-0-9827757-0-7

Library of Congress Cataloging-in-Publication Data

Smith, Grant F.
Divert! : NUMEC, Zalman Shapiro and the diversion of US weapons grade uranium into the Israeli nuclear weapons program / by Grant F. Smith.
p. cm.
Includes bibliographical references.
ISBN 978-0-9827757-0-7 (alk. paper)
1. Nuclear weapons--Israel. 2. Shapiro, Zalman M. (Zalman Mordecai), 1920- 3. Nuclear Materials and Equipment Corp. 4. U.S. Atomic Energy Commission. 5. Highly enriched uranium--Israel. 6. Highly enriched uranium--United States. 7. Diversion of goods--Israel. 8. Diversion of goods--United States. 9. Espionage, Israeli--United States. I. Title.
UA853.I8S58 2012
623.4'5119095694--dc23
     2011050232

In April 1976 Attorney General Edward Levi ordered the FBI to reopen an investigation into Dr. Zalman Mordecai Shapiro and the Nuclear Materials and Equipment Corporation (NUMEC). Since the mid-1960s enforcement and regulatory agencies suspected NUMEC had illegally diverted U.S. government-owned weapons-grade nuclear material into Israel's clandestine nuclear weapons program.

The newly reopened investigation looked beyond violations of the Atomic Energy Act. The FBI's new mandate was to also uncover "any attempt by anyone in the executive branch to prevent or impede an investigation into this alleged diversion, or to withhold any information regarding this alleged diversion from any investigative body."

The FBI code-named its investigation "DIVERT." After interrogating high-level officials from U.S. government agencies and NUMEC employee eyewitnesses to nuclear diversion, the investigation was suddenly terminated in 1981. Why?

Table of Contents

Table of Figures

About the Author

Grant F. Smith is director of the Washington, DC-based Institute for Research: Middle Eastern Policy (IRmep). IRmep is an independent nonprofit that studies U.S. policy formulation toward the Middle East. Smith's research and analysis about trade, law enforcement, and opportunity costs have appeared in *Antiwar.com*, the *Financial Times of London, Inc. Magazine, Arab News, Kiplinger, Gannet, The Wall Street Journal, The Washington Post, The Jewish Daily Forward, Al-Eqtisadiah, Khaleej Times, The New York Times, The Minneapolis Star Tribune, The Daily Star*, the Associated Press, Reuters, *The Washington Report on Middle East Affairs*, and specialty publications such as the U.S. State Department's *Washington File*. Smith has been a frequent guest analyst on Voice of America (VOA) television and Radio France Internationale. He has also appeared on the BBC, C-SPAN, Al Jazeera, CNN, PressTV, and numerous public radio programs. In 2003, Smith launched IRmep at a policy symposium in the Rayburn House Office Building on Capitol Hill warning about the grave dangers of adopting a 1996 neoconservative plan for involving the U.S. military more deeply across the Middle East.

Smith's research assignments have taken him to more than 40 countries. Before joining IRmep, he was a senior analyst and later research program manager at the Boston-based Yankee Group Research, Inc. Smith taught undergraduate, graduate, and executive education courses in finance, research, and marketing for five years at CESA in Bogotá, one of Colombia's top-ranked business schools. Before that, Smith was a marketing manager at the Minneapolis-based Investors Diversified Services (IDS), now Ameriprise Financial Advisors. Smith completed a BA in International Relations from the University of Minnesota and a master's in International Management from the University of St. Thomas in St. Paul, Minnesota. Jeff Stein of the *Washington Post* named Smith "a Washington, D.C. author who has made a career out of writing critical books on Israeli spying and lobbying." *Mondoweiss.com* blogger Philip Weiss claimed "the best investigative work is being done by Grant Smith at IRmep..."

Smith receiving a declassified NUMEC GAO report in 2010.

Introduction

In the early 1960s Israeli covert operatives and their U.S. allies began clandestinely diverting highly enriched uranium to Israel from a nuclear reprocessing facility in Pennsylvania. Their mandate was clear. Israel's leadership decided Israel should become a nuclear weapons power almost as soon as the state was established in 1948. Although Israeli operatives pursued nuclear know-how, materials and funding from many corners of the earth, stealing large quantities of weapons-grade uranium from the world's largest production stream required more than just inside help. Accidents, mistakes or detection by law enforcement could be disastrous. Yet the payoff was potentially enormous. Israel could immediately assemble a deployable stockpile of gun-type weapons—bypassing the most complicated and capital-intensive link in the production chain. The site of the diversion was a front company called the Nuclear Materials and Equipment Corporation launched at Apollo, Pennsylvania. NUMEC start-up funding was gathered by a mysterious venture capitalist with a secret history of maritime smuggling between the U.S., Europe and Israel.

Atomic Energy Agency Commissioners and Department of Defense overseers who had a "need to know" understood that NUMEC was reprocessing enriched material into fuel for America's growing fleet of nuclear-powered submarines and ships. NUMEC was also contracting to win secret projects such as nuclear jet engine fuel and power plant fuel for advanced space satellites. NUMEC received over 25 tons of U.S. uranium from the U.S. Navy and Westinghouse during its span of operations.

Despite growing circumstantial evidence of unacceptably high losses of nuclear material, regulators fought allegations of diversion. They feared a backlash against the Atomic Energy Commission and its conflicting roles to promote private nuclear industry while safeguarding the most dangerous military material on earth. Yet AEC commissioners and congressional overseers knew something was fishy. NUMEC charged into the capital-intensive nuclear reprocessing industry as a three-person corporation with a few thousand dollars. Subsequent infusions of start-up capital were organized by David Lowenthal, who's illustrious past raised questions of motivation. Lowenthal fought in Israel's war of independence and was well-connected with high-level Israeli intelligence officers and top political leaders. NUMEC and its parent corporation, Apollo Industries, were tiny and severely undercapitalized players compared to reprocessing industry rivals. New evidence reveals that subsequent NUMEC buyouts by Atlantic Richfield Company and Babcock & Wilcox were orchestrated by the AEC to precipitate a change in NUMEC's management. After Shapiro and Lowenthal left, unaccounted material dropped to industry norms. But in the long run, both companies assumed unsustainable liabilities for worker risks and heavy pollution that predated their acquisitions of NUMEC.

Today, with hindsight and modern research tools unavailable to FBI and CIA investigators in the 1960s, hidden connections are now observable. NUMEC emerges from background noise as yet another in a line of disposable, single-purpose, self-financing smuggling front companies set up by Israel's nationalists in the United States. But few of Israel's conventional arms smuggling fronts ever produced so much toxic fallout.

NUMEC founder and President Dr. Zalman Shapiro, a genius inventor, was as aloof to his employees' well-being as he was to the long-term environmental impact of NUMEC's operations. His presence at strange hours in the plant and loading dock shortly before mysterious shipments raced away on truck beds for direct loading onto Israeli flagged cargo ships or El Al airliners were recounted to law enforcement officials by a few distraught plant employees. But this was more than a decade after the employees had left NUMEC and gradually begun to fall ill to radiation sickness. It was much too late for criminal prosecutions. Many former employees felt compelled to hold their tongues until NUMEC could no longer damage their economic future in hardscrabble Apollo, a one-company town. Only after it was far too late were NUMEC's mysterious guests claiming to be scientists positively identified as top Israeli covert intelligence operatives and nuclear weapons production coordinators working undercover.

Shapiro, fellow executives and his closest partners in NUMEC not only had the means, motives and opportunities to divert uranium into Israel's nuclear weapons program during a critical time period. They nearly all shared an unbending ideological commitment to Israel that to them justified criminal activity. Shapiro met David Lowenthal during "organizing work"[1] in such influential and interconnected lobbying and fundraising organizations such as the Zionist Organization of America and United Jewish Appeal.[i] Shapiro's many elite relationships—including legal experts from an Israeli government-funded U.S. law firm—and his leadership role in Israel lobbying organizations made him an all but impossible target for law enforcement officials.

Local stories and lore of weapons-grade nuclear material diversions to Israel—among the earliest from a free supermarket advertisement newspaper—have blended into a confusing smokescreen of self-serving tell-all biographies by Israeli spies directly involved in the heist and purposeful disinformation periodically emitted by Shapiro and his allies. Not only are there "smoking guns" and indisputable evidence of wrongdoing, the trail of spent cartridge casings leads directly from Apollo to Dimona and back to Washington. Demands that somebody finally do something about NUMEC trickled up to U.S. regulators, members of Congress and law enforcement officials for decades. Questions over what to do about NUMEC dogged a string of American presidents until the case was finally buried during the transition from the Carter to Reagan administration.

Vast quantities of "materials unaccounted for" won NUMEC a notorious place in the nuclear industry record books. Department of Energy statistics published in 2001 reveal that NUMEC had almost twice the rate of "materials unaccounted for" as cohort reprocessing facilities during the time it operated under Shapiro

---

i Formed in 1939 and later merged with the Council of Jewish Federations and United Israel Appeal to form United Jewish Communities in 1999.

and Lowenthal. Losses peaked during the visit of Israeli intelligence officer Rafael Eitan, who later ran spy Jonathan Pollard against the U.S.

The political, human and financial toll on America has been disastrous and growing. One early CIA assessment speculated that Israel would become ever less willing to negotiate peace with its neighbors after acquiring a nuclear arsenal. From the first day, Israel's weapons were squarely — though probably not literally — pointed at America. Nuclear "coercion" of the United States for conventional weapons and unconditional diplomatic support — called "nuclear coercion" by realists and "blackmail" by the less charitable — has long been perceivable to the observant.

The Kennedy administration was dead-set against a nuclear-armed Israel and did everything in its power to stop it up to the very moment of JFK's assassination. The Johnson administration, deeply indebted to Israel's nuclear funding coordinator Abraham Feinberg, successfully delayed bona fide investigations into the matter. As the statute of limitations slowly ran out, U.S. presidents began a shameful policy of lying to Americans and the rest of the world about Israel's arsenal. The CIA's early assessment of how a nuclear-tipped Israel would behave has now proven largely accurate. America has lost a great deal of sovereignty, wealth, credibility and leverage in order for Israel to gain a secret nuclear arsenal — all because it simply could not (and in some cases, would not) keep Israel's agents and helpers in the U.S. from carting away all the nuclear know-how, material, and related technologies they could carry.

In historical perspective, it is entirely unsurprising that the founders and support network of a country that stole enormous quantities of conventional American WWII surplus weapons, munitions, weapons-making machinery and supplies through the vast Jewish Agency/Haganah theft and smuggling network front companies[ii] would within a decade also establish another entity for systematically diverting American nuclear technologies and know-how. Thanks to newly declassified documents, the means by which Israel and its American helpers stole nuclear material is now opening up to serious study.

These new revelations pose grave policy questions. Why did U.S. presidents subvert rule of law and criminal prosecutions? Why has the establishment news media consistently omitted key data while getting so many important facts about the NUMEC diversion story wrong? How are U.S. government agencies, particularly the FBI, CIA and NSA, still keeping many files about NUMEC and other nuclear diversions to Israel secret from the American people? Why has no single U.S. President ever delivered a straight answer about Israeli nuclear weapons while sitting in office? Why were all of the most expedient and sensible solutions to the NUMEC diversion problem ignored when they could have done the most good?

Although much of the NUMEC story centers on the 1960s, the crisis is far from over. A series of political, courtroom and corporate maneuvers has quietly shifted the cost of NUMEC's $170 million toxic dump cleanup onto unwitting U.S. taxpayers. Pointed questions about this cost-shifting loom. What are the current total costs of NUMEC to America, not only in strategic and domestic

---

[ii] Also known as the Sonneborn smuggling network.

political terms, but in the cost of misdirected litigation, a serious public health crisis in Pennsylvania and the ballooning U.S. Army Corps of Engineers cleanup? How has the ever-present influence of Israel lobby campaign contributions, behind-the-scenes maneuvering and political appointee cross-promotion kept U.S. government information and due redress over NUMEC bottled up? How has the same Israel lobby influence network on the periphery of NUMEC sought to quietly restore Zalman Shapiro's tarnished reputation by secretly pressuring the highest levels of government, as it already has done with conventional arms smugglers such as Charles Winters, Adolph Schwimmer and Herman Greenspun, and nuclear technology smuggler Arnon Milchan?   Perhaps the single most important question posed by NUMEC is also the timeliest: Is there anything the Israeli government and its vast network of operatives and elite political contributors in the U.S. would not do to expand Israel's power and influence? With sound of war drums for America to attack Iran growing louder, *Divert* provides the terrifying answer.

1. Israel builds a U.S. smuggling network

Jewish Agency chairman David Ben-Gurion visited New York accompanied by Executive Secretary Theodore Kolleck in November of 1946 to establish a vast arms and people smuggling network. Under intense lobbying by Zionist organizations, President Harry S Truman asked that Britain allow 100,000 Holocaust refugees to emigrate to Palestine in November of 1945. Great Britain's Foreign Secretary, Ernest Bevin blocked this effort, ratcheting the permitted number down to 1,500 per month. Ben-Gurion was determined to fight back.

Operating parallel to legitimate charitable relief fund-raising organizations and counting on a generally supportive Americans, the Haganah smuggling network worked to create and fortify the state of Israel through highly illegal activities. Though frequently detected by U.S. law enforcement and intelligence agencies, it presented a confusing, politically sensitive target to prosecutors. The Haganah smuggling network's elite supporters, such as former Secretary of Treasury Henry Morgenthau Jr.,[iii] political fundraiser Abraham Feinberg, New York lawyer Nahum Bernstein, and businessman Rudolph Sonneborn were able to intervene in coordination with the Israeli Prime Minister to prevent the warranted prosecution of smuggling network members and financiers. Highly effective in quashing investigations and prosecutions while pressuring the FBI to back off, the Haganah smuggling network provided the model for how Israel could utilize political and financial power in a way that undermined rule of law in the U.S. On direct orders from Israel, Feinberg pressured the Justice Department to drop proposed prosecutions of almost a hundred lower-level Haganah smugglers swept up in a dragnet.[2] Nahum Amber Bernstein, Feinberg's associate and a key Haganah leader prized for his skill setting up front companies, also transcended U.S. law enforcement.[3]

Nahum Bernstein[iv] was a well-connected New York lawyer and later became the founding chairman of the Jerusalem Foundation. After graduating from law school and a stint doing investigations for a law firm, Bernstein went to work as an Office of Strategic Services[v] operative training U.S. agents for "overseas operations in enemy territory" during WWII. In 1943 Bernstein listed his qualifications as "police methods including surveillance, cover, wire-tapping, Dictaphone, sound recordings, agent's reports, records, statistics, witness interrogation and personnel."

Bernstein began applying his skills in service to Israel from his New York law office (with a branch in Tel Aviv) which was placed in charge of $750,000 in Jewish Agency funding and contracted to represent the organization.

---

iii The FBI categorized all pages of its hefty counterespionage file on Henry Morgenthau Jr—suspected of financing smuggling operations, foreign agency, espionage and nuclear weapons funding—as "unreleasable" in 2011.
iv Also known as Ned Bernay, AKA Meyers, AKA, Corman
v Predecessor to the CIA

Bernstein was involved with smuggling front companies Foundry Associates, Materials for Palestine and also incorporated Americans for Haganah led by Abraham Feinberg and David Wahl. He defended *Haganah Speak's* successor publication *Israel Speaks* against Justice Department demands that it register as an Israeli foreign agent under a public disclosure law.

The FBI opened a file detailing Bernstein's "professed knowledge with regard to establishing and directing business fronts for covert operations." Bernstein was instrumental in providing the funding for Haganah smuggling network operative Adolph "Al" Schwimmer needed to illegally purchase surplus WWII transport aircraft for Israel. Schwimmer was later convicted of a felony. Bernstein, called in as a defense witness during Schwimmer's Neutrality Act trial in January of 1950, confessed that he had personally disbursed the Jewish Agency funding that purchased the aircraft. In July of 1950 Assistant U.S. Attorney Hershel E. Champlin claimed that if the U.S. government "had known of Bernstein's participation in the conspiracy involving Schwimmer et al prior to his admission of it in court, he [Bernstein] would have been indicted along with the others." But U.S. Attorneys in Los Angeles, Miami and New York later declined to prosecute. Bernstein was simply too well-connected to prosecute. He also performed legitimate undercover services.

Bernstein's law firm created a front company called "Yardley Fabrics" to aid in a wire tapping-operation conducted by a New York district attorney investigating insurance fraud. However the FBI later linked Bernstein to sophisticated technical surveillance operations targeting Arab officials participating in UN functions, Arab consular office break-ins and classified information leaks from the U.S. State Department. But the Justice Department declined all FBI suggestions that Bernstein be prosecuted or at least be made to openly register as a foreign agent of the Israeli government. Like Abraham Feinberg, despite numerous alleged violations of important U.S. laws, Bernstein enjoyed complete prosecutorial immunity in the United States. It did not hurt that his law partner was later elected district attorney and New York police officers migrated onto his payroll.[4]

When the U.S. demobilized after WWII, peace unleashed dangerous quantities of war materiel onto the market. The War Assets Administration (WAA) oversaw sales of enormous stocks of highly specialized machinery and military equipment. WAA mandated this surplus had to be converted to civilian use or decommissioned, rendered inoperable and sold as scrap. The Haganah drive to build a self-reliant military-industrial capacity began when Ben-Gurion sent engineer Haim Slavin to New York to research modern ammunition and arms production. Slavin operated under the truism that it is faster and cheaper to acquire the technology of others than to develop the same capability oneself. He began researching modern arms production (with no formal authorization by the Department of War) while commissioning the design of an entirely new weapon code named "the gun" for the Haganah while searching for highly specialized WWII surplus production machinery throughout the United States.

The Sonneborn network front companies, named after organizer Rudolph Sonneborn, bore innocuous names such as "Machinery Processing and Converting Company." They acquired, stored, packaged, disguised, and illegally exported U.S. military goods. Their first purchase included six tons of machinery from the Remington Arms plant in Bridgeport, CT for manufacturing .303 caliber ammunition for "the gun." The network could acquire state-of-the-art ammunition-making equipment worth hundreds of thousands of dollars at the price of $70 per ton only by promising it would be completely decommissioned. Another WAA deal routed through a friendly entrepreneur's corporation secured 200 tons of M-3 demolition explosive at the price of 10 cents per 2.25-pound block, just as the U.S. Department of State declared an embargo on all arms shipments to the Middle East.

The network's core competencies involved hiding contraband until it could be shipped offshore. Sophisticated military-industrial gear was disassembled, catalogued, and disguised as civilian machinery and innocent-looking components that would make it past U.S. customs inspectors for shipment to Palestine. Ammunition and firearms were welded into the centers of giant boilers or generators, while TNT crates were stenciled with innocuous labels.

The Haganah network was also highly active in manpower exports. NUMEC financier David Lowenthal hired on to smuggle and acquire many black skills and contacts. Haganah friends inside and outside the U.S. government compiled and provided timely intelligence for military personnel recruitment operations. One front, Materials and Manpower for Palestine, surreptitiously obtained the entire U.S. armed service chaplains denominational database which allowed the Haganah to direct targeted appeals to Jewish veterans in the United States during its drive to recruit volunteers to fight in Palestine.

The network operatives thought big. Even after the U.S. State Department declared its embargo on arms shipments to the Middle East, smugglers purchased a flat-top aircraft carrier from the WAA for $125,000. The plan was to ferry arms and displaced persons to Palestine on the *U.S.S. Attu* and later fully restore the carrier for air attacks.[5]

Nathan Liff, who acquired a WAA contract for decommissioning surplus arms, owned a Honolulu scrap yard that was the site of a typical arms theft operation. A dedicated Zionist, Liff notified his Haganah contacts during a visit to New York about his access to surplus war planes. Al Schwimmer a wartime TWA flight engineer who worked in an aircraft reconditioning and air freight business in Burbank, sent Haganah West Coast coordinator Hank Greenspun to Hawaii to look over Liff's inventory and procure functioning surplus aircraft engines.[6]

Greenspun noticed brand-new crated .30 and .50 caliber machine guns in a military section of the yard full of stock that had not been rendered inoperable. The crates were not only still owned by the military, but actively patrolled by U.S. Marines. Greenspun observed the sentries' timetable and used a forklift to steal 58 crates containing 500 machine guns. He carefully replaced the new stock with crates of guns already rendered inoperable from Liff's side of the yard.[7] Greenspun transported the guns to Los Angeles for transshipment to Palestine via Mexico. He almost lost the 35 tons of machine guns out of San Pedro harbor while employing a civilian yacht for the Los Angeles-to-Acapulco leg of the smuggling operation. The machine guns arrived for deployment in Israel by October of 1948.[8]

Theodore Kolleck also established front operations with Latin American dictators, including Anastasio Somoza in Nicaragua. Somoza bought operable WAA stock from the U.S. government as a sovereign allied state, which he reshipped to Palestine in exchange for a 3.5 percent kickback. Haganah operatives also coordinated with gangster boss Sam Kay to traffic arms through Cuba and Panama. [9] This foreign front-company model would later be used by Israel to divert significant quantities of yellow-cake uranium from Europe, purchase heavy water from the United States Atomic Energy Commission, and ship large quantities of U-235 to France, which was secretly building Israel's Dimona plutonium production facility.

Israel's air transportation industry began when Adolph "Al" Schwimmer purchased three surplus military Lockheed Constellations from the WAA for $45,000. The sticker price for new commercial service models, depending on the equipment configuration, was $685,000 to $720,000. The airplanes were capable of flying 300 miles per hour, had a service ceiling of 16,000 feet, and could carry 100 passengers or 10 tons of cargo. Schwimmer used another $20,000 of the network's funds to rent space at the Lockheed Air Terminal, where he added 10 smaller surplus C-46 Commando cargo planes under the name of Schwimmer Aviation. But Schwimmer was not actually going into business for himself; it was just another Israeli government front.

Schwimmer made a proposal to an out-of-luck Florida cargo entrepreneur, Charles Winters, who had purchased two B-17 bombers and converted them for civilian use. Each was capable of carrying seven tons of bombs and had cost taxpayers $204,370 to manufacture. When Winters' Caribbean fruit cargo business failed to prosper, Schwimmer asked if he was interested in flying the bombers to "somewhere in Europe." Winters navigated the bombers across the Atlantic to Czechoslovakia, where they were refitted for war and used to attack Egypt. He would later pay for this by a stint in prison. He was the only convicted Haganah smuggler to serve meaningful time in prison.

Schwimmer's air fleet, at times tracked and impounded by the FBI, left the United States for Panama by registering under a shell corporation as a Panamanian airline in order to evade U.S. export controls. It soon departed Panama and went into service in Europe, ferrying military supplies between Czechoslovakia and Tel Aviv. The U.S. Central Intelligence Agency detected the activity and filed a report titled *Clandestine Air Transport Operations* on May 28, 1948. The report's cover letter advised that "U.S. National Security is unfavorably affected by these developments and that it could be seriously jeopardized by continued illicit traffic in the 'implements of war.'" The CIA saw it as a false flag operation. Schwimmer's crews operating in Europe "dressed in U.S. Army uniforms without insignia" which deceived airport authorities in sovereign nations such as Switzerland into believing Schwimmer's air transport smuggling ring was really a "U.S. Air Force Operation."[10] The use of such pseudo official cover would be repeated in other high-stakes gambits.

Arab nations attacked the newly founded Israel in 1948 after the United Nations decision to partition the British-controlled territory of Palestine into Jewish and Arab states. Jewish forces, extremely well-armed by the Haganah network, prevailed and seized territories far beyond those authorized at the

United Nations. The U.S. smugglers were largely immunized by Israel's victory. Sonneborn smuggling organizations handling "black" goods gradually "went legit" after Israel won independence. The Supply Mission of the State of Israel in New York absorbed Machinery and Metals Company to manage military acquisitions. Materials for Palestine became Materials for Israel and stopped handling military equipment in favor of basic civilian goods for immigrants, including medical supplies, clothing, footwear, and vehicles. Land and Labor for Israel quietly shut down for less centralized recruiting efforts. All of the entities that could have been criminally prosecuted simply disappeared or reconstituted into other corporate forms.

The FBI, like the CIA in Europe, was alerted early on to the massive smuggling activities taking place across the United States, but took little effective action. In 1949, Charles Winters pled guilty to illegally exporting airplanes and was sentenced to 18 months in prison. Schwimmer of Service Airways was charged with conspiracy to violate the Neutrality Act and was found guilty along with Leo Gardner, Rey Selk. The captured smugglers were ordered to pay fines of $10,000.

But none of the leaders of the Haganah arms smuggling network were ever indicted. Henry Montor, the United Jewish Appeal leader who organized the first Haganah meeting, became founder of the Israel Bond Organization, which successfully floated its first issue of $52 million in 1951. He was never prosecuted for his smuggling network fundraising efforts operating in tandem with the UJA. Montor left the U.S. to live in Rome and Jerusalem in 1957.[11] Rudolf Sonneborn retired quietly as director of Witco Chemical Company and died in 1986. William Levitt was celebrated for postwar American mass production housing such as his "Levittown" development. Levitt provided a $1 million loan at no interest for the purchase of 15 Messerschmitt ME-109 fighter aircraft from Czechoslovakia for the Haganah, but never faced legal consequences for violating the Neutrality Act.[12]

Al Schwimmer prospered after becoming managing director of Israel Aircraft Industries (later Israel Aerospace Industries) after Israel's war of independence.[vi] With the backing of Ben-Gurion and Shimon Peres (Director General of the Ministry of Defense), Schwimmer worked to make IAI an indispensable vendor to the Israeli Air Force in the 1950s. The ambitious IAI attempted to manufacture modern fighter jets suitable for domestic military use and export. Later, recognizing necessity of economies of scale and industrial shortcomings, IAI settled into a more specialized role as an advanced modification, upgrade, and improvement vendor for existing fighters, commercial aircraft, and helicopter airframes, as well as manufacturing engines and electronics systems.

Most organizations and individuals in the Haganah network had plausible cover stories and connection to elites across U.S. politics, business, law and government. This made them difficult targets for law enforcement. An unusual meeting was held after network members were arrested in Canada smuggling prototype assault rifle components across the border in 1947. Leaders of the network traveled with a high-level Jewish Agency representative to Washington,

---

[vi] Known by Palestinian Arab as "al Nakba" or "the disaster."

DC and met with Robert R. Nathan, who had led the U.S. industrial mobilization in WWII, becoming the War Production Board's chairman in 1942.[13]

Nathan brokered a summit with FBI Director J. Edgar Hoover. The Royal Canadian Mounted Police had already "asked the FBI to cooperate in tracking down the sources and personnel involved and maybe prosecuting." This law enforcement initiative presented a major threat to the smuggling network and exposed the Jewish Agency's leadership role. Nathan flatly told the FBI director that the network's activities were not "anything damaging to the United States. But it is not straight up and aboveboard. Some prominent people and some important organizations could be hurt." Nathan assured Hoover that none of the weapons involved in the smuggling ring would ever be used in or against the United States, and left the meeting feeling that the FBI director was "sympathetic," but with no indication that he would "cooperate."[14]

Over time, all of the criminal records of Haganah smugglers were expunged, and even the reputations of the "little fish" convicted in court were carefully rehabilitated to hero status. In 1950, Nathan Liff offered compelling testimony in a Los Angeles courtroom during the trial over Greenspun and Schwimmer's violations of the Neutrality and Export Control Acts. Liff explained to jurors that he gave guns to "young Jewish boys who went to the door of Hitler's ovens" to bring Holocaust survivors to Palestine.[15] John F. Kennedy pardoned Hank Greenspun in 1961 after winning Israel lobby support in his presidential election campaign, though this relationship quickly soured. Bill Clinton pardoned Al Schwimmer in the year 2000, even though Schwimmer never personally applied for a pardon or expressed any contrition for his actions.[vii] Israel's U.S. supporters, led by Hank Greenspun's son, filed on his behalf. Schwimmer felt pardon requests demanded he "fill out all sorts of papers asking for forgiveness, telling the Justice Department you're sorry, you did wrong, and you regret it, and you won't do it again. I didn't feel that way, and I still don't. I didn't feel I had done anything wrong, so I never applied."[16]

Charles Winters, the only network member to actually serve a meaningful prison sentence, was posthumously pardoned in December of 2008 by President George W. Bush after intense lobbying by Steven Spielberg and other prominent American Jews eager to repair the historical record. Although no American, with the exception of Richard Kelley Smyth, has ever been convicted of smuggling nuclear-related items to Israel, NUMEC founder Zalman Shapiro—with the aid of Israeli government-paid lobbyists—has sought rehabilitation in the form of a formal professional recognition from President Barak Obama.

Former CIA Tel Aviv Station Chief John Hadden once claimed that the Nuclear Materials and Equipment Corporation was an "Israeli operation from the beginning." To understand this claim without benefit of the full CIA file on NUMEC, one must trace the career trajectory of Apollo Industries and David Lowenthal. David Lowenthal had a foreign intelligence and military career that was little known outside of his circle of close friends, which included his neighbor Zalman Shapiro. More than any other, Lowenthal created NUMEC

[vii] These are normally required in filings with the U.S. Department of Justice Office of the Pardon Attorney.

through the acquisition of the Apollo Steel Company in Pennsylvania, providing facilities for NUMEC to incubate while he raised additional capital that would allow the reprocessor to win a steady flow of government contracts and special nuclear material.

David and both of his parents, Markus and Sarah Latke, were born in Dobromil, Poland. David Luzer Lowenthal was born September 14, 1921. The family arrived in the U.S. onboard the SS Pulaski on August 2, 1934.[17] David was awarded derivative citizenship on August 2, 1938 on the basis of his father's naturalization. [18] He served in the United States armed forces during World War II and was fully naturalized on July 27, 1945 in the U.S. District Court of Pittsburgh, Pennsylvania.[19]

Lowenthal was soon recruited into the underground and fought during Israel's 1948 war for independence under Meir Amit. Amit later rose to become head of Israeli intelligence. Lowenthal was a close friend of David Ben-Gurion who became the first prime minister of Israel. FBI sources were impressed after seeing desert photographs of Lowenthal with Moshe Dayan and David Ben-Gurion.

According Israel nuclear program researcher Avner Cohen, David Ben-Gurion was virtually alone in launching Israel's nuclear weapons program in the mid-1950s. In his year 2010 book *The Worst Kept Secret: Israel's Bargain with the Bomb,* Cohen writes, "Initiating a national nuclear project, with an eye on to the bomb, had been high on Ben-Gurion's agenda even before he returned to power in early 1955 from a year-long self-imposed exile. By that time he already was convinced that Israel should embark on a nuclear project, but the practical concerns about the project's feasibility, concerns he had entertained earlier, remained. Could Israel pursue a nuclear weapons project on its own? If not, was there a foreign supplier who could provide the required technology? Could that foreign supplier be trusted, politically and technologically, to do so secretly and reliably?"[20]

Israel's secret contract with France to build a plutonium-based nuclear facility at Dimona while deceiving the Kennedy administration's non-proliferation efforts and inspector visits to Dimona under various pretexts has now been well-documented. The mechanics of NUMEC's diversions of bomb-making material, which allowed Israel to assemble weapons (and acquire feeder fuel) even before Dimona was fully operational is not.

Lowenthal not only traveled to Israel during the key milestones of Israel's nuclear weapons development program, he was also uniquely well-positioned to divert tax-deductible funding raised by the United Jewish Appeal back into the United States and into NUMEC. During the late 1950s and 1960s similar mass diversions of tens of millions of dollars in charitable relief funds were laundered through the Jewish Agency's American Section back into American lobbying and public relations campaigns, ultimately leading to a secret war with the JFK Justice Department and Senate Foreign Relations Committee.[viii]

According to the FBI investigation of NUMEC, Lowenthal visited Israel in 1956, 1960 and 1965. In 1956, when applying for his passport, Lowenthal

---

[viii] See the book "America's Defense Line: The Justice Department's Battle to Register the Israel Lobby as Agents of a Foreign Government"

indicated he would attend the World Zionist Congress in Israel for one month. In 1960, Lowenthal revealed he would be traveling in Italy, France and Israel on a UJA [United Jewish Appeal] Study Mission.[21] Lowenthal had easy access to millions in UJA funds for projects deemed vital by Israel's leadership. Acquiring a nuclear arsenal was Israel's very top priority.

Lowenthal's visit to Israel in 1956 coincided with the formal Israeli decision to obtain both yellowcake and weapons-grade uranium. Presumably it was during this trip that Lowenthal received the go-ahead to create yet another Israeli front company — this time designed not to steal or smuggle surplus WWII arms — but to obtain and clandestinely divert useful quantities of U.S. weapons-grade uranium and plutonium. It would require a more substantial investment and appearance of legitimacy than typical Haganah smuggling fronts. Lowenthal already had the experience and contacts to establish and fund such front companies from his service a decade earlier aboard the *1947 Exodus*. As before, established organizations such as the Zionist Organization of America would be tapped for board members and political support.

Lowenthal's 1960s United Jewish Appeal visits to Israel and France overlap with the period during which the Dimona reactor was being readied to launch, while 1965 was the year Israel's nuclear bomb builders at RAFAEL were soliciting guidance from Shimon Peres about precisely what type of weapons should be assembled from materials at hand, long before Dimona could produce the necessary weapons-grade material.[22]

Shapiro and Lowenthal both enjoyed rock-star status from government officials during visits to Israel. In the U.S. the pair appeared to be legitimate businessmen. After WWII, Lowenthal labored in middle management within the private sector. He married on December 23, 1951, while working as a manager at the Federal Paper Company.[23] Lowenthal then joined the Columbus Pipe and Equipment Company of Ohio, the parent company of Mount Vernon Bridge Company in 1954. Partner Morton Chatkin and David Lowenthal were entering joint business ventures by 1953, acquiring control of the Mount Vernon Bridge Company with other investors. Lowenthal stayed on as Vice President until 1955, when he sold his interest and acquired the defunct Apollo Steel plant.[24]

After a visit to Israel in 1956, Lowenthal executed an unnecessarily complex merger between three corporations which more than anything else brought together top Zionist leaders within a plausible smokescreen of defunct, bankrupt recently acquired corporations with long histories. They then built up Apollo Industries as a holding company, launching Raychord Steel and acquiring other small companies for tax write-offs, while nurturing the fledgling NUMEC until it was no longer an asset to Israel's nuclear weapons program. FBI wiretaps revealed NUMEC's utter disregard for worker and environmental safety as the founders successfully saddled successor parent companies and U.S. taxpayers with the burden of cleaning up after NUMEC.

2. AEC, USS Nautilus and the birth of NUMEC

Zalman Shapiro liked to drive fast. As he blazed along the 30 miles of roadway between his comfortable home in Pittsburgh and the Nuclear Materials and Equipment Corporation at Apollo, Pennsylvania, he seemed the very picture of youthful entrepreneurial vigor. Shapiro's reprocessing facility fueled the Navy's growing fleet of nuclear powered submarines and surface ships under the illustrious command of the, prickly, no-nonsense Admiral Hyman Rickover. NUMEC's contracts to produce fuel for Nautilus class submarines necessitated a seemingly limitless government supply of highly enriched uranium to NUMEC, totaling 22 tons between 1957 and 1967.[25] NUMEC was clearly on the new frontier of atomic private enterprise. Yet NUMEC was as deceptive as Shapiro's youthful appearance. An FBI surveillance team tracking Shapiro's every move duly noted his age was difficult to determine since he "dyed his hair." Apollo's creaky and aged infrastructure was wholly inappropriate and undercapitalized for the toxic work NUMEC undertook, necessitating Shapiro's attempts to abscond with loaned equipment owned by a U.S. government national laboratory under the pretext that it was "contaminated".

Zalman Mordecai Shapiro was born on May 12, 1920 in Canton, Ohio. His entrepreneurial drive was infused with a powerful devotion to serve Israel. Son of an orthodox rabbi from Lithuania, Shapiro studied hard and became valedictorian of his high school class. Having lost relatives to Nazi violence, and claiming to have endured anti-Semitic insults as a child, he felt "strongly about the need for an independent Jewish state."[26] Like many of his fellow Zionists, he pronounced himself incapable of distinguishing between advancing U.S. interests and those of Israel. Such advancements, claimed Shapiro to any who would listen, were all one and the same.

Shapiro's nuclear career began at Westinghouse Research Laboratory in August of 1948, shortly after receiving his PhD in chemistry from Johns Hopkins University. This marked the same period Shapiro became active in the Zionist Organization of America and the American Technion Society, which raised funding and provided equipment and guidance to Technion Institute of Technology at Haifa.[ix] Over a span of 60 days he submitted six patent ideas which were accepted and won him recognition and rewards from Westinghouse. In 1948 the U.S. Navy awarded Westinghouse a contract to design and build the first pressurized water reactors to power a new type of submarine. The contract required two reactors. The first was for a land-based prototype in Idaho. The second was destined for New London, Connecticut where General Dynamics would build the USS Nautilus. The talented Shapiro was invited to transfer to the Atomic Energy Commission Naval Nuclear Power Laboratory operated by Westinghouse as part of the reactor team when the high-stakes project was launched in February of 1949.

[ix] Described as "Israel's MIT."

The timeline was tight and reactor assembly was on the project's critical path. Any delay in putting together a safe, smoothly functioning reactor would jeopardize the entire project. This revolutionary propulsion system for a prototype submarine had to be ready by 1953 when reactor core installation was scheduled. When daunting engineering challenges threatened to derail the timeline, Shapiro helped provide the most innovative solutions.

One nagging reactor design problem was how to prevent high pressure heated water from reacting with uranium within the fuel elements. Radioactive fission elements could not be allowed to enter the submarine's coolant stream. Initial designs written up by project physicists called for cladding U-235 fuel with artificial zirconium metal which had both a high melting point and the desired low neutron-capture in addition to corrosion resistance.

During the early phase of the Nautilus project, not enough was known about the metallurgic and chemical properties of man-made zirconium metal. The thousand-dollar-per-pound metal available to the project was of inconsistent purity. It often reacted with water to form a disruptive white zirconium oxide powder. Shapiro worked twelve hour shifts, seven days a week to determine the cause of impurities and design industrial processes that would deliver consistently high quality materials. A special facility was built to produce zirconium under Shapiro's specifications in just three months in order to keep the Nautilus program on track. When control rods for the Idaho reactor were also found to be "metallurgically faulty," Shapiro was again brought in to design a production process that could replace faulty rods with corrosion resistant, ductile hafnium replacements. This major improvement to power control was vital since hafnium had attractive neutron absorption properties. When the Idaho core was finally shipped and successfully tested, Admiral Hyman G. Rickover cited Zalman Shapiro as one of four individuals most responsible for the success of the Nautilus program.

USS Nautilus (designated SSN-571) signaled its historic message "underway on nuclear power" on January 17, 1955 and began conducting sea trials that revolutionized naval operations. Nuclear propulsion allowed greater depths and submerged durations beyond the capability of any conventionally powered submarine. This meant that all of the submarine warfare rules written in WWII had to be tossed out the window. By 1958 USS Nautilus became the first submarine to transit the North Pole underwater to demonstrate the future deterrent capability of submarine-launched ballistic missiles.

During his period at Bettis from 1949-1957, Shapiro rose through the ranks as senior engineer, manager, physical chemistry manager, to become the assistant division manager of the Pressurized Water Reactor Division. [27] Shapiro's rising star, deepening relationship with Rickover, and the Atomic Energy Agency's simultaneous drive to promote private nuclear industry in the United States all converged to enable Shapiro's own business plan—reprocess growing supplies of government-owned and leased uranium, plutonium and other radioactive sources into value-added products for government projects and civil industry.

Meanwhile, across the ocean, Israel was faring poorly. In an attempt to keep colonial powers from pulling out of the Suez Canal Zone, Israel launched false-flag terrorist attacks against the U.S. and British targets in the summer of 1954.

Israel failed in its attempt to frame "local nationalists" in a way that would draw in U.S. troops. Israel instead saw its agents arrested and sentenced to prison in Egypt. Notes taken at a July 23, 1954 National Security Council meeting revealed Eisenhower's plummeting opinion of Israel. "The President commented on a conversation which he had recently had with a visitor from Israel, who had stated to him that the government in Israel was thoroughly unreligious and materialistic. The President said he had been astounded by such a statement, since he had been of the opinion that a good many members of the Israeli government were religious fanatics."

After Egyptian President Gamal Abdul Nasser nationalized the Suez Canal, Israel, France and the United Kingdom attacked on October 29, 1956. The U.S. and USSR persuaded the belligerents to withdraw from Egypt. This turbocharged Israel and its global support network's scramble for a new military edge that would allow the tiny country to cast greater influence over a world dominated by two nuclear-tipped superpowers. But Israel could now tap France for support at Dimona as payment for its services attempting to roll back Egypt.

In 2009 author Stephanie Cooke wrote a derogatory but serviceable description of the Atomic Energy Commission. "The AEC had become an oligarchy controlling all facets of the military and civilian sides of nuclear energy, promoting them and at the same time attempting to regulate them, and it had fallen down on the regulatory side...a growing legion of critics saw too many inbuilt conflicts of interest"[28] The AEC's handling of NUMEC displays all of its most egregious flaws, self-serving tendencies and abuses of power. NUMEC was a major factor in the AEC's ultimate demise under a growing chorus of public suspicion and criticism from Congress—which included the powerful Joint Committee on Atomic Energy which oversaw the AEC.

The Atomic Energy Commission was created by the McMahon Atomic Energy Act on August 1946. This law transferred control of atomic energy from the Manhattan Project's weapons-oriented military oversight to civilian control on January 1, 1947. It was an awesome responsibility. Under the guiding vision of giving AEC the lead role promoting atomic energy for public welfare and private enterprise, the McMahon Act explicitly outlawed nuclear technology transfers between the United States and foreign countries, requiring FBI background checks of any contractors and scientists seeking access to AEC facilities. It also gave the government an absolute monopoly on the production of fissionable materials. The AEC took over weapons oversight from the Manhattan Project and management of the Los Alamos Scientific Laboratory and other major infrastructure including the sprawling Hanford plutonium production site in Washington state. In 1954 the U.S. Atomic Energy Act was amended to permit nuclear technology and material exports to countries that promised not to develop nuclear weapons. This freed U.S. companies to sell technology, resell AEC licensed materials and advance Eisenhower's program of Atoms for Peace, announced to the UN General Assembly in December of 1953.

The U.S. significantly increased funding to build civilian power reactors to jumpstart the domestic nuclear industry, while offering foreign aid and information programs to other countries interested in U.S. technology. This included nuclear training, technical information and help building small research reactors. In March of 1955 Eisenhower directed the AEC to provide "limited amounts of raw and fissionable materials" to "free world" nations. Absent a credible international safeguards regime, it was a recipe for proliferation quickly taken advantage of by Israel, India and Pakistan.

NUMEC was incorporated on December 31, 1956 with $4,500 in property and three directors (Zalman Shapiro, Frederick Forscher, and Leonard Pepkowitz).[29] It became the youngest, most-cutting edge company to be incubated by Apollo Industries. Zalman Shapiro was still working at the Atomic Energy Commission's Bettis plant operating in the suburbs of Pittsburgh by the Westinghouse Electric Corporation when he incorporated NUMEC. Although Hyman Rickover was allegedly angered by Shapiro's decision to leave, he was soon contracting for fuel with NUMEC.

NUMEC commenced operations on February 1, 1957 with just 14 employees.[30] NUMEC lab workers labored in a wood-beamed room precariously lit by strung fluorescent lights. Even Shapiro recognized the substandard conditions at Apollo as workers raced to fill their first production order fabricated from uranium oxide for the Pennsylvania advanced reactor. "We literally worked nights and days in pots and pans to fill that one."[31] NUMEC's clandestine purpose manifested itself within the plant in more subtle ways. Shapiro ordered that all proceeds from vending machine sales in the plant be donated to Israel. Workers later earned brownie points by stuffing nylon stockings and cigarettes into shipments bound for Israel.

A steadily increasing flow of government contracts allowed NUMEC to quickly expand. NUMEC shared a facility with Lowenthal's Raychord Steel production facility in Apollo. NUMEC then opened a new 20,000 square foot building five miles away in Parks Township on a site with 58 acres of raw land. Shapiro tapped other investors. Kiski Valley Enterprises backed the State of Pennsylvania's 100 percent loan guarantee financing system to support NUMEC's expansion with four banks putting up 50 percent of the capital needed for expansion. The State of Pennsylvania contributed 30 percent and Apollo Industries pitched in another 20 percent of the capital. In 1960 Apollo's workforce blossomed to 230 as the company beat out the bids of 38 competitors to win a plutonium contract valued at a healthy $2.6 million[x] over the next three and a half years. [32]

By 1968 NUMEC's workforce reached 1,000 employees, most working within 246,000 square feet of fabrication facilities in and around Apollo, and a smaller group producing Boron-10 at a Niagara facility in Lewiston, New York.[33] The Atomic Energy Commission, which supplied all uranium and other special nuclear materials on a strict contractual basis while also regulating NUMEC, became aware of unacceptably high levels of missing nuclear material in the early 1960s.

The first alarm bells were sounded by the meticulous and uncompromising Admiral Hyman Rickover. The head of the Navy's nuclear ship program was NUMEC's biggest customer by 1962. Rickover was so upset about NUMEC's lax security and stream of foreign visitors and foreign employees, he directed an

---

[x] Worth $19.4 million in today's dollars.

aide to send a critical letter addressed to Zalman Shapiro. It expressed the Admiral's anger about of NUMEC's "apathy" toward security and the inherent risks of the company's growing ties to foreign countries. A dozen countries were sending 50-60 aliens to visit every year and NUMEC itself employed six aliens, among them an Israeli metallurgist assigned to highly sensitive plutonium work. The Israeli was forced to resign shortly after Rickover lodged his formal complaint.[34]

The Atomic Energy Commission was also gradually becoming uneasy about NUMEC. A January 13, 1962 report concluded that "numerous deficiencies were found in NUMEC's overall security program." A follow-up January 22 report threatened that if security violations "continue to develop, classified weapons work may be withheld from NUMEC."[35] This may have referred to NUMEC's secretive production contracts at the Lewiston facility.[xi]

NUMEC's main facility at Apollo was licensed to possess U.S. government-owned and enriched uranium to manufacture fuel, recover precious radioactive materials from scrap and conduct nuclear research and development beginning in December of 1957.[36] NUMEC principally engaged in small-scale production of low and high enriched uranium and thorium fuel until 1962. By 1963 most of the Apollo facility was dedicated to the continuous production of uranium fuel. NUMEC's key "value-added" industrial process was converting low-enriched uranium hexafluoride (Hex) gas delivered in steel containers from the gaseous diffusion plants into uranium dioxide used in the fuel rods of nuclear reactors. From a valuation perspective, the only real comparative advantage NUMEC ever developed was its ability to maintain a steady stream of large U.S. government contracts.

For Apollo workers, NUMEC was quite literally the only game in town. According to Harold Cupec, a former NUMEC employee "You needed a job. You didn't make a whole lot of money, a buck and a half an hour, but you had $90 every week coming in, you know? So that, that helped out a lot, you know. Just because you had a steady job. I came from Allegheny [Steel], but I was laid off all the time. Just stayed there for ten years. Then they were going to sell out. And they told us, we could get another job, you know, though they hated to see us go, but they didn't have a buyer yet...Ended up in there [NUMEC] for 23 years."[37]

1963 marked a turning point for business expansion when NUMEC added a new production line of highly-enriched uranium fuel for U.S. Navy propulsion reactors. But NUMEC later expanded into almost seventy related goods and service offerings including moderating and control materials, analytical laboratories, scrap recovery, uranium storage, and research and development projects.[38]

Apollo Borough began to suffer the consequences of undercapitalized, mismanaged nuclear fuel production long after NUMEC's founders departed.

[xi] All declassified government documents focus on nuclear materials losses that occurred at NUMEC facilities located in Apollo and Parks Township. There currently is no publicly available information about NUMEC's Boron 10 production facility at Lewiston, within a vast facility that worked on secret weapons programs.

Lawrence Frain, who lived on Armstrong Avenue from 1959 to 1963 next to the Apollo plant, did not worry about soot and pollution spewing into the air from NUMEC's smokestacks. It seemed only a minor annoyance as it left a grayish-white film on his 1960 Ford. But such ignorance was later recognized as deadly. "I remember a guy walking around with a {Geiger] meter. Sometimes he'd say 'They let a lot out last night.' Neighbors thought he was a little off, but maybe he knew something." [39] Frain's wife, sister and niece all died of cancer, while Frain contracted melanoma. All around him, many neighbors and friends gradually became sick and died of cancer.

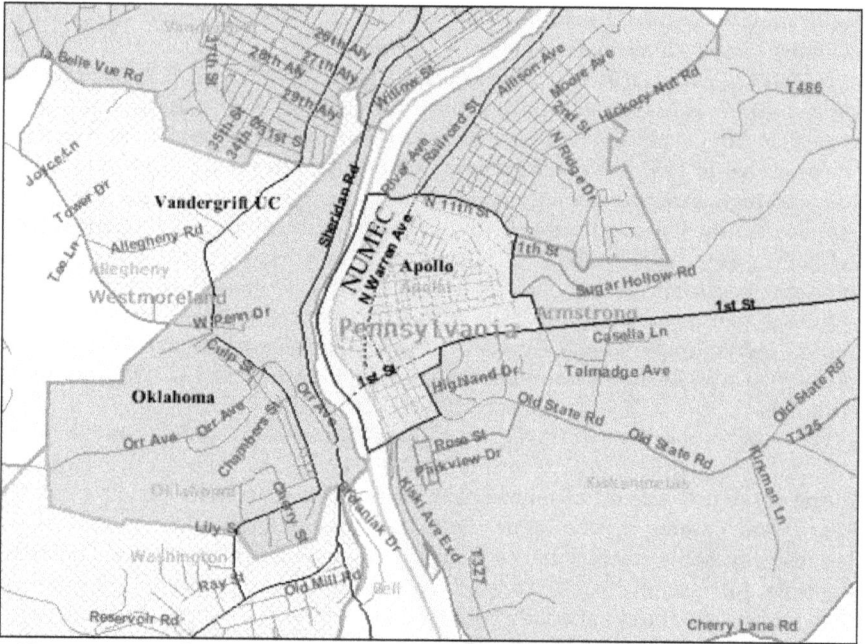

Figure 1 Map of Apollo Borough, PA. (2.8 miles across)

The center of Apollo Borough was a staggeringly inappropriate site for a nuclear fuel reprocessing plant. Only two decades earlier, Apollo and other towns along major river tributaries had suffered deadly floods that devastated Pittsburgh as the Ohio River crested 46 feet above its banks.[40] The Apollo steel works were situated on a hundred-year floodplain alongside the Kiskiminetas River[xii]. Patricia Ameno, an Apollo resident turned environmental activist, noted the threat to the river. "NUMEC was established in the old [steel] mill which was located directly in the corporate confines of the town, on the main street (route 66) , up against the Kiskiminetas River which is and has been a major source of drinking water, and across the street from several homes, including my parent's house and deli where I grew up. Dr. Shapiro had the mill's dirt floors covered

---

[xii] A 27 mile long tributary of the Allegheny River.

with cement, applied fresh paint, built his production lines and then opened for business." [41]

Figure 1 NUMEC facility in the center of Apollo, Pennsylvania

NUMEC's reprocessing operations in the center Apollo had a major limitation — waste disposal. NUMEC built a new facility five miles up Route 66 in Parks Township to specialize in plutonium processing. The waste problem seemed to go away, at least from Shapiro and Lowenthal's perspective. Like the Apollo facility, Parks Township was also in proximity to homes and small businesses, including the Farmers Delight Dairy Farm which supplied milk to local schools. Parks Township soon became not only NUMEC's AEC-authorized nuclear waste dump, but NUMEC also offered waste burial services to any plant operating nationwide that was willing to truck toxic material into Parks.

Few residents initially understood the hazards of the unfenced Parks Township dumping site. Ameno's memory is clear. "The burials were made in an extended abandoned mine area that sits at an uphill gradient from the Kiski River with an open and unfenced access to the burial field, where children played, adults hunted and family pets would roam." [42] Not all of NUMEC's waste dumped at Parks was even stored in stainless steel barrels; some atmospheric chamber (glove-box) waste was buried in cardboard boxes. [43] NUMEC experienced avoidable radioactive spills [44] that were a result of cost-cutting and compliance violations of the already lax rules of the time.

But in the 1960s, working at the new Parks Township plutonium facility seemed to be major career advancement to the workers transferred over from NUMEC's Apollo facility. Long-time NUMEC employee Skip McGuire compared the two facilities. "It [Apollo facility] weren't nothing but an old steel mill. They'd come in, remodeled it, but it still had the dirt floor, and it was colder than heck in there, in winter time, and as soon as we left the front entrance through the place we ended up in lab coats and booties, covers over our shoes to walk through the place. The place was saturated with acid. Half the guys would come out with their shoes burned off of them, after a couple months of work. Their shoes would be eaten off. Then I was transferred down to the plutonium plants in Parks Township, and I thought I was walking into a palace. It was clean, and nice and everything, but we still didn't know the danger of the place."[45]

Figure 2 NUMEC facility at Parks Township.

However, looks were deceiving. NUMEC's Parks plutonium facility was not a safe work site. According to Larry Giunta, a former NUMEC employee "In 1965 working at NUMEC we had predominately plutonium and uranium that we worked with. The plutonium was contained in glove boxes, and the uranium was in the open atmosphere. There were no controls on the workers. We didn't understand what we were dealing with at that time. We had—again—things in the open air, and we worked in glove boxes. A glove box was a plastic, Plexiglas cube with rubber gloves in it. And you handled the toxic materials and plutonium with the rubber gloves like you were working with it on the shop bench. So you know the hazards of a shop bench. There were pointy things, there were sharp things, there were jagged things. And quite often, on a daily basis, little holes would perforate the rubber gloves you were wearing. And you would come in and out of the glove box many times during the morning, and the only time they monitored was when you were leaving the facility. They had a little hand monitor there, when you were leaving the fab[rication] area. And many times at that area, people were caught with contaminated hands, contaminated forearms, contaminated shoes, then they were sent to the decon[tamination]. But during the morning, when you were working, you were working with that decontamination."[46]

Tiny amounts of plutonium could be deadly. In March of 1969, Shapiro recorded that an employee who accidentally pricked his finger with a "sharp instrument where plutonium [was] involved and the finger was amputated."[47] Despite the risks, NUMEC could count on the depressed economic environment to keep wages low and mouths shut. One NUMEC employee started at the plant in 1959 receiving $1.50 an hour quit 30 years later making only $13 an hour. "They never warned us. Early on, there was a taped line on the floor that divided the contaminated area from the side that wasn't contaminated...but it was in the air..."[48]

Figure 3 NUMEC's Parks Township plutonium facility.

Jack Zimmerman worked at NUMEC from 1959 to 1964 handling uranium dust. Shortly after he was hired at age 18, a company executive callously told him to "drink a lot of beer to flush your system out." Zimmerman felt it was a condescending, self-serving way for executives, who chose to live in Pittsburgh, to cover up what they already knew. "Shapiro and the big shots knew...The engineers knew...How do you forgive somebody that possibly ruined your life?"[49] By 2008 he had undergone a dozen skin cancer operations and was scheduled for two more.

Beer was also not a sufficient remedy for Garry Walker, who worked in the plant for 30 years after joining in 1959. Interviewed by the *Pittsburgh Tribune* in 2008, Walker stated that "after 1983 my kidneys quit, and I was on dialysis for 2.5 years, then I got a transplant, after the transplant, I got cancer, and I'm on dialysis again, I've been on it for three years, three times a week, Monday,

Wednesday, Friday, six in the morning until 11 and I'm waiting for a kidney right now."

Figure 4 Scanning an employee for radiation.

Radiation detection and decontamination were dicey processes. NUMEC. work rules left employees exposed for long periods of time. Larry Giunta, former NUMEC employee gave details about how glove box decontamination functioned. He "was contaminated many times. The glove would get perforated, sometimes you knew that the glove got perforated, you pulled your hand out and now you knew that your hand was contaminated, so you went over where they had a little room, and you went over and you washed your hand. If that didn't get it off you contacted health safety, and they put potassium permanganate on your hand, it would dissolve potassium permanganate crystals and make a liquid out of it, you would dip your hand in this potassium permanganate, and it would turn your hand purple. You would leave it on your hand for 4 or 5-minutes, I think two minutes is the max recommendation now, but at NUMEC, there wasn't a guideline. After it was on, what it would do is oxidize a layer of your skin, then you would dip your hand in sodium bisulfate which would neutralize the potassium permanganate, turning your hand back to its normal color again, also taking off a couple layers of skin. If that didn't work, then you'd put a rubber glove on, tape it at the wrist, and sweat the remaining plutonium out of your pores." [50]

3. Highly enriched uranium goes missing

Regulators closely observed and documented how NUMEC was administered in the early 1960s—before the MUF issue erupted into a threat to the AEC and terminal financial crisis for NUMEC. One insightful perspective comes from Charles A. Keller, the U.S. Department of Energy's assistant manager for manufacturing and support who worked out of Oak Ridge. In 1978 Keller made his personal diary entries available to the FBI—which promptly classified them as "top secret" until 1985. Keller's almost clinical observations provide a glimpse into just how inadequate NUMEC's infrastructure and location was for the type of government contracting it pursued. Keller also documented how the AEC initially reacted to reports of MUF, only to develop a self-severing and exculpatory theory about NUMEC's missing material that had no empirical basis. Keller's meticulous notes also expose the regulated industry—and how NUMEC frequently disregarded the rules by soliciting preferential treatment or hiring away insiders.

On June 20, 1962, Keller recorded that NUMEC's "files are a mess" when the New York Operations Office ordered him review NUMEC's "contracts, legal and finance" division. Subsequent regulators and investigators would conclude that NUMEC was purposely haphazard about record production and retention as an operational tactic of a broader deception scheme. To Keller, NUMEC seemed to be enjoying preferential treatment when it received $30-$35,000 from the AEC as a "progress payment" even though "Not a single contract has been completely closed."[51]

In mid-July of 1962 communications with NUMEC's chief financial officer, Frederick Forscher, Keller threatened to hold up a shipment of 645 kilograms of uranium hexafluoride until AEC approval was given on a feasibility report. Known as "hex" to industry insiders, the material can be processed into to both fuel nuclear reactors and to produce material for nuclear weapons. NUMEC was at the time reprocessing the material for the NERVA (Nuclear Engine for Rocket Vehicle Application) experimental nuclear rocket engine. [52]

Not until September 24, 1962 were the 11 cylinders of hex finally released and shipped to NUMEC for NERVA work. NUMEC's role in the supply chain was converting the gas to uranium dioxide, a black crystalline radioactive powder used to make nuclear fuel rods. NUMEC was supposed to delay processing the batch until the AEC formally approved NUMEC's feasibility report, to be presented to inspectors visiting the plant on the 29th of September. On October 3, 1962 Keller finally informed NUMEC the feasibility report was approved and that the material would be released. [53] Such seemingly rigid AEC control over special nuclear materials would ultimately lead to changes in how materials could be handled, eventually leading to private SNM ownership by reprocessors. In reality the "release" decisions were almost entirely trust-based and not frequently subject to verification. The AEC did not even bother to verify reports of a major fire that allegedly resulted in radioactive materials loss at NUMEC.

CFO Fred Forscher called Keller from NUMEC at 10pm on February 10, 1963 to report that a small fire had erupted in one of NUMEC's "special storage" vaults. Plant personnel heard a popping noise, like "a large firecracker in the vault." Wearing face masks while dispersing Metalex powder, NUMEC personnel finally extinguished the blaze. NUMEC reported the cause was uranium dicarbide (UC_2), a nuclear reactor fuel, which had been improperly vented into the plant's atmosphere. "As soon as UC_2 exposed to air, it ignited — small firecracker like explosions...the bottles broke spilling the contents...in all 5 bottles containing 8.8 kgs of FE U (Iron-Uranium) involved, all station material."[54]

This was NUMEC's first, but far from last, documented major discharge of radioactivity into Apollo's environment due to lax materials handling. Details of a far more nefarious incident a few years later would not become public until the year 2011. During 1978 testimony to members of Congress, Zalman Shapiro revealed there may even have been plant records stored in the same storage vaults as SNM. He conveniently blamed fires and labor strikes for periodic record destruction — records that would have provided vital additional insights into how NUMEC was really operated — if only they had survived.

Keller calmly noted there was apparently no radiation hazard and that NUMEC was washing down and cleaning up the affected area of the accident without any disruption to the rest of the plant. NUMEC officials did not call in any outside fire assistance from Apollo or Armstrong County. Certainly neither would have been adequately equipped to deal with nuclear fire hazards at that time. NUMEC plant employees gamely pitched in, cleaning up the mess while dressed in their regular work clothes which were later "picked up for decontamination." Forscher was apparently trying to be cooperative as he told Keller that he wanted "to know what reports are required." Keller told him he would check and advise. While casually discussing the fire, Forscher spun yet another tale to Keller about a NUMEC incident involving a large plutonium shipment. [55]

NUMEC had received a contract from the New York AEC operations center to develop a technique for decontaminating and recovering precious metals. After soliciting samples, NUMEC received 14 crates of material but reported opening only two. [56] NUMEC charged the AEC $150,000 to develop a recovery process. Years later, NUMEC asked the New York AEC operations office to pick up the rest of the unopened crates in order to free up plant storage space. NUMEC steel-banded the crates and shipped them back on January 4, 1963. [57]

Handled several times between NUMEC and its final destination at Brookhaven National Laboratory,[xiii] a plutonium leak from a 15 gallon drum was discovered when the shipment entered Jersey City. NUMEC claimed they had not known the shipment even contained plutonium since "they hadn't opened the package." Keller cautiously noted "no one has pointed a finger at NUMEC on this...looks like the original shipper did a poor job of packaging....Forscher said after incident they learned several millions of dollars of precious metal was in the

---

[xiii] A U.S. national laboratory established in 1947 located in Upton, New York on Long Island.

shipment." [58] NUMEC was such a secure site, Forscher seemed to imply, that large quantities of precious plutonium could safely reside there undisturbed for years.

Keller continued to diligently investigate the NUMEC vault fire to see if the required toxicity surveys were being run and whether any hazardous material had gone airborne and vented outside the plant. [59] On February 11, Forscher confided that NUMEC had not issued any public press release about the incident to Apollo residents. An old dumbwaiter shaft to the decrepit steel plant's second floor had vented the toxic smoke out of the storage vault and up through a roof ventilator. NUMEC's roof was contaminated. "Roof around ventilator read 600 cts/min sq....checked outside building—no contamination." [60]

All NUMEC personnel were subsequently ordered to have a urinalysis. The NUMEC vault, roof around the ventilator, hallways, and an upstairs lab area had to be decontaminated, though at the time of Keller's report the total area in need of decontamination was "not known, still checking." Keller spread word about the fire to other AEC officials as NUMEC shut down operations to "keep from tracking stuff around until they get decontamination done." [61]

But NUMEC's management team was preoccupied with a far more pressing problem—cash flow. Forscher constantly complained about not being paid for reprocessing contracts until after shipments left the plant, which he continued to characterize as the epitome of security. On May 9, 1963 Forscher advised Keller that the Pittsburgh Naval Reactors Operations Office, along with naval people from Schenectady, were visiting NUMEC for a "nuclear safety review on handling of Navy core materials." Forscher suggested that AEC New York office people might want to participate in the inspection. Keller demurred but indicated he'd be happy to see any derivative report. [62] Only much later did such invitations and guidance lead one inspector to tell FBI investigators he felt "steered" by NUMEC's management during mandatory onsite inspections.

On March 11, 1963 NUMEC contacted Keller about payment for a contract bid at $32,275. Keller said he'd check on the status, noticing that the AEC was only authorized to pay $27,768 for NUMEC's completed work. On the 14th Forscher claimed that all plant workers checked out okay on the urinalysis tests and that NUMEC would deduct losses of the material it claimed to have lost in the fire equally between Astronuclear and Bettis contract work. [63] The AEC agreed to this claimed materials loss without ever having sent out an inspector to NUMEC to verify whether material had actually been lost. Like NUMEC, the AEC also never announced the potential radiation dangers to the public, it simply took Forscher's word that the entire fire incident had been resolved.

Under the not-so-watchful eye of the AEC, NUMEC ramped up its business with Israel. Under the U.S.-Israel Agreement for Cooperation, the AEC authorized NUMEC to "fabricate four plutonium-beryllium neutron sources" ostensibly for use by the "Department of Nuclear Science, Israel Institute for Technology." NUMEC was allowed to ship 320 grams of plutonium valued at $12,600 with the irradiators. The shipment left U.S. shores for Israel on June 30, 1963 via a 600 pound El Al Airlines shipment container covered with warning labels.[64]

In May and June of 1963, Keller worried that Bettis was urgently requesting reports on NUMEC's materials handling practices. On July 18, 1963, after Forscher called in to complain about "dispersion of responsibility" in the AEC, Keller advised him that Doug George at the AEC's Division of Nuclear Materials Management "has said he is only interested in the security of the material, behind this is a feeling that he needs essentially all the info in a feasibility report to determine that [NUMEC] security is adequate." In August, Keller called Forscher about problems with three separate contracts and "asked that he look into their claims and that we discuss further. He agreed to do." Three days later Forscher suddenly stalled wanting to "hold up discussing questionable lots until visit." The two agreed that this could take place on August 24.[65]

On March 31, 1964 Keller began to tackle even more serious problems over a Westinghouse Astronuclear Lab (WANL) purchase order[xiv] contract won by NUMEC. NUMEC had claimed that "technical difficulties" were creating a "lot of scrap" which required that the amount of uranium supplied for the successful completion of contract would now have to be more than doubled from 400 kilograms to 1,086 kilograms. NUMEC was taking on 100 percent of the financial responsibility for the added inventory. [66] As documented later in interviews with a national laboratory staffer, NUMEC was not at all unwilling to give out false information in order to obtain such valuable government property.

At the time, NUMEC was forced to pay an "inventory use charge" of 4.75% of the value of HEU delivered under contract beginning 90 days after delivery of the last finished product. October 30, 1964 was the date of the last NUMEC WANL processed shipment, but WANL received only 762 kilograms of uranium product back from NUMEC, which also returned 70 kilograms of material as "recovered scrap" to the AEC. Keller nervously scribbled down his concerns at the discrepancy. "253 kgs U [uranium] still outstanding (book balance). Purchase order says all scrap to be recovered within 180 days after final delivery. April 28, 1965 is the crucial date..."[67] The clock was ticking. Could NUMEC return all of the HEU it had been given for contract fulfillment and pay the inventory use charge?

Keller noted cautiously that "NUMEC says they can meet the April 28 date..." However NUMEC now wanted to "transfer material to Supply Agreement." Keller verified to WANL that this meant NUMEC would take material from another government contract — signed with the Space Nuclear Propulsion Office of Cleveland — in order to cover all of the HEU owed to WANL. NUMEC would then theoretically recover HEU from waste produced during the WANL contract and return it later. "Advised Yates that we could take material under Supply Agreement if SNPO-C decided in best interest of government to do." While Keller did not at the time consider whether NUMEC may have been diverting material out of the country, he did suspect they were at least trying to make more profit. "I pointed out that they [SNPO-C] should consider this avenue carefully since it appeared that NUMEC could get a windfall by delaying payment of losses. (Expected losses plus material unaccounted for equals approximately 35 Kgs at moment). Told Yates I thought a good physical

---

[xiv] Purchase order number 9/62

inventory should be made before material put under Supply Agreement in order to determine what losses have occurred to date. Transfer Book Inventory and then adjust to physical with payment on difference to prevent any windfall...."[68] NUMEC would not fare well under a bona fide audit of its material, even in such a flexible regulatory environment.

In the 1960s, "Supply Agreements" represented a loosening of AEC rules requiring the strict segregation of special nuclear material by specific contract. In order to avoid the red tape of written AEC and supplier approvals needed to comingle materials, the Supply Agreement provided more flexibility by allowing facilities to comingle materials from various government contracts. Companies such as NUMEC would acquire Supply Agreement materials under AEC license but would then pay "use charges." These charges were then offset through the issuance of credit vouchers in amounts agreed to under specific government contracts in order to control the material. The company was still financially responsible for any losses. When waste was recovered, credit vouchers would also be issued to offset the Supply Agreement inventory value in whole or part.[69] But years of investigations and surveys would reveal that NUMEC was not simply shifting HEU between contracts in order to gain additional time to recover vast amounts of material from scrap. NUMEC would never recover or return massive quantities missing HEU.

In the early and mid-1960s, licensed recipients of U.S. government-supplied HEU had to pay full value of any lost material. Unfortunately for NUMEC this payment was not based on a company's own "book" estimates but rather an annual physical AEC inventory (called a survey) with payment for any missing material due immediately.[70] In reality, although $10 per gram seemed like a high price, the material itself could be priceless to any country unable to invest billions of dollars to build gaseous diffusion plants. On April 6, 1965, Doug George from the AEC Division of Nuclear Materials Management informed Keller that the AEC wanted to send in some AEC staff to be present for the NUMEC inventory "to assist us and also provide firsthand knowledge" of the situation. The names of AEC officials Lovett and Solem were floated. Keller told George "this would give us no pain."[71]

On April 15, 1965 Walt Scheib from NUMEC called Keller about the WANL "job and shift of residual material to Supply Agreement." The AEC headquarters agreed to allow NUMEC to substitute materials from the Supply Agreement, noting that it "Will transfer what is determined on hand at time of survey. Will pay for losses to WANL." NUMEC temporarily delayed its day of reckoning and mandatory fines for a few more months until the AEC's physical inventory was completed.[72] Not until 1980 would a NUMEC employee interviewed by the FBI detail how exactly Shapiro and unknown helpers smuggled the HEU from the U.S. to Israel in the spring of 1965.

The Space Nuclear Propulsion Office-Cleveland was not pleased about this cozy arrangement over the MUF or that NUMEC was continuing to win new government contracts. It wanted this displeasure to be felt at NUMEC. Keller jotted down that "SCPO-C somewhat unhappy about recent award of scrap contract to NUMEC (WANL Scrap) due to fact they were having collection problems with NUMEC. Wanted to know if we could hold up award. Told him we had no basis for such actions. Someone would have to get NUMEC on disqualified bidders list and this might be extremely difficult...."[73] In practice, later investigations found the AEC had far more coercive regulatory power than it ever exercised, whether demanding materials accountability from contractors or warranted cancellation of security clearances over suspected breaches, the AEC usually found excuses for not exercising its statutory authority over NUMEC when warranted.

By June 25, 1965 the AEC documented the estimated amount of lost material at NUMEC and was ready to fine the company. "RE NUMEC – WANL Contract. Sending a bill based on our survey numbers. Will modify bill if we change our position on amount of material in filters [material trapped, but not yet extracted from disposable processing filters at NUMEC]."[74] On June 28, the Keller revealed the enormity of the fine. "Bill is $765,000 on old WANL contract." While SCPO-C finally agreed to lift its opposition to the transfer of its material to the Supply Agreement, the AEC Division of Nuclear Materials Management decided the enormous loss was serious enough that it was now time to impose "continuous surveillance" over NUMEC's operations to prevent further losses.[75]

The AEC tried to be helpful by raising its official estimate of how much material had been trapped in NUMEC's conversion filters used during the WANL job, fudging the numbers in NUMEC's favor. "Ed Marshall [of the Nuclear Materials Accountability Branch of Oak Ridge operations] proposed increasing [estimated] filter content..35 grams to 48 grams based on recovery data." But even this unrealistically high filter estimate could not account for the vast amount of "unaccounted for" material which NUMEC claimed would be found in a future scrap recovery. The AEC nevertheless pitched the higher filter loss theory with "NUMEC (Les Weber) answering our letter to effect that filters contain all losses."[76] Official AEC surveys of the filters conducted by a disinterested third party would soon prove otherwise.

On July 6, 1965 the AEC readied a revised NUMEC bill for "approx. $750,000. P.J. Haycock of the AEC Division of Nuclear Materials Management discussed 'witnessing' the actual recovery of material from scrap at NUMEC. Keller advised 'we could probably do with help and having my people work up a plan.' We would follow only through dissolution and sampling which Haycock agrees is enough, including sampling of residues."[77]

Haycock also solicited Keller's views on the financial impact of billing the tiny NUMEC for so much lost material. "Told him not matter of my concern, but that I thought if a billing is in order they should submit. This could bankrupt NUMEC per Shapiro, so getting [AEC staffer Edward J.] Bloch on board as there may be political pressure."[78] Bloch was the program's point person, often briefing other AEC officials on the status of NERVA.[79] Keller apparently felt Bloch would try to defend NUMEC, and he was right, though Bloch was far from the only AEC official willing to bend the rules to avoid accountability.

Bloch ordered that the bill not to be sent to NUMEC until Howard Brown, the assistant general manager for administration at the AEC headquarters, had carefully reviewed it. Bloch also recommended an informational brief be prepared for AEC commissioners about the enormous fine. At the time, any invoice was unwelcome news at NUMEC. AEC accounts revealed that the cash-strapped NUMEC already had outstanding unpaid invoices of $53,748.33 plus interest for being more than thirty days delinquent. [80]

Bloch wanted harder numbers. This would delay the fine, but at least he was "not sympathetic to a book inventory only on NUMEC." But like many others, Bloch was unwilling to trust NUMEC's lofty estimates of material that could be recovered from waste in some distant future. According to Keller's notes, because the AEC "can't ignore physical inventory that we've taken" Bloch pushed for counting and then discounting the value of NUMEC scrap material that was "moving in for processing" which would help lower the massive fine. Douglas George also wanted a more scientific approach that did not at all rely on any of NUMEC's own filter recovery estimates. George proposed that a neutral third party sample NUMEC's filters "on a random basis," recovering material offsite to get absolutely reliable data. [81]

On July 12, the situation escalated further during a meeting between SNPO-C, the Counsel Controller and AEC staff gathered to discuss NUMEC. Keller heard "Lots of comments regarding inventory Supply Agreement, etc." Keller visited NUMEC again on July 21, 1965 and by July 22 AEC headquarters drafted a letter to both WANL and NUMEC confirming an accommodating approach after new "agreements made by Shapiro." [82]

Shapiro promised to augment a July 2, 1965 letter from Les Weber "stating precisely what added inventory exists that should be included in inventory, i.e. including mine 2 [an AEC approved NUMEC toxic waste disposal trench, see figure 21]." Shapiro promised to further specify in writing how NUMEC would verify that much of the missing highly enriched uranium was in fact located in the waste dump. Shapiro was also to confirm to SNPO HQ that he agreed to a 8.3 kilogram loss and that NUMEC would pay degradation charges on material returned to date totaling $145,000, but not for scrap material allegedly still in NUMEC's possession that would someday be returned. [83]

While Shapiro probably hoped to dodge a bullet by lowering AEC fines even as he promised to recover HEU from NUMEC's waste pits, concerns were now moving up the chain of command at the AEC. Assistant General Manager Howard Brown asked whether the AEC was withholding any payments to NUMEC and requested a written summation of outstanding claims that were in dispute. He also wanted to know all about NUMEC's past record handling U.S. government furnished special nuclear materials. [84]

AEC Assistant General Manager for Administration John W. Vinciguerra fired off a terse February 14 memo demanding to know the AEC's policy toward applying contractual financial responsibility to NUMEC and "what loss factors did we allow?" [85] Meanwhile, NUMEC continued to stammer, stall and change the subject. On July 26 CFO Fred Forscher brashly told Keller that NUMEC had invested a great deal in uncompensated upgrades to improve NUMEC processing equipment. He boasted NUMEC was getting NERVA fuel dust production up from 2 kilograms per day to 5-6 kilograms a day. "Forscher feels NUMEC being cheated by not being given credit for upgrading even though in conflict with contract terms and conditions." Keller openly recognized NUMEC's production accomplishments. But the AEC was not distracted. Keller knew that the filter samples, which NUMEC claimed had trapped unrealistically high quantities of U-235 from the WANL job, were finally being sent offsite to the government-owned New Brunswick Laboratory for testing. [86]

On August 2nd Congress finally became involved, making it clear that the AEC should not be coddling NUMEC. The Joint Committee on Atomic Energy (JCAE) was most concerned about questions neither the AEC nor NUMEC seemed to want to address "where did material get to, i.e. safeguards end of business." Keller thought that they were "Not just interested in getting $, need to know what happened to material... JCAE staff isn't going to intervene on behalf of NUMEC. Take best action we can to protect the government without being too vindictive..."[87]

On August 2, Keller's staff met to discuss the NUMEC crisis and float a new proposal to Shapiro, stating flatly that a mere $145,000 in fines was simply not sufficient. AEC's harsher new position—shored up by the JCAE—was that it had already been very patient and reasonable with Shapiro, but since AEC now seriously questioned whether NUMEC even possessed the nuclear material, it would have to submit a complete bill for the lost WANL U-235. Graciously, the AEC offered to transfer any material recovered from WANL scrap in the future to NUMEC's Supply Agreement based on verifiable physical inventories. The AEC would no longer permit NUMEC to include any inflated "book" estimates of recoverable material allegedly buried in its toxic waste pits. On August 6, Keller received a much more ominous information request from far up the AEC chain of command, signaling the first inklings that NUMEC diverted material off shore. "[AEC Commissioner James T.] Ramey wants [documentation of] shipments from NUMEC to foreign countries. "[88]

As if sensing a brewing storm at the AEC, Shapiro quickly fired off a salvo of self-congratulatory claims to the AEC, claiming that an Oak Ridge Operations Office survey team found special nuclear materials control at "NUMEC was greatly improved and that we [the AEC] were happy." This contrasting sharply with Keller's missive that AEC was "not satisfied." Moreover headquarters was now demanding that Keller supply "statements from our personnel as to what they told NUMEC at close of Survey. Statements are to be best recollection." There were to be no crossed channels or leniency as NUMEC played one AEC official against another. But AEC also remorsefully pondered whether it should have excluded NUMEC's inflated buried waste estimates. "No one brought up buried inventory because 'someone' from Oak Ridge told them not to report."

On August 25 Les Weber from NUMEC contacted Keller about the "accountability survey report" requesting a detailed discussion with Keller's team on "what is needed" to resolve the missing uranium issue. Keller told him that the AEC's problem was "in great part due to [the] fact you can't follow material from one job to another internally." NUMEC didn't want to be fined for material it now ardently claimed could be recovered from its waste dump. [89] But NUMEC wasn't about to attempt a comprehensive recovery effort. On September 13, 1965 NUMEC claimed it wouldn't open a toxic pit filled in 1961 since it didn't believe it would be economical or worthwhile. When NUMEC opened a burial pit dug in 1963 with renewed hope for recovery...it instead revealed evidence of a future environmental disaster in waiting. Keller documented the AEC's horrified observation. "Digging of 1963 pit required removal of 30 feet of overburden (at least 40 feet deep to bottom of drums). The [19]63 pit has at least 2 active springs. This pit has a lot of WANL residues." [90]

Shapiro stridently claimed that "10-20 kilograms [of HEU] may have been buried in carbon filters that shouldn't have been buried. These used in connection with plasma spraying operation. NUMEC has good set of records established with internal control documents that are being used. Mechanism now exists. [Inventory Control] manual about 50% written. To be complete on January 1, 1966." [91] But NUMEC's new corporate handbook for future inventory control obviously wouldn't rectify any past materials handling issues.

In addition to the hopeful new internal processes and promised recovery from filters buried in the waste pit, Shapiro nervously opened a NUMEC waste pit dug in 1962. AEC noted NUMEC's slow, laborious work. "About one-half done on hand picking the [1962] pit." But then progress slowed and finally stopped. AEC staffer Charles L. Marshall allegedly told "Shapiro to effect that no need to do further recovery of buried material." By September 24, this cross-jurisdictional snafu was cleared up with NUMEC having won another delay on paying the WANL fine. "Shapiro met with Brown and Tremmel on 9/24. Shapiro agreed to open the [19]63 pit. Nothing will be done on billing on WANL order until we get more data..." [92]

Figure 5 NUMEC workers digging in Pit #1 at Parks Township in 1966.

In October, NUMEC started slowly digging again. Water gushed from the face of the toxic pit at two gallons per minute. Troubled by the environmental implications, on October 6 the Division of Nuclear Materials Management suggested "putting someone at NUMEC to witness the excavation of buried material." But Keller pointed out the danger of that since "this appears to be a contract matter" suggesting that WANL should cover it. The liability issues were high over the "chances of our people being claimed as accessory to the fact before, during and after if there is a labor dispute on how material was handled, i.e. because we failed to suggest any changes in procedure or action we considered questionable." [93] In 2011, the U.S. Army Corps of Engineers would have to halt a sophisticated cleanup of NUMEC waste over similar high risk materials handling issues.

Keller thought "this might be prejudicial to our doing inventory of facility at [a] later date. Pointed out we believe that once they have recovered material that complete physical inventory is necessary to be sure they haven't robbed Peter to pay Paul." The question of scheduling any direct AEC observations of NUMEC materials processing was still also outstanding. The AEC Division of Nuclear Materials Management agreed to try to get a big NUMEC stakeholder, the Pittsburgh Naval Reactor, to cover staffing needs of the observation since "there is an interest there too." [94]

Concerns at the Joint Atomic Energy Congressional committee were building as the AEC struggled to answer even their most basic questions. Keller lamented "[AEC Assistant General Manager] Brown wants overall loss figure on NUMEC compared with AEC operations if possible...If they don't find 52 KGs in pit, what do we do safeguards wise?" At an AEC meeting on NUMEC on October 26, 1965, it was finally time for NUMEC to "fish or cut bait." [95]

A complete listing of NUMEC's foreign shipments compiled in 1965 raised many more questions than it answered.[96] NUMEC officially claimed that it had shipped 8,800 kilos of uranium to overseas customers, which included 425 kilos of highly enriched uranium-235. According to NUMEC's own figures, France was the destination for 69% of the itemized shipments, followed by Netherlands (13%) and Germany (5.3%). The AEC recorded that its examination of the records backing 32 of the foreign shipments found 26 were "incomplete, inaccurate or missing."[97] NUMEC's data included no record of HEU shipments to Israel.

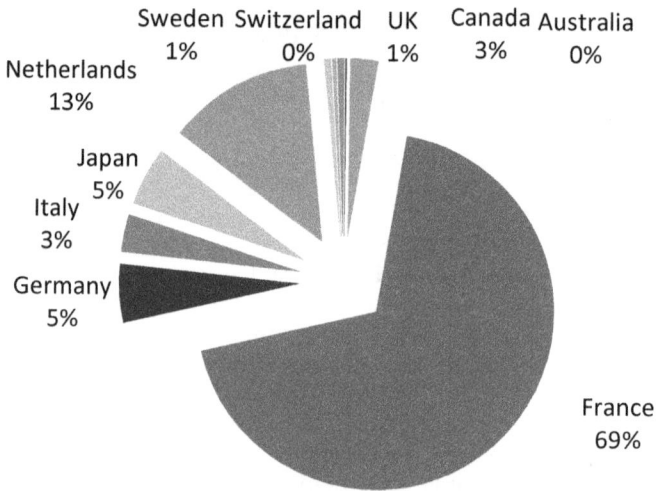

Figure 6 NUMEC reported U-235 exports 1957-1965.

It is worthwhile to juxtapose NUMEC's timeline of claimed shipments to the period of French covert assistance constructing Israel's Dimona nuclear weapons material plant. NUMEC's total shipments during the one and a half final years of the Eisenhower administration were robust (4,300 kilos). Construction of Dimona started late 1957 or early 1958 and was completed in 1962. During the JFK administration which fell just short of three years, NUMEC claimed shipments dwindled to only 504 kilos, most from a single shipment (300 kilos) delivered in May of 1963. After completion of the Dimona reactor and LBJ's assumption of the

U.S. presidency on November 22, 1963, NUMEC's French shipments revived to 1,300 kilograms through mid-1965.[98][xv]

Despite investigations conducted in the 1960s and 1970s and unreliable NUMEC records, no definitive audit following the trail of NUMEC's overseas shipments of licensed SNM to an identifiable European nuclear power facility and then on to waste disposal or back to the U.S. for reprocessing has ever been publicly released. This is unsurprising since determining what happened to most SNM shipments overseas authorized by the U.S. is a challenge that goes far beyond NUMEC. A 2011 Government Accountability Office report revealed that "DOE, NRC, and State are not able to fully account for U.S. nuclear material overseas that is subject to nuclear cooperation agreement terms because the Agreements do not stipulate systematic reporting of such information, and there is no U.S. policy to pursue or obtain such information." In a 1993 examination of 17,500 kilograms shipped overseas, the U.S. was only able to fully account for 1,160 kilograms. In 2011, flaws embedded in the Atomic Energy Act continue to haunt the U.S. since "No U.S. Law or policy directs U.S. agencies to obtain information regarding the location and disposition of U.S. nuclear material at overseas facilities."[99]

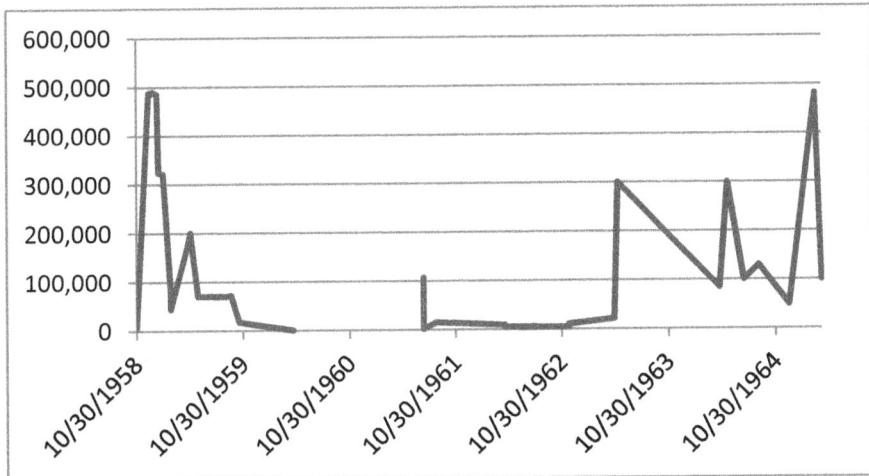

Figure 7 NUMEC U (low enriched) exports to France 1957-1965 (Grams).

But despite the availability of shortages, AEC investigators still had little context for how the NUMEC piece might have fit into the puzzle of David Lowenthal's earlier French smuggling career and ongoing connections with Israeli intelligence operatives. Despite the JCAE's early suspicion of diversion, by November 1, 1965 the AEC was actively quashing such theories, which some

---

[xv] More simply, NUMEC's transfers to France while the French were secretly colluding to build Dimona were 3,000 KG/year under Eisenhower, 178 KG/year under JFK, and 641 KG/year under LBJ.

staffers shared, and instead worked to exact financial "accountability for WANL residues at NUMEC in such a fashion as to prevent any windfalls to NUMEC and control of shifting material to other accountability, etc." On December 7, the AEC also began probing basic operational issues of NUMEC accountability and peripheral safeguard matters as materials transit. Keller admitted "Shipment to NUMEC on which safety problem exists. Navy order. Part of basic problem of who is responsible in transit..."[100]

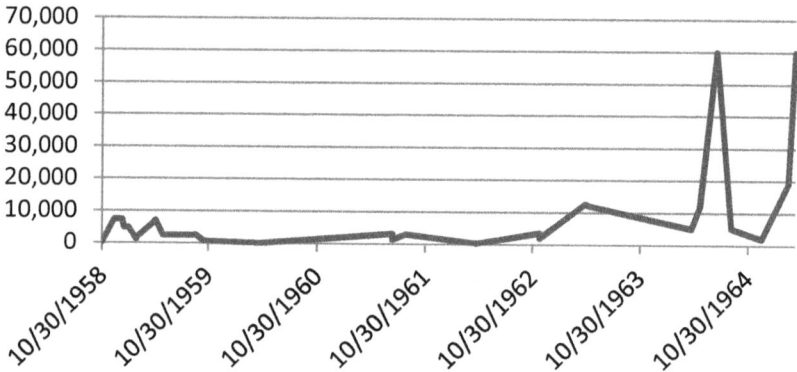

Figure 8 NUMEC U-235 (HEU) exports to France 1957-1965 (Grams).

On November 12, 1965 Douglas George and James E. Lovett met with Zalman Shapiro and Jack Newman to discuss survey findings and upbraid Shapiro over his failure to place a technically competent person in charge of inventory. NUMEC simply had to correct the situation because George "personally did not intend to go through the experience of the previous few months with NUMEC again." Shapiro countered that he had tried to hire away a competent specialist from the 3M Corporation, with no luck. Did George have any suggestions?

In exasperation, George rattled off a list of people with the required qualifications, including Lovett and other members of AEC staff. Though George later claimed it merely demonstrated the existence of a pool of available professionals that could be hired after a bona fide search, the suggestion would soon become a point of controversy. Later that day, S.A. Weber from NUMEC contacted Lovett. The very next day Shapiro offered him a job. Lovett resigned from the AEC on January 1, reporting for work at NUMEC on January 3, 1966. By February 3, Lovett was representing NUMEC from the far end of the table at an AEC meeting.

Although the AEC's general counsel investigated the Lovett hiring matter as a potential conflict of interest and referred it to the Justice Department, no action was taken. George later chided Shapiro for neutralizing the highly experienced and valued Lovett as an AEC survey team member at such a sensitive time. Shapiro offered to withdraw the offer, but George sourly told him it would not affect the impact of Lovett's loss to the AEC.[101] The AEC forged ahead without Lovett.

By February 6, 1966 Keller provided Brown with a list of all industrial processing companies holding special nuclear materials and descriptions of their conversion processes, lists of overseas shipments, and the total volume of SNM handled by processors, both per annum and on a cumulative basis. [102] But the AEC would never adequately resolve the JCAE's concerns about NUMEC. As the crisis grew, the AEC descended into crisis mode over NUMEC as a matter of self-preservation.

On February 14, 1966 the AEC convened an emergency meeting attended by four Commissioners, including AEC Chairman Glenn T. Seaborg, 25 staffers, the agency's legal counsel and two managers. Although the written meeting agenda was blandly titled "Safeguards and Domestic Material Accountability" the AEC's copious meeting minutes reveal the gathering's sole objective to develop a common, defensible and unified agency position to sell to the JCAE, FBI and the American public if news of NUMEC's extreme losses ever leaked out. The St. Valentine's Day gathering had the medium-term result of securing AEC's Congressional funding by presenting NUMEC's missing SNM as merely processing loss. The AEC's processing loss theory shielded NUMEC from a timely and warranted FBI atomic material diversion investigation mandated under the statutes of the Atomic Energy Act.

At the onset of the meeting, Assistant General Manager Howard Brown urged AEC overseers not to make NUMEC a "whipping boy" for the entire domestic material accountability system as it then existed. Brown urged AEC staff to consider that NUMEC's losses were instead "traceable to features [of] that system" though it was admittedly a system based entirely on the "presumption of honesty." While that system was "reasonably good" argued Brown, the AEC was now forced to coldly address how the system worked when "the presumption[s] of honesty were removed."

Brown updated attendees that the AEC's official figures for total NUMEC plant losses over eight years of operations now stood at 178 kilograms, of which 84.2 were officially considered by the AEC to be "known" processing losses. NUMEC reported losses of 149 kilograms during these eight years, however AEC staff determined in its November 1965 survey that NUMEC's underreported losses totaled another 29 kilograms, bringing reported losses up to 178 kilograms. He also confirmed that the SNM NUMEC had received for the WANL contract was particularly potent, having been weapons-grade "93 percent U-235." Since the AEC's congressional budget authorization was fast approaching, anticipating tough questioning from JCAE quickly became the order of the day.

Commissioner Ramey solicited all staff present at the meeting whether they had any information about the amount of material NUMEC had shipped abroad over the past eight years. Brown said staff did have data, but it was exclusively based on faulty NUMEC records, with no independent verification.[xvi] Brown delivered the bad news that "if collusion between a shipper and a foreign government were assumed, it would be theoretically possible to ship material abroad in excess of the amounts indicated in the company's records." The AEC's

---

[xvi] NUMEC did not itemize any SNM shipments to Israel since it claimed all shipments were for irradiators and since "source" material shipped, such as Cobalt-60, was not fissile material.

accountability system, based on presumptions of honesty "might not reveal a deliberate and systematic attempt to divert material in this manner." Perhaps suddenly realizing the inexcusably flimsy nature of the accountability system he was describing, Brown attempted to backtrack, claiming "it was important to bear in mind that the presumption of honesty was not a mindless assumption. Specifically, the Atomic Energy Act provided severe criminal penalties for violation of accountability procedures. The deterrent value of these penalties had been considered fundamental to the entire system of domestic safeguards. Analogously, the international safeguards system relied upon formal sovereign guarantees of foreign governments."

Brown was optimistic that NUMEC apparently shipped relatively small amounts of HEU abroad—in contrast to others such as Westinghouse and General Electric. NUMEC also seemed mainly to ship "slightly enriched material." Commissioner Ramey emphasized that according to available information, most of NUMEC's export material had gone directly to France, Japan and Australia.

Myron Kratzer, who was intensely interested in promoting proper packaging for the growing volumes of international SNM shipments, broke in trying to outline how a diversion conspiracy would have to function. "In order for domestic industry successfully to divert material, substantial numbers of presumably loyal citizens would have to be deceived. Moreover, collusion between a U.S. and a foreign firm would also normally require collusion between the foreign plant management and the foreign government." Perhaps without realizing it, Kratzer was outlining precisely the situation at NUMEC. Shapiro had formed a clandestine corporate partnership with the Israeli Atomic Energy Commission, which was in charge of nuclear weapons development, though the AEC did not yet know the IAEC's full mandate. NUMEC's top HEU importer was France, which was secretly building Dimona, and could serve as Nicaragua had in the Haganah days as a transshipment point. Lowenthal was also connected to the upper tiers of Israeli intelligence, while Shapiro in constant contact with Israel's U.S. based covert operatives. The base of NUMEC's venture capital network were mostly dedicated Zionists. Many were active leaders in the Zionist Organization of America.

However none of this background was taken into consideration by the AEC. Under Brown's careful guidance, the AEC St. Valentine's Day meeting soon turned toward strategies for testifying before the JCAE. "The basic Commission position should be that AEC had no evidence or suspicion that diversion occurred; neither could the Commission say unequivocally that the material had not been diverted. Staff did, though, have a theory[xvii] to support its lack of suspicion. Specifically, staff had determined during its two surveys of NUMEC that the company had consistently underestimated its actual process losses. Additionally, the difference between actual and estimated losses appeared to have been passed on from completed jobs to new jobs. Thus losses attributable to

---

[xvii] Although AEC meeting notes clearly refer to processing losses as merely a theory, from this point on Commissioner Glenn T. Seaborg and Zalman Shapiro would claim it was actually the AEC's formal "finding."

the WANL contract probably included an accumulation of deferred losses over an eight-year period."

The AEC's "Peter robbing to pay Paul" theory sounded good at the time, even though it had no more empirical basis than a diversion theory. All pursuits of processing losses in NUMEC waste pits and filters had come up with MUF far below what Shapiro carried on his books. Shapiro's serial red-herrings and dishonesty should all have pointed to diversion. A nuclear arsenal was growing in Israel at the same time NUMEC's HEU inventory was dwindling. But processing loss was a theory that served the AEC's interests far more than diversion. But even as it publicly proclaimed processing losses, the AEC began to tackle how to safeguard SNM from diversion—rather than working to reduce industry processing losses—as the most urgent course of action precipitated by the NUMEC crisis.

In the St. Valentine meeting, the AEC went to great lengths to explain away NUMECs losses through the most favorable possible interpretation, including the spinning the final independent analysis of NUMEC filters. But in reality that independent survey only proved NUMEC had again misled the AEC by claiming the filters contained much of the MUF. Incredibly, lack of filter HEU recovery, in the opinion of AEC, now only meant that material must have eluded the filters ended up somewhere else.

Willful self-delusion spread. "For example, NUMEC reflected in its inventory estimates of approximately 31 grams of U-235 per filter. Gamma spectrometry of over 700 such filters, verified by chemical analysis of samples, supported only an average of about 12 grams per filter. NUMEC estimated that more than 50 kilograms of U-235 were contained in equipment and various combustible wastes which had been discarded in burial grounds. In connection with staff examination of the burial pits NUMEC incinerated and analyzed representative samples and concluded that only 5-6 kilograms would be recovered from these burial pits. Independent analysis by AEC confirmed this lower estimate. Additionally, the consistently high rate of return on scrap recovery contracts contributed to the theory that NUMEC did not take full account of losses as they occurred and compounded them through successive contracts." Brown became NUMEC's advocate stressing the high complexity and "extremely difficult" contract duties for WANL that NUMEC had performed, categorizing it as "indicative" of the type of high loss jobs NUMEC bid on. Shapiro would later adopt the same argument during testimony to Congressional investigators.

However, if the presumption of good faith were subtracted from the AEC's materials accountability analysis, the results provided more evidence for diversion than processing losses. The fact that NUMEC's filters did not have the higher estimated HEU lodged in them were as much a verification that SNM may never have passed through the filters at all, much less into Apollo's air and water. Final recovery of SNM from the walls and ducting of NUMEC's facilities would never yield all of the MUF. Shapiro's assurances to the AEC that his burial pits accounted for MUF managed to once again stall the AEC. But such delay tactics also pointed to diversion.

AEC Chairman Glenn Seaborg suggested a comparison be made between NUMEC's total losses and losses "typically experienced in AEC plants." Brown claimed that NUMEC's losses were approximately 1.2 percent, compared to AEC experience of process losses of less than 1 percent. "In other words, NUMEC's losses over the eight year period were high, but not exorbitantly high. The percentage loss under the WANL contract was of course substantially higher. It was, however, staff's theory that this contract had become the 'banker' for the other losses." In 2001 Department of Energy research would later accurately peg NUMEC's losses at 2%, almost twice the industry norm and one hundred times the losses the operation would experience after Shapiro and his management team finally left NUMEC.

The AEC grappled with how it would explain to the JCAE how it "'permitted' such a condition to exist at NUMEC.'" The best argument was that AEC had such a "plethora" of nuclear material and there was such a concerted institutional drive to accelerate the development of peaceful uses of nuclear energy in the U.S. and overseas. The Atoms for Peace program was a vital element of U.S. foreign policy, while countering nuclear proliferation had simply "not been a decisive consideration." Evidently, this compromised situation had been detected by Israeli intelligence and exploited by its covert operatives who could freely send scientists abroad to covertly penetrate secure facilities.

The AEC candidly admitted that its own accountability procedures for counter-proliferation were largely "ambivalent." While AEC shipped HEU under armed escort, it had not required the same level of security from its subcontractors that were also moving government-owned material. AEC admitted it had over-relied from the beginning on financial accountability as a sole deterrent to subcontractor materials loss. Such an honors-based system was as tempting to Israeli covert operatives as the cache of WWII surplus weapons two decades earlier.

When the topic of interviewing NUMEC employees in charge of material accountability surfaced, Brown was at first reluctant, noting that "a number of individuals had over the years performed this function at NUMEC. All but one or two had left the company's employ... Brown's personal belief was that the yield from such interviews would be low." Brown was right. The commissioners nevertheless approved NUMEC employee interviews, assuming that even marginal information could be "helpful." It would only be apparent much later just how deficient the AEC's hastily contrived survey and amateurish interviewers were, compared to the FBI's more exhaustive and meticulous debriefings.

The politically savvy AEC commissioner Glenn Seaborg chimed in again emphasizing "the desirability of stressing, to the JCAE, [AEC] staff's theory in support of the belief that no diversion had taken place at NUMEC." General Manager Hollingsworth seconded Seaborg's approach. This largely evidence-free decision offered the most protection to the AEC's future. It would remain Seaborg's unaltered public position for the rest of his life. Only in private would Seaborg pen an obvious question to his biographer, Benjamin Loeb "...am I trying too hard...to argue NUMEC's case?"[103] History reveals that indeed he was.

Brown wanted to cut to the chase by asking NUMEC for permission to examine its confidential financial records, proposing "such an examination would give staff some degree of additional confidence that diversion had not occurred." AEC staff immediately lodged their objections. Advancing the pretext that somehow the AEC wasn't the center of what was essentially a vast state-run industry, they felt "the impact on both NUMEC itself and the nuclear industry would, to say the least, be traumatic." But NUMEC was so tiny compared to industry cohorts and battling so hard to save its own neck that it was in no position to refuse. Brown received formal approval to contact Shapiro and ask if NUMEC would voluntarily make the requested financial records available.

Brown summarized AEC staff views and institutional consensus on the theory of non-diversion, which would make it "unnecessary to involve the FBI formally in the matter." Based on experience with NUMEC and huge flaws in materials accountability oversight, the AEC further committed to "study measures to tighten the system and make recommendations to the Commission." This was unavoidable since with "increasing numbers of reactors" under construction and the prospect of "private ownership" of special nuclear materials was just around the corner. AEC would have to at least appear to resolve the diversion threat.

General Manager Hollingsworth frankly noted that AEC had huge credibility problems. Its mandates were contradictory, having a direct financial interest in nuclear material, another role to promote public health and safety, and a responsibility to prevent the diversion threats to U.S. national security. "He believed it could be fairly argued that AEC's present system was not completely responsive to the latter interest." But like the others, Hollingsworth then circled back to turf protection, arguing in favor of not letting the AEC be "forced into hasty or ill-considered action on the basis of the NUMEC situation alone." The AEC would soon be forced to move away from its "sole" emphasis on contractor "financial responsibility" for inventory control. But AEC threw away its single best opportunity for survival by continuing to provide cover for NUMEC rather than credibly pursue diversion.

During the St. Valentine's Day meeting, AEC staffers strategized about which JCAE committee members would ask what questions. "On both the international and domestic fronts the Commission was vulnerable to the criticism that there was substantial disparity between the provisions of its accountability systems and procedures in practice of direct AEC operations." Before adjourning, AEC staff fretted over what do if there were a "premature leak of the NUMEC situation" that could "lead to sensational and probably inaccurate press reports." Commissioner Seaborg recommended preparing talking points in advance. Luckily, the JCAE had already imposed "strict limitations" on all communications about NUMEC. While future hearings about NUMEC would take place in closed session, the threat of publicity could not be ruled out. Any JCAE member could raise NUMEC in a public hearing and then the story could explode.[104] Seaborg was right to worry. Word about NUMEC losses would not be kept from the public indefinitely. When they finally broke, the AEC's carefully sealed Valentine to NUMEC would be clawed open by more skeptical and less captive investigators.

On February 17, 1966, AEC assistant manager for operations Howard Brown belatedly notified the AEC's FBI liaison of the formal findings of "sloppy management of nuclear materials by NUMEC, an AEC subcontractor." Brown reported that since April of 1965 the AEC had been making "extensive internal technical checks" at NUMEC which were also reported to the Joint Committee on Atomic Energy in Congress. The AEC told the JCAE that based on an absence of evidence, or suspected violation of the law, that AEC on February 14, 1966 "determined that inquiry by FBI was not then warranted." But when the JCAE dutifully prodded the AEC to find out what had actually happened to the HEU, rather than only protecting the U.S. government from financial loss, AEC had decided it should interview past and present NUMEC employees, the kind of investigatory work normally done by the FBI. Brown asked if the FBI would like to take the investigative lead. Unfortunately, the FBI chose the more attractive option that "under the circumstances, it is not felt we should do so."

An internal FBI memorandum to the head of domestic intelligence dated February 18, 1966 summarized the AEC's pitch at the time. "NUMEC received 1,012 kilograms of uranium-235 from AEC to process into fuel elements for nuclear reactors for space propulsion. This subcontract was completed 10/31/1964. In April, 1965, an AEC inventory indicated a loss, fixed by later AEC check in November, 1965 at 61 kilograms, valued at $764,000. Prior to this determination, AEC had many meetings and discussions with NUMEC officials and conducted extensive internal investigation. Relatively unproductive technical searches of NUMEC premises and waste disposal pits were also made. In addition, a survey of the plant's operations since 1957 revealed that NUMEC had had a total cumulative loss, on all AEC subcontracts, of 178 kilograms, all but 61 of which AEC considers properly accounted for by normal processing losses. While it cannot say unequivocally that theft or diversion of the 61 kilograms has not taken place, AEC believes that NUMEC consistently underestimated its processing losses and that the loss (61 kilograms) being charged to the latest subcontract actually represents an accumulation of losses over an 8-year period. However, because the NUMEC records system was not then so set up, it could not be determined when the various losses occurred or whether material provided for the latest subcontract was used, knowingly or inadvertently, to offset losses on other contracts." Brown then assured the FBI that at very least there would be no financial loss to the U.S. government.[105] xviii

On March 1, 1966, a memo from FBI Director John Edgar Hoover approved the FBI's hands-off approach of letting the AEC take the investigative lead through NUMEC employee interviews, and by extension treating the loss as a purely "an administrative" matter. [106] It was a move many regulators and law enforcement officials would later criticize as the AEC carefully structured a questionnaire designed to produce the very answers it most wanted to hear.

During March of 1966 the AEC interviewed 37 current and former employees of NUMEC, including executives such as Zalman Shapiro. Many of the interviews took place in a private office in the NUMEC administration building at Apollo, Pennsylvania. Former employees were interviewed either at home or their current places of business. Its mandate for a clear outcome predetermined, the AEC officially reported that no diversion took place at NUMEC. Documented internal deliberations and subsequent FBI interviews of the same employees reveal it was clearly not a bona fide discovery effort. The AEC also

---

xviii Also notes possible conflict of interest over NUMEC's hiring Lovett.

failed to act on highly disturbing findings of health hazards and NUMEC's treatment of its workforce.

In an April 6, 1966 report to Howard Brown, a three-member team drawn from the AEC Divisions of Inspection, Security and Materials Management summarized interview results. "All knowledgeable interviewees believed the losses were sustained during processing operations, due in part to (1) poorly constructed, inadequate or obsolete equipment, (2) discarding of dismantled piping and allied equipment containing enriched uranium (3) discarding of high level wastes to burial; (4) mixing of wash waters of various isotopic enrichments which may have resulted in mixtures uneconomical to recover..(5) to diluting the contents of liquids in waste storage tanks having relatively high U-235 concentration in order to meet maximum permissible discard limits, rather than determining the economic recoverability of such high level wastes prior to dilution. A few accidental spills were recounted but these did not appear to be greater or more significant than similar incidents in other comparable plants. Former employees attributed the low pay at NUMEC to high turnover and a less conscientious workforce more prone to waste."

Not a single interviewee claimed knowledge of "enriched uranium being stolen or otherwise diverted from the plant." Even employees who were "resentful" of their former employer—and there were plenty who were—could not identify any motive for wanting to steal such material, citing the health risk, criminal punishment and most importantly "lack of a market for such material." While "many persons felt that thefts of minor quantities of material were possible" none revealed any firsthand knowledge of such thefts.

Management was just as unequivocal, but also undercut the AEC's "robbing Peter to pay Paul" theory. "NUMEC management has stated that to the best of their knowledge there have been no instances of deliberate commingling of enriched uranium assigned to different contracts." The AEC survey team did not explain how this unhelpful admission would allow their "robbing Peter to pay Paul" theory to continue. NUMEC staff seemed almost adamant to kill off that theory, unaware that it formed the primary basis for the AEC's exculpatory testimony to the JCAE. "None of the interviewees knew of any instructions to commingle such material, nor were they aware of the transfer of enriched uranium from one specific contract to another."[107] Zalman Shapiro, testifying before Congress in 1978, would later recant these denials, instead heralding NUMEC as a commingling innovator that righteously forced the AEC to adopt more realistic material control protocols with contractors.

The AEC survey team, having no truly independent investigative mandate, effectively telegraphed what they did (and did not) want to hear through a formal opening statement read to all NUMEC personnel interviewed. "The AEC is in the process of determining to the extent that technical and other means permit, the exact disposition of certain losses of government-owned special nuclear material (enriched uranium) reported by NUMEC in the course of the company's operations over the past eight years. We want to make clear that NUMEC has taken steps to satisfy its financial responsibility to the government for this material. Moreover, we have no reason to believe from our inquiries to date that these losses are attributable to other than operational considerations.

Nevertheless, because of the nature of the material and the AEC's responsibilities in the area of national security, it is incumbent upon us to obtain all relevant information bearing on NUMEC's use of and control over this material. The information being collected is intended for internal government use. We have, however, informed the company of our interest and the company has offered to cooperate."[108]

The AEC also advised interviewees that they were not obligated to answer questions, and could have their own legal counsel present, and that while it was not a criminal investigation "certain questions asked may bear upon the compliance with applicable laws and AEC regulations." There is no record of any employee retaining legal counsel for the interviews since few could likely have afforded it. Exposed, lacking guidance, and with no economic incentive for rocking the boat, employee interviews differed radically from those later obtained by the FBI—when it was far too late to staunch diversions or prosecute the perpetrators.

The AEC's first interview question appeared to cut to the chase. "Do you know of any instances where enriched uranium was ever removed from the plant without proper authority and documentation?" [109] However the question missed the possibility that SNM was diverted under the cover of proper documentation and the watchful eye of Zalman Shapiro. None of the other seven questions were tailored to probe a clandestine diversion perpetrated by NUMEC's own management. The AEC's institutionalized "presumption of honesty" was endemic throughout questionnaire, which sought to uncover unforeseen waste, low-level worker removal of material, or samples accidently shipped overseas rather than a subtle ongoing diversion conducted by NUMEC's own top officials and mysterious outside visitors other employees could not identify.[xix]

[xix] Approved AEC questions for NUMEC employees were:

Do you know of any instances where enriched uranium was ever removed from the plant without proper authority and documentation?

Are you aware of any specific enriched uranium items or containers that were missing and could not be relocated? If so, what actions were taken?

On the basis of your experience and observations, where do you think the greatest enriched uranium losses occur at NUMEC?

Do you know of any enriched uranium losses (spills, leaky valves, or of enriched uranium accidentally discarded, e.g. down the drain) that were never recovered, reported or estimates made of the losses?

Have you ever brought any control matters to the attention of the management? Circumstances? Result?

Have you ever been instructed by company management or supervisors to use, or do you know of any use of enriched uranium from one specific contract to another (e.g. from WANL contract, Job 1231)?

What has been your experience with respect to company practice in assessing and reporting losses under the AEC lease agreement and under individual AEC contracts?

In your job, identify your role and company practices regarding material shipped off-site, foreign and domestic?

A close examination of individual responses to the AEC's interviews reveals the investigators should have been looking for the "means, motives and opportunities" outlined in Marvin Kratzer's St. Valentine's Day conspiracy theory since many employees verified that no viable market existed for lower-level employees to successfully fence nuclear material for cash. Some laborers, like management, also directly contradicted the prevailing AEC theory that the WANL contract was the "banker" for an eight-year cumulative loss while floating banal claims the AEC already knew to be false.

Robert H. Moore, NUMEC's foreman of material accountability, had been employed since the beginning of 1961 and maintained all of the company's material accountability documents. Moore claimed the largest portion of NUMECs losses actually were attributable to the WANL contract since it was recycled so many times, making a process loss of 1% per cycle multiplied by "four or five." He said that although the missing SNM "is referred to as being lost, much of this has collected on the ceiling and superstructure of the plant, the walls, in various parts of the plant." Investigative journalist Seymour Hersh would later use almost the same phrasing to exonerate NUMEC in his book *The Samson Option*. Moore then baldly made a claim the AEC had already debunked.

Although the AEC's own independent sampling of recoverable U-235 per filter was only 12 grams of material versus NUMEC's claimed 31, Moore claimed NUMEC was "in the process of reclaiming material from several hundred filters which have been stored for several years and he said he feels that quite a bit of U-235 will turn up in this way." Like the other employees, Moore added that "even if anyone would steal material, he is at a loss to understand what they would do with it because they certainly could not sell it domestically."

NUMEC's Manager of Analytical Chemistry James Scott also pointed out the lack of any wholesale market for stolen SNM. "There is no market for it except to persons actively engaged in the fuel material business and any such persons are able to get all the material they need legitimately. He added that he cannot conceive of any legitimate user of SNM running the risk of being involved in purchasing SNM in some underhanded way."

Floyd Joyner was the 34th employee ever hired by NUMEC but had been fired on August 5, 1964 after being told by he was "not doing his job." Joyner was bitter toward both Shapiro and NUMEC and told the interviewers that he had been made into a "fall guy" over the AEC's "criticism of NUMEC's unsatisfactory accountability function." Joyner criticized NUMEC's accountability as "poor" because NUMEC lacked management depth and had "too many hats for the few people." Interviewers happily noted that although Joyner was "obviously bitter" toward NUMEC "he states he believed the losses were all due to processing conditions and supported the loyalty of NUMEC employees, management, and officers." The investigators did not choose to question why NUMEC had such a thin management layer or obvious follow-on questions about its financial structure and venture capitalist David Lowenthal. However, being disgruntled did not necessarily make the witnesses more or less credible. Another disgruntled employee would finally reveal crushing details of SNM diversion 25 years later in a way that could directly imperil his own and other worker liability claims for radiation sickness.

The AEC had evidence it needed to validate the theory that no diversion occurred at NUMEC. No former or current NUMEC employee interviewed by the AEC admitted seeing or participating in diversion. FBI interviews later documented that NUMEC's corporate culture viewed the AEC as a regulator out to shut down the plant. No solid employee leads would surface until the 1980s when the FBI again interviewed many of the same NUMEC employees long after their fears of job loss, economic retribution and armed threats from NUMEC's pistol-toting loading dock guards had somewhat faded.

Howard Brown eagerly trumpeted the NUMEC interview findings to the executive director of the Joint Committee on Atomic Energy on May 18, 1966. He also sent a final copy of NUMEC's newly penned "Control of Enriched Uranium" report on shipping procedures and practices along with internal financial records related to foreign shipments. Brown concluded that "in view of NUMEC's relationships with foreign interests in the peaceful uses of atomic energy and our concurrent efforts to determine the disposition of special nuclear materials at NUMEC" were successful. The AEC's director of the division of security met with Shapiro on March 8, 1966 to make sure that AEC recommendations about security and materials management were being taken seriously. Brown certified that "these investigative actions have disclosed no evidence of theft or diversion" and that the company was taking satisfactory steps to improve procedures. Brown also noted that the AEC was responding to the JCAE's concerns by reviewing safeguards policies and procedures agency-wide, including sticky questions about how to verify overseas shipments. The security threat presented by NUMEC was being massaged into the framework of an industry-wide opportunity for improvement and a learning moment. But no U.S. regulator would ever resolve the overseas safeguards problem.

The AEC's next NUMEC materials survey was scheduled for September of 1966, but the organization eagerly considered doing it "sooner mainly to see whether they are taking action on recommendations." This enthusiasm waned when the AEC controller's office caught wind of a rumor that NUMEC wanted to transfer even more material, this time from a New York contract, into its Supply Agreement to cover WANL losses. On May 25, the AEC's Dick Yates questioned whether—given its heavy losses—NUMEC would even be allowed to bid on new contracts for the NERVA. These contracts started at a hefty 800 kilograms of NERVA fuel "plus 4 options taking up to 1,600 kilograms." Such agreements required written AEC headquarters verification that contractors could meet "financial" wherewithal tests for any materials losses. It wasn't clear after the WANL financial penalty that NUMEC had such financial resilience, so Keller quietly asked that his division be advised if NUMEC were a successful bidder.[110]

4. Zalman Shapiro investigated as an Israeli foreign agent

Despite the AEC's initially successful efforts to sell its processing loss theory, NUMEC's president was not yet out of the woods. The FBI interviewed Zalman Shapiro over his possible obligation to register as an Israeli foreign agent. This occurred after the chief of the research branch of the AEC's division of security notified the AEC's FBI liaison that Shapiro was negotiating with the Israeli government to set up a joint venture. Clem Palazzolo notified the FBI that NUMEC wished to export Cobalt 60 sources to Israel under an innovative citrus irradiation plan.[111] Another AEC inspection file from a July 2, 1964 audit alarmingly revealed that "NUMEC is a sales agent for the Government of Israel through its Minister of Defense, Division of Supplies" in New York City. The AEC inspection revealed NUMEC had yet another joint venture food irradiation program with Chemical and Phosphates Limited, which listed its address as "PO Box 1428, Haifa, Israel." [112] Under the 1938 Foreign Agents Registration Act, all agents active in the U.S. must openly declare their foreign principals and disclose financial, public relations and political activities every six months in a public office of the U.S. Department of Justice. Was Shapiro violating the law?

On June 15, 1966 two FBI special agents interviewed Zalman Shapiro. He disclosed that one year earlier he had formed a company called the Israel NUMEC Isotopes and Radiation Enterprises Limited (ISORAD) on a "partnership" basis with the Israeli government. Under the agreement NUMEC contributed a third of the equity value in ISORAD while receiving 50 per cent of the company's stock. ISORAD's eight-member board of directors was composed of four people nominated by NUMEC and four from Israel. Shapiro served as chairman of the board alongside steel magnate and American Jewish Joint Distribution Committee official Leon Falk, and President of the Pittsburgh re-development center Benjamin Rosen, a prominent Pittsburgh realtor. Phillip Powers, past president of the Western Pennsylvania Power Company served as assistant chairman of the ISORAD board.

Shapiro claimed the four Board members named by Israel included the head of the Israel Atomic Energy Commission, the head of the research branch of the Israeli Atomic Energy Commission, the head of the Israeli government's citrus board and an unnamed "prominent Israeli banker." The last ISORAD board meeting was held during October, 1965, attended by all the American board members. According to Shapiro, NUMEC and the Israeli government wanted to invest in "radiation research" to reduce spoilage in citrus fruits by eradicating the bacteria which causes rot and eliminating live larvae of the Mediterranean fruit fly before transportation. In addition to citrus, ISORAD was also engaged in research projects involving the irradiation of other agricultural products.xx

---

xx Shapiro would later change his story, claiming to Seymour Hersh that his numerous clandestine meetings with Israeli intelligence, government officials

Shapiro mentioned that just before forming ISORAD, he and NUMEC representatives discussed their plans with the AEC, including the security division, and a contact at the U.S. Embassy in Israel the FBI assumed to be the U.S. ambassador.

Shapiro's position in ISORAD made him an associate of Joseph Eyal, the Science Attaché of the Embassy of Israel in Washington and the purchasing Commission of the Israeli Government in New York City. Shapiro told the FBI that NUMEC had already sold "manufactured products" to the Government of Israel after securing a license from the Export Control Division of the U.S. Department of Commerce. Shapiro emphasized that he made every effort to comply with the laws of the United States in connection with this business venture. Neither he nor NUMEC were registered as Israeli foreign agents with the U.S. Department of Justice since NUMEC's association with the Israeli government was "a bona fide business venture." The FBI noted "In his opinion, he has never operated or served as an agent of the Israeli Government or any other government...His records of shipments with the Department of Commerce are open to inspection."

According to Shapiro, ISORAD was purely a "business venture for the sole purpose of making a profit for NUMEC." Shapiro said he had also attempted to form similar ventures with Spain and Germany without success. Shapiro told the FBI he would immediately review the 1938 Foreign Agent Registration Act statutes "for the purpose of determining if he or his company were violating the law in any manner whatsoever...He pointed out that he had never considered the possibility that he or his company were violating the law."[113]

Agents interviewing Shapiro felt he appeared to be "extremely cooperative and friendly to Bureau Agents." But at the time the FBI had no means to evaluate the substance or veracity of any of the information divulged by Shapiro, or juxtapose how large irradiator shipments through ISORAD aligned with MUF estimates from the AEC. There was much more the FBI did not know.

According to information published in 2010 by author Avner Cohen, the Israeli Atomic Energy Commission was the primary cover organization for Israel's clandestine nuclear weapons development program. Shapiro's business partner, Ernst David Bergmann, chaired the Israeli Atomic Energy Commission from 1954 to 1966. Early on Bergmann kept even the existence of the organization a state secret. Observations of Shapiro during his many visits to Israel noted high Israeli government officials exhibit deference to Shapiro far in excess of his status, indicating much more was going on in Israel than destroying the DNA of fruit and airborne pests in order to boost citrus exports.[114]

According to Cohen, the IAEC was created by David Ben-Gurion's secret executive order on June 13, 1952. It was established as a civilian agency only as an "external identity." In public it tried to mimic the U.S. Atomic Energy Commission's role by regulating the development of atomic research, energy and science. This allowed Israeli nuclear scientists to apply for and receive accreditation in international scientific bodies. But perhaps not everybody was

---

and top figures in Israel's nuclear weapons development program observed by the FBI all involved work on a secret project to protect Israel's water supply.

fooled. AEC Commissioner Glenn Seaborg refused to permit the organization to claim his official 1966 visit to Israel was hosted by the IAEC.

The Israeli Atomic Energy Commission's primary activity was entirely secret: laying the foundation for the development of Israel's nuclear arsenal. This secret IAEC nuclear research center, dubbed Machon 4 in the mid 1950's "was operated, funded and functioned as the research unit of the Ministry of Defense's Research and Planning Division (EMET)." By the mid 1960's as NUMEC entered into its partnership the IAEC was assuming overall command of both Dimona nuclear material production and RAFAEL (the Armaments Development Authority). NUMEC was indirectly partnered with the two principal pillars of Israel's nuclear weapons program.[115]

Shapiro's other prominent Israeli "business" partner, Joseph Eyal, served as attaché of the section of the Israeli embassy out of which operated the LAKAM (Bureau of Scientific Relations) an Israeli intelligence and covert operations agency that collected scientific and technical intelligence from abroad. LAKAM was formed in 1957 by Benyamin Blumberg to play key roles in stealing nuclear weapons components and enriched fuel for the Dimona nuclear reactor. This, like the IAEC's 'dual roles,' required the management of "deniable" front companies. Israel later claimed that LAKAM was closed down after Israeli spy Jonathan Pollard was captured in 1985 after stealing more than a million highly classified documents from the United States.

During the reign of NUMEC, ISORAD was perfectly "deniable" as an inexpensive and disposable front company that could generate a steady stream of overseas shipments with no questions asked. A December 18, 1968 FBI interview with a NUMEC informant revealed the small investment and short duration of the joint venture. According to the source "$25,000 was invested by NUMEC and $25,000 was invested by the State of Israel in order to organize ISORAD. NUMEC got back most of the $25,000 by selling to ISORAD a radiator developed by NUMEC. He added that ISORAD never accomplished anything beyond the experimental stage and experiments were conducted in Israel for the above stated purposes [inspection of the shelf life of oranges in Israel]..." After Atlantic Richfield acquired NUMEC in 1967, it simply gave NUMEC's ISORAD stock to the Israeli partners.

The insider noted the near impossibility of an FBI investigation since, "ISORAD had no personnel assigned to duty in the United States and no building sites in the United States. Certain Israeli persons would have titles in ISORAD and conducted the above stated experiments in Israel."[116] In hindsight, ISORAD accomplished a great deal as a credible cover for shipping sealed equipment with radioactive material hidden inside directly to Israel. However, conclusive eyewitness accounts of this diversion scheme wouldn't be documented by the FBI until the 1980s. In 1966 the FBI had no way of knowing what Shapiro's Israeli "business associates" were truly up to, or that LAKAM's key operative, master spy Rafael "Rafi' Eitan would be paying an undercover visit to NUMEC within two years.

After the FBI's friendly visit Assistant Attorney General of the Internal Security Division J. Walter Yeagley privately ruled on the Shapiro FARA matter.xxi On September 14, 1966 Yeagley advised the FBI that because NUMEC had acquired an export license from the Department of Commerce on its own behalf (rather than on that of the Israeli government) it did not appear that Shapiro or NUMEC were operating as foreign agents subject to FARA. This thin rationale seemed to be made without the benefit of FBI observations of Shapiro's movements, all of which strongly suggested foreign agency. Yeagley's failure to secure Shapiro's registration under FARA might have seemed at the time to be an honest mistake, given the cloak of secrecy surrounding the Israeli Atomic Energy Commission and LAKAM. Yet even when foreign agency was blatantly obvious, Yeagley usually failed to obtain compliance — most especially when the foreign principal resided in Israel.

In 1951 the FARA office ordered Isaiah "Si" Kenen to continue registering as a foreign agent after he resigned from his post at the Israeli Ministry of Foreign Affair's New York "Office of Information." Kenen, who had pledged to continue working as a representative of the Israeli government in America simply ignored the order. In 1962 Yeagley, under direct authority of Attorney General Robert F. Kennedy, ordered the American Zionist Council to register under FARA after it was discovered clandestinely receiving direction and millions of dollars to lobby the U.S. government on behalf of the Jewish Agency. The AZC fought the order, and ultimately obtained the Justice Department's permission to file only a nonstandard "summary" registration covering a mere three months of activity. Even that registration was kept secret until 2008. Kenen, who ran the AZC's unincorporated lobbying division and received Jewish Agency funding and guidance, reconstituted the AZC as the American Israel Public Affairs Committee (AIPAC) just six weeks after the AZC FARA order — and never registered the new corporation under FARA.[117] Yeagley and future leaders of the FARA unit would soon grow accustomed to being ignored or circumvented by such groups. Back at the AEC, things were not proceeding smoothly.

Even by July 29, 1966 NUMEC had still not paid its fine for the WANL losses, and was again insistently raising questions over whether its account could be credited for alleged equipment upgrades. Keller was somewhat sympathetic on the issue. "Personally I saw point of credit for upgrading since for every downgrading there was upgrading. We got this recognized in contracts after we took over, but couldn't do anything about early NYO contracts. Told him he was free to again raise question, but I doubted anything would be gained." Presumably, if the cash-strapped NUMEC could obtain some AEC credit vouchers for equipment upgrades, it could offset the looming fines. NUMEC was risking a great deal by raising the issue of equipment. It was later discovered trying to abscond with valuable equipment borrowed from national laboratories by reporting them as "contaminated."[118]

---

xxi The FARA division does not normally release records revealing its deliberative process over who should register. This has recently led to a lawsuit since in many cases decisions appear to be arbitrary and politicized.

On August 15, the AEC division of contracts temporarily placed a three-year hold on processing further NUMEC contract extensions. The AEC survey of NUMEC in 1966 was designed to be a comprehensive effort and was scheduled for the week of October 8. The AEC itself was being audited by Congress's own investigative division, the Government Accounting Office.[xxii] Keller worried that "GAO may be here week of 9/12 to go over work papers of previous NUMEC [surveys]."[xxiii] AEC officials arranged with Keller to take a first look at the NUMEC survey before the GAO arrived. But for the AEC, time ran out. The news media had finally gotten wind of NUMEC.

"U.S. Plant loses 100 kilos of U-235" screamed the headline. Few Americans felt immediate cause for alarm since *The New York Times* article bearing that disturbing headline was published in the Paris edition on September 18, 1966. The earlier U.S. headline, dateline Washington on September 17 by John W. Finney was much more tranquil. It dutifully reported talking points developed during the AEC St. Valentine's Day meeting. More sedately titled "Uranium Losses Spur Drive for Tighter U.S. Control of Fissionable Materials," it was the first establishment media reporting of the NUMEC SNM loss in America. The article downplayed most aspects of the crisis signaling diversion and easily discoverable details that might have given better insight into what was happening at Apollo. Indeed, it made no mention at all of NUMEC or Zalman Shapiro.

The strained message was surreally contradictory. Given the undeniable losses, the AEC would refocus on materials accountability in the name of avoiding diversion into nuclear weapons programs. But the *Times* reported that no diversion actually occurred. Missing material ended up in dust, drains and sea. "The Atomic Energy recently discovered that one of its industrial contractors had lost more than 100 kilograms of highly enriched uranium—enough to fabricate six atomic bombs. After an extensive investigation, it was concluded that the fissionable material had been lost in the normal process of fabricating reactor fuel rods and had not been diverted clandestinely to the manufacture of nuclear weapons. The discovery of the million dollar loss, however, so disturbed the commission that it has quietly begun an effort to strengthen what it now acknowledges are its inadequate controls to assure that some of the growing quantities of fissionable materials in the hands of domestic and foreign companies is not being diverted into the secret manufacture of atomic weapons."

The *New York Times* emphasized the lost 90% enriched U-235 was enough material to fabricate "half a dozen" bombs. "By the probably conservative estimate made by Arnold Kramish, a scientist for the RAND Corporation, 16 kilograms of enriched uranium (and six kilograms of plutonium) are required to manufacture an atomic bomb. The lost uranium would be sufficient, therefore, to make at least six small atomic bombs of the size dropped on Hiroshima in 1945." In the absence of any independent investigation of its own, the *Times* calmly assured readers that "The lost uranium, however, was not all together or even in

---

xxii Now the Government Accountability Office
xxiii Of three separate GAO reports on NUMEC, only one has been publicly released, in heavily redacted form at the insistence of the CIA.

substantial pieces. Rather it had been scattered to the winds. Some of it had been lost as scrap during the machine tooling and been swept up and buried; some of it disappeared as dust that was caught on filters, and some presumably had been washed down drains and carried away to the sea."

The AEC St. Valentine's Day caveat, that diversion couldn't definitively be ruled out—made no appearance in the *New York Times* exposé which credited JCAE Executive Director John T. Conway with disclosing the explosive story during a speech urging stricter domestic and international SNM control protocols. With the benefit of hindsight, the entire article seems deceptive. If all of the mysterious plant's losses could be attributed to process waste, why wasn't the AEC working to protect the environment by increasing recovery rates and reducing production line waste? Why wasn't environmental protection made a priority at Apollo where the plant resided in the midst of the population center? What about Parks Township where the NUMEC dump also threatened nearby residents? The *Times* struggled to cover diversion. "The other development is the potential spread of atomic weapons to other nations. As long as only small, research quantities of fissionable materials were involved, controls were needed more in principal than in fact. However, now that other nations are on the threshold of acquiring and producing large amounts of fissionable materials, there is a somewhat belated recognition that strict safeguards against diversion are a necessary corollary of atomic nonproliferation measures."

In breaking the NUMEC affair the *New York Times* clearly failed ask or answer the warranted questions. Who founded and ran the mysterious plant? Did they, by any chance, have a history of smuggling anything to foreign countries interested in building nuclear weapons? How did losses really compare with industry cohorts? Who were the mysterious plant's customers and business partners? Was it shipping much product overseas? What about following the money? On October 10, 1966 Seaborg noted that he had sent a letter to the editor "protesting" allegations about NUMEC after an editorial finally named the mysterious plant.[119]

Given the relatively small number of AEC nuclear contractors in Pennsylvania, the *Times* could have publicly identified NUMEC via process of elimination or a cursory review of its own business news archives—if it did not already know NUMEC's identity when it first published the story. A review of Shapiro and mezzanine-level NUMEC investors or consultation of *Times* back issues would have set off at least some alarm bells. When more relevant information finally began circulating, a wealth of more penetrating—and much less charitable— investigative reporting from second-tier publications would drive widespread popular demands for redress. But fortunately for Shapiro and NUMEC, most of that uncompromised, data-driven investigative reporting was still a decade away. But with Congress and its GAO in hot pursuit, the AEC had an incentive to bury NUMEC forever, inside another large AEC contractor.

On October 27, Keller met with Dean Crowther and Tom Stewart of the GAO and "discussed the background on nuclear materials management problems at NUMEC. GAO plans to issue statement of facts to JCAE with copies to AEC on 11/7, and have a close out meeting with NUMEC on November 10." Keller broke with the AEC party line in his candid assessment to the GAO. "I pointed out situation not improving and only way I thought it could be resolved quickly was to stop flows of material, but this would probably bankrupt company. GAO personnel noted same deficiencies in NUMEC inventory as we did. I pointed out that any action would have to be overall basis with full AEC HQS backing."[xxiv] The GAO was scheduled to meet with NUMEC on November 18 to "present a statement of facts on NUMEC."[120]

On November 14 Keller summed up the high costs AEC incurred testing NUMEC's waste for recovery potential at New Brunswick Laboratory. "Will take 160 man hours of overtime to get by that date (December 1). Earliest 12/15 without overtime. Take 40-50 hours overtime to do by 12-9. Estimate 10 man hours/sample. Have only 3 people that can do." Meanwhile, NUMEC was falling behind on its shipments of finished product to SNPO-C. Keller worked through the options with Dick Yates, reading through the relevant supply agreements. "Re return of material at lower assay than furnished. Should it be cut off when there is a U (uranium) balance rather than U-235 balance? Read him clauses of the Supply Agreement covering blending (Art. 5). Told him I thought one must look at contract provisions re blending. Generally I feel that the contractor should be held to a U Balance. In NUMEC case, however, we have a special situation..."[121]

Keller recognized the uranium shortage at NUMEC and proposed two options. "Let NUMEC ship back against 1231 (contract) based on u-235 originally furnished with charges for degradation (cut off 11/22/66 per Yates). This recognizes fact that 1231 was dumping ground for all services." But was it? No clear evidence for this theory was ever publicly documented. The AEC's own survey revealed a string of employees all claiming that there had been no mixing between contracts. The other alternative explored by Keller "is to base on U-235 balance with charges for degradation and loss. This would complicate handling off {excess] i.e. what to assign to it." The financial impact to the government, rather than diversion, seemed to be the AEC's top concern. "The first of above, while perhaps arguable by lawyers, is probably best operationally and would get AEC the most $$."

The AEC approved overtime on NUMEC filter sampling toward the goal of the GAO "report getting out." Meanwhile, NUMEC continued to shuffle uranium between contracts in order to meet tight production deadlines. On November 17, Lovett sought permission to transfer 16.6 kilograms of enriched uranium held in a station account to use in work for Argonne National Laboratory and Union Carbide jobs. Keller obliged his former co-worker and "told him I saw no reason why we couldn't transfer and to submit a transfer document."

On December 7, the AEC's John Vinciguerra spread bad news that the JCAE requested billing for use charges on the uranium held at NUMEC. This would drastically increase the amount of capital NUMEC needed to continue its shoestring operations. Congressional overseers also wanted NUMEC to pay up immediately on its WANL fine. Two days later Dean Crowther presented GAO's

---

[xxiv] The FBI would later make the very same recommendation to cut off material flows to NUMEC. After the recommendation was ignored, the FBI largely shut down its investigation of NUMEC and Zalman Shapiro between 1969 and 1976.

final statement of facts on NUMEC. After reading through them, Keller admitted he had "no quarrel with facts." The AEC then prodded NUMEC to complete the survey of materials unaccounted for and pay the financial penalty for losses on the WANL contract.[122]

5. CIA demands a second FBI investigation with surveillance

In 1967 the AEC established its "Office of Safeguards and Materials Management" and a new "Division of Nuclear Safeguards." AEC did this "with a view to preventing future losses such as occurred at NUMEC."[123] Meanwhile covert U.S. environmental sampling around Israel's Dimona nuclear weapons facility detected traces of HEU bearing the same Portsmouth, Ohio signature as the U.S. naval reactor fuel NUMEC processed at Apollo.[124] The possibility that Israel itself had enriched the uranium was discounted at the time by the AEC. CIA Director Richard Helms wrote a classified letter to Attorney General Ramsey Clark that HEU "processed at Apollo might have ended up at Dimona" and requested that the FBI investigate NUMEC. Helms also informed President Lyndon Johnson about Israel's nuclear weapons program, to which LBJ famously responded, "Don't tell anyone else, even [Secretary of State] Dean Rusk and [Defense Secretary] Robert McNamara."[125]

In May of 1968 the FBI advised Attorney General Ramsey Clark that it was now initiating a wide-ranging but "discreet investigation" of Zalman Shapiro. Because of the sensitive nature of the investigation, argued the FBI "No grand jury subpoena is being used in this instance."[126] Investigating agents advised "offices who have not had previous communications on this matter" that "during the period from 1957 to 1966, NUMEC reported losses of approximately 572 pounds of U-235, the fissionable ingredient of uranium which can be utilized in military explosives. Further, during the period 8/1958 to 10/1965 NUMEC shipped approximately 935 pounds of U-235 overseas to various sections of the world under some 28 different contracts."[127]

Although NUMEC was located in Apollo, Shapiro lived with his wife Evelyn and three children on the same leafy lane in Pittsburgh — Bartlett Street — as venture capitalist David Lowenthal. FBI special agents in charge of surveillance wanted to set up a fixed observation point in Shapiro's neighborhood, but it was a strictly a market for homeowners. The FBI noted it was "not possible at the present time because there are no suitable rental faculties in the immediate area..."[128]

Portions of the FBI files deal frankly with Shapiro and the Israeli nuclear program. The Bureau classified all of its surveillance files at the "Secret" level on the basis that disclosure could be "detrimental to the national security interests of the United States." The FBI's "discreet and carefully planned" observations recorded Shapiro typically leaving his residence between 7:40 and 8:11 AM in a 1967 four-door dark blue Buick Electra or a bronze 1967 Ford sedan. "He travels to his place of employment in Apollo, PA, to date, nonstop using the same route. He is known to generally travel at very high rate of speed."[129]

Further complicating the FBI's surveillance of Shapiro was the ever-widening range of his travels. The fines for the uranium loss and increasing Congressional scrutiny meant that the company could no longer operate without a major capital infusion. Shapiro and Lowenthal agreed to sell NUMEC to a large petroleum corporation — Atlantic Richfield Company — in April 1967. Atlantic Richfield was the deep-pocketed parent company that Apollo Industries could never be. With annual sales of $1.6 billion and a net income of $130 million, it referred to its acquisition of NUMEC in a small footnote of its 1967 Annual Report as a "minor subsidiary" requiring the issuance of only 84,500 Atlantic Richfield shares to purchase.[130] Shapiro had responsibilities at an Atlantic Richfield office in Philadelphia as well as the NUMEC administrative office in Apollo. There most of Shapiro's corporate administrative duties were carried out from "a large old home converted into office space" on Second Avenue in Apollo, four blocks from the main plant.[131]

Agents from FBI field offices began tailing Shapiro as he visited Atlantic Richfield offices in New York and Pennsylvania, but treated him with extreme caution. Due to the "sensitivity of this surveillance"[132] agents frequently broke away to avoid detection. In one instance, agents reported that Shapiro "walked to 500 Fifth Avenue NYC, and took an elevator which services floors 38 through 58. It is noted that 500 Fifth Avenue is a very large office building. Surveillance was discontinued, inasmuch as the subject pushed his way onto a crowded elevator, and there was no room for another person..."[133]

On September 6, 1968 Clem Palazzolo, the AEC Security Chief who had first notified the FBI of Shapiro's suspicious joint ventures with the Israeli government in 1966, forwarded a suspicious permission request from NUMEC to the AEC. Four Israelis were seeking permission from the AEC's New York City operations division for a September 10 visit to NUMEC.

NUMEC's formal written requests from NUMEC's manager of security listed Avraham Hermoni, Scientific Counselor from the Israeli embassy, Dr. Ephraim Biegun, from the Department of Electronics, Avraham Bendor, Department of Electronics, and Raphael Eitan, Chemist of the Ministry of Defense. The reason given for the official visit was to discuss "thermoelectric devices" which the letter carefully noted was an "unclassified" subject area. The only NUMEC personnel with whom the four Israelis initially sought an audience was Zalman Shapiro, also the NUMEC executive officially sponsoring the visit. Barry Walsh of the AEC Security Division issued a formal approval letter on September 20, 1968 ten days after the visit took placed.

On September 27, NUMEC's security manager Bruce Rice advised Palazzolo after the visit had taken place[134] that two other NUMEC employees from the "Energy Conversion Department" also met with the Israeli "thermoelectric generator specialists."[135] Shapiro was flying high. Just days before meeting with the Israeli covert operations team, Shapiro announced a $5 million NUMEC facilities expansion for administrative and fabrication infrastructure arm-in-arm with Atlantic Richfield's president. NUMEC was also announcing a joint venture with the AEC and National Heart Institute to produce a nuclear-powered cardiac pacemaker.[136]

NUMEC's security director assured the AEC that only unclassified topics were discussed during the Israeli visit. "Discussion with the Israeli nationals concerned the possibility of developing plutonium fueled thermo-electric generator systems in the 5 to 50 milliwatt power level. Specifically, they were interested in 10 generators in the 5 milliwatt range. Each of which would be fueled with about 2 grams of plutonium. The 50 milliwatt generator is considered a remote possibility, but would use approximately 20 grams of plutonium. The generators are of the terrestrial type."

NUMEC's report reassured the AEC security office that "We are proceeding to make a proposal to these gentlemen for this work using, of course, only unclassified information which is already in the public domain. It is also our understanding that these same gentlemen have visited several of the major nuclear organizations in the United States to develop proposals from them on these items."[137] CIA Tel Aviv Station Chief John Hadden would later repeat NUMEC's characterization of the Israeli covert operation team as "gentlemen" during a revealing BBC interview about nuclear diversion.

Neither the AEC nor FBI had any idea at the time the true identities of the Israeli visitors. Shapiro was meeting with Israel's elite nuclear weapons development officials and top covert operative, none of whom had any expertise whatsoever in "thermo electric devices." Ephraim Biegun was head of the Israeli technical department of Israel's Secret Services from 1960-1970. Avraham Hermoni was technical director of Israel's nuclear bomb project at RAFAEL and had a leadership role in the highly decentralized Dimona project. Bendor was a long-time Shin Bet operative and Eitan's "right-hand man" on overseas operations. Rafael Eitan was a long-time Mossad and LAKAM operative who later directed Jonathan Pollard's massive espionage program against the United States. In June of 1986 Middle East analyst Anthony Cordesman assessed the significance of Eitan's presence at NUMEC. "There is no conceivable reason for Eitan to have gone but for the nuclear material."[138]

In the early 1990s when author Avner Cohen was writing his groundbreaking book *Israel and the Bomb,* Avraham Hermoni confided details about some key decision points of the Israeli nuclear weapons program on the strict condition that they never be detailed back to him. Only in 2010 did Cohen finally reveal how Hermoni sought guidance from Deputy Minister of Defense Shimon Peres[xxv] over what types of weapons Israel should actually assemble. "During the early to mid-1960's Hermoni was a technical director in RAFAEL, one of the three individuals serving as the technological eyes and ears for RAFAEL's boss, Munya Mardor. Hermoni's area of responsibility was overseeing RAFAEL's role in the nuclear project, which was under the overall responsibility of Shimon Peres, then deputy minister of defense....Sometime around 1964/1965...he wrote a memo to Shimon Peres, the project's overall chief executive, asking him for guidance. Specifically, Hermoni listed three technological options, with each describing a particular technical product that the project could work toward. Hermoni remembered that memo vividly because he considered it one of the most important he had ever written. Although Hermoni refused in 1992 to be too specific as to what those options were, he left me with the understanding that they ranged from a crude nuclear explosive device to a fully deliverable weapons

---

xxv According to secret apartheid-era documents released to author Sasha Polakow-Suransky, Peres offered to sell nuclear-tipped Jericho missiles to South Africa.

system (a bomb). His question to Peres was, in essence, How far should Israel go with its nuclear option? What should the developers ultimately aim for?"[139] Although Peres did not provide specific guidance "The project leaders knew how to proceed, and their superiors had no problems with the path they chose." [140]

The Washington Field Office tried to conduct "discreet" surveillance during Shapiro's attendance of the 1968 Atomic Industrial Forum running concurrent to the American Nuclear Society and Atom Fair in cooperation with the Atomic Energy Commission. But full-time observations of both the Shoreham and Sheraton Park hotels during the event "failed to spot the subject." The Washington Field Office reported its intention to "conduct spot checks in the above hotels during the duration of the above sessions and maintain contact with established sources in an effort to determine whether the subject is engaged in any unusual activity…"[141]

October 6, 1968 a confidential FBI source disclosed that Shapiro planned to visit Israel "during the latter part of November, 1968 for a period of approximately two weeks." The source had no idea why Shapiro was visiting but was of the opinion that he "owned property in Israel" based on a comment made by Shapiro's sister Zipporah Schefrin. Like Zalman, Zipporah was a dedicated Zionist, working as a volunteer at the Jewish Museum in New York City. She was a member of Hadassah (the Women's Zionist Organization) and chairman of the women's division of the United Jewish Appeal, as well as a board member of the Jewish Family Service and a case worker for the National Council of Jewish Women.[142]

Zipporah's husband Dr. Alex Schefrin was also planning to leave New York for Israel on October 10, 1968 accompanied by 63 others.[143] During this sensitive period, the Israeli government officials and the White House were battling over whether a proposed U.S. sale of Phantom fighter jets should be linked to Israel's entry into the Nuclear Non-Proliferation Treaty and be subject to verification inspections of its Dimona nuclear material facility.[144] A fourth FBI source disagreed, saying that Shapiro was not going to travel to Israel, implying that it was unnecessary. "This source stated that David Lowenthal, President of Raychord Corporation, Apollo, PA, and a long-time personal friend of Shapiro, made two recent trips to Israel, but that Lowenthal has not discussed the purpose of these visits to Israel."[145]

Shapiro spoke again with Hermoni and Beigun while visiting Washington, D.C. on September 30, chatting tantalizingly over some urgent matter that "although there were problems, both were anxious to move ahead." Shapiro explained to the Israelis that "they must first determine if the materials are available for exportation." Shapiro then set off alarm bells when he sent an October 8 letter to AEC Chairman Glenn Seaborg seeking clarification of the parameters for obtaining and exporting plutonium. Rather than mention that his inquiry was based on Israeli demands for U.S. plutonium, Shapiro ambiguously cited "domestic and foreign applications" and NUMEC's desire to "do development work on their own" as the impetus for his inquiry.[146] Shapiro did not mention that Dimona's nuclear weapons makers might have wanted to compare the quality of Israel's newly initiated Dimona plutonium output with America's.

Seaborg's November 20 response opened with a breezy and familiar "Dear Zal." Seaborg made it clear that "the 500 grams of plutonium 238 made available on April 1968 may be used for foreign applications as well as domestic applications. Overseas distribution within the free world is made on a non-discriminatory basis..." Seaborg clarified that there would be no restrictions on the amount of plutonium 238 for a single thermoelectric generator, provided it was below "criticality." After responding to all of Shapiro's points, Seaborg ended with a friendly "I hope this information has been helpful to you and that I have answered all your questions concerning the use of plutonium 238 in milliwatt and microwatt thermoelectric generators. I would be happy to have you discuss these matters further with Commission personnel if you so desire." Seaborg then named AEC personnel who would "be glad to arrange to meet with you."[147]

On November 22, 1968, the FBI uncovered grave mistrust of NUMEC within lower levels of the AEC during an interview with a source at Lawrence Radiation Laboratory. The source revealed how in 1965 NUMEC was handling U.S. government-owned plutonium. About three years earlier the interviewee "had partial responsibility for a project in the Special Materials group of the laboratory involving a research and development contract with subject's firm...for processing of nuclear material...the Special Materials Group, handled details of the contract; and...only met the subject briefly when the subject was at the Laboratory on business."

The interviewee "had no knowledge of any connections between the subject with the Israeli government. The subject and his firm had been a source of concern...while the contract was in existence. Under the contract, the Laboratory furnished Special Nuclear Material to NUMEC for processing. Accountability for such material is strict because of its cost and toxicity poisonous. Past experiences had established average percentages of loss of material during process of the type done by NUMEC and the loss by the firm seemed excessive..."[148]

The FBI's source at Lawrence Livermore also testified to Shapiro's approach to the government. "The equipment used by NUMEC was government-owned and furnished by the Laboratory. Upon completion of the contract, the firm tried to persuade the Laboratory to leave the equipment where it was on the grounds that it would be cheaper to leave it than return it to Livermore and decontaminate it. In support of this argument, the firm furnished lists of equipment it claimed had been contaminated. Subsequent inspections...showed that some of these items had never been used at all." This testimony led the FBI source to "suspect NUMEC of trying to get the equipment free and created further distrust of the firm."[149] An assistant to the Deputy Director testified in 1976 that the value of the equipment NUMEC tried to obtain for free was $130,000.[150] In October 17, 1968 the FBI recorded that it was continuing to "follow and report subject's activities" in Pittsburgh, Pennsylvania.[151] But the FBI's surveillance capabilities stopped at the U.S. shoreline, and Shapiro was again headed overseas.

Zalman Shapiro's trip to Israel in the fall of 1968 was either a restful vacation, or a hectic series of high level business and pressured government meetings — depending on who asked him about it. On November 29, 1968 Shapiro boarded an El Al flight for Tel Aviv from JFK airport in New York, returning December

12, 1968. An FBI source knowledgeable about Shapiro's long time friend and business associate Joseph Swartz claimed Shapiro "informed Swartz that his trip to Israel was satisfactory but very hectic and that he obtained no rest during his visit." Shapiro claimed he was conducting NUMEC and ISORAD business at the "same time."[152]

Shapiro then surprised Swartz with the news that he was seriously considering whether or not to settle in Israel within the next half decade. Though already offered a professorship, Shapiro was unsure what type of work he would pursue if he moved. The "Israeli government is presently very interested in converting their technology in the nuclear field into commercial enterprises" but Shapiro would first have to raise a million dollars to invest. [153] Shapiro told NUMEC vice president Oscar Gray that the Israeli government had pitched several business ventures including laboratories of the Pittsburgh Testing Laboratory and Battelle laboratory type.[xxvi] For these military-related ventures, the Israeli government was willing to guarantee a $2.5 million investment if Shapiro could raise half a million dollars to invest. Shapiro did not think it would be easy, but signaled he was "going to attempt to raise the money."[154] Shapiro believed that he could secure contracts from both the industrial and military branches of the Israeli government "to insure their use of the laboratory." According to Shapiro, "anyone with a little money and desire could accomplish plenty in Israel at this time."

Shapiro closely followed the proposed sales of American-made Phantom F-4 fighter jets, confidently boasting that "Israel is capable to actually build their own fighter planes as far as airframes and electronics are concerned, but that they lack the knowledge for designing the planes, which necessitates new technologies."[xxvii] Israel was "anxious to convert their technologies in nuclear centers to commercial business" but Shapiro said he was not interested in devoting all of his time to this work. During a 1978 Congressional interview Shapiro would deny knowing anything more than what was published in newspapers about which "nuclear centers" in Israel were so robust as be able to feed "commercial business."

Shapiro was also interested in a new Israeli technique for drilling holes in diamonds with the use of a laser, an application with "numerous commercial possibilities."[155] He claimed ISORAD was trying to procure an irradiator and that he brought back "a couple of bids to review and to determine if the basis for their analysis is realistic."[156] But Shapiro was much more guarded and dismissive in his conversations with another NUMEC employee who inquired about his visit, characterizing it as a "two week vacation" that gave him a chance to "catch up on some overdue rest."[157]

[xxvi] Batelle was an AEC contractor instrumental in developing new technologies with commercial applications. Batelle was also involved in the Manhattan project and DoD biological and chemical weapons contracts.

[xxvii] Israel soon overcame this problem by stealing French Mirage jet fighter plans from a Swiss contractor, aided by an insider Swiss engineer named Albert Fraunknecht.

By late 1968 NUMEC had already sent numerous irradiators to Israel. NUMEC's smaller three foot high versions of this equipment allowed the neutron bombardment of any specimen of animal, plant or seed that could fit into the 12.5 liter "irradiation tank" of the device. NUMEC's prices for differing configurations depended on the amount of plutonium that would fuel the Neutron-Pac. The NUMEC-AA fueled with 16 grams cost $665, while the 80 gram NUMEC-H was $1,065. In its brochure, NUMEC carefully noted that "Prices do not include the contained plutonium which must be leased from the USAEC by domestic customers and purchased from the USAEC by non-domestic customers.[158]

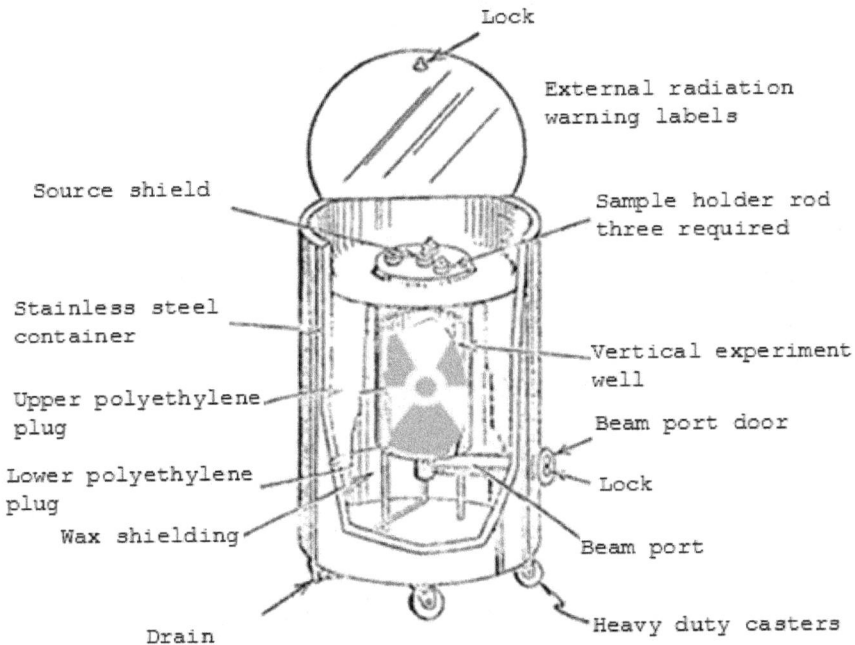

Figure 9 Sketch of NUMEC's Neutron-Pac® irradiator.

In October of 1968 another FBI source revealed that "about the same period of time that NUMEC sustained the unaccountable losses of U-235, subject [Shapiro] was involved in the development and manufacture of food irradiators for Israel. Source advised that at least one large irradiator was manufactured, and a number of small units called 'howitzers' were manufactured and sent to Israel. Source was of the opinion that had U-235 or any other nuclear material been available for shipment to Israel, it would have been a simple matter of placing large quantities of the material in these food irradiator units and shipping to Israel with no questions asked."[159]

On December 21, 1968 the FBI recorded that Shapiro's wife Evelyn had asked Joseph Swartz's wife Reva to act as a courier to deliver a package to an Israel Atomic Energy Agency contact during an upcoming visit to Israel. After observing this social network, on January 20, 1969 the scope of the FBI's investigation expanded from its original mandate of revealing "the nature and extent of Zalman Shapiro's relationship with the Government of Israel" to also investigating other persons "connected with or in addition to, or as a result of, his association with Israeli officials and sympathizers in the United States."[160] The FBI concluded that Shapiro "has stated that he seeks new employees for NUMEC, who in addition to their technical skills, may have contacts in the AEC or other government agencies. Source has indicated subject does this type of thing many times in a cover or background manner which reflects a character trait of ruthlessness..."[161]

1964	1965	1966	1967
Boron plant contract	Food irradiators R&D	Food irradiation	Medical electric equipment
Fish irradiators	Hawaiian fruit	Heart pacemaker	survey
Hanford isotopes	irradiator	Scrap recovery	Bids for Hanford
plant	Indian Point	contracts	contract
Hexaflouride	Israel, cobalt-60	SNAP generators	Irradiated food
Israel project	shipment to	Uranium	Survey
Sefor fuel contract	plutonium fuel	contract, Sohio	
Shippingport fuel contract	SNAP-25		

Figure 10 NUMEC claimed projects 1964-1967 to *Nuclear Industry Magazine*.

On February 14, 1969 the FBI sent Secretary of State William P. Rogers, the attorney general and National Security Advisor Henry Kissinger a top secret report about the suspected diversions of HEU from NUMEC to Israel. It informed the leaders that "in view of the sensitive positions of our sources in this matter, it is requested that the information contained in the attachment be handled on a need-to-know basis."[162] The contents of this communication have never been fully declassified, but it generated a February 28 recommendation that two special agents of the Pittsburgh FBI field office be given incentive awards for their "discreet investigation" resulting in "ascertaining his [Shapiro's] nefarious activities" documented in the 56 page attorney general report.[163] On March 5, 1969 the special agent in charge of the New Haven, CT office also forwarded damning surveillance photographs of Zalman Shapiro and promised more of "companions" that "will be forwarded when developed."[164]

On February 18, 1969 the FBI sent the summary report of its Zalman Shapiro investigation which was shared with the Assistant Attorney General Walter Yeagley, still in charge of foreign agent registration through the internal security division, and Director of the Division of Security at the Atomic Energy Commission William Riley. The FBI documented Shapiro's trip to Israel and that although the Atomic Energy Agency had formally disapproved, NUMEC was once again employing an Israeli nuclear scientist. "Subject, [claimed] acting on an unofficial approval of AEC inspectors, hired an Israeli nuclear scientist to work at NUMEC for one year. Official AEC position was one of disapproval. AEC has advised the aforementioned scientist was Baruch Cinai....Subject later sought AEC permission to have other foreign scientists to work in NUMEC laboratories. AEC stated that they disapprove of this procedure..."[165] Cinai's mysterious presence at NUMEC in the early and late 1960s before large documented losses were not further explored by the FBI.

By early 1969, the incoming Nixon administration clearly saw the NUMEC situation as a matter of possible ongoing espionage, rather than just nuclear diversion. AEC commissioner Glenn T. Seaborg was a long-time Nixon associate and fellow Californian who first met the future president in 1948. Nixon invited Seaborg to "stick together" with him and even named Seaborg to the press as part of his "brain trust" in order to boost his scientific bona fides during the 1960 presidential campaign. This move later caused trouble for Seaborg's nomination to head the AEC under JFK as he struggled to prove his Democratic Party bona fides. The record reveals Seaborg endlessly lobbied the administration that it was counterproductive to pursue the diversion angle.[166] The entire upper echelon of the Nixon administration would soon become embroiled in the politics of what to do about Shapiro as he deftly switched employers to work at a nuclear weapons builder and while seeking the highest level U.S. security clearance.

This successful derailment of the diversion focus was unfortunate. According to Victor Gilinsky and Roger Mattson "In June 1969, Shapiro had a rushed meeting at Pittsburgh International Airport with another Israeli scientific attaché, Jeruham Kafkafi, which Shapiro again had difficulty explaining. Interviewed by AEC security two months later, Shapiro, surprised that the AEC knew about the meeting, first claimed it was about an overdue invoice. When the AEC interviewers did not buy his story, he called back to say Kafkafi asked about an individual at Oak Ridge National Laboratory, whose name he could not remember. (A possible explanation: Because the meeting came right before what was to be the last U.S. inspection of Dimona, Kafkafi might have sought information about U.S. team members who were assigned to the inspection.) Despite such dissembling, Shapiro did not lose Seaborg's backing or that of the AEC commissioners."[167] The AEC and Justice Department explored the rushed meeting and Shapiro's evasive responses in great depth. But lacking context for why Shapiro's dissembling was so important, the AEC later simply let it go. "In its report on the Shapiro interview to JCAE Chairman Chet Holifield, the AEC admitted that Shapiro was 'less than completely candid' in discussing his relationship with Israeli officials, but that the [Atomic Energy] commission 'does not contemplate further action in this matter at this time.'"[168]

As the FBI continued to trail Shapiro on visits to the National Institutes of Health where he was pitching a nuclear-fueled cardiac pacemaker, it also unearthed more information about his financial affairs. "Subject's mother-in-law, who died in approximately September, 1968, left estate valued at $150,000."[169] "Subject would be receiving a salary at NUMEC as of April 1, 1969 of $50,000, which would be an increase of $5,000 over his 1968 salary."[170] But the FBI apparently missed the huge payoff Shapiro was entitled to receive for NUMEC's sale to Atlantic Richfield. Because of this, Shapiro's move to Kawecki-Berylco appeared from the outside to be an economically-driven career move rather than an attempt to gain access to advanced hydrogen bomb designs which would be at least as valuable to Israel as Shapiro's access to SNM at NUMEC.

At the very same time the FBI was tracking him, Shapiro was tracking the careers of allies at the AEC. When told that a high-ranking official was resigning from NUMEC, an informant claimed "this disclosure obviously disturbed the subject, who replied that if such were the case, they would be without a friend in AEC."[171] Although the FBI transcript censored which AEC employee was allegedly resigning, the fact that the transcription indicates difficulty spelling the name, with "PH" in brackets, the letter count of the censored name, and the fact that that the Commissioner resigned in 1971, reveals that NUMEC's "friend" was probably none other than Glenn T. Seaborg.

Seaborg refused to cooperate with subsequent FBI interviews investigating NUMEC in the late 1970s. Seaborg had long been tracked by another organization, the Weizmann Institute and Abraham Feinberg, Israel's nuclear weapons funding coordinator. The organization had brashly solicited an AEC "research" grant from Seaborg in early 1967, which the Commissioner passed down to staff for further consideration.[172]

The FBI documented NUMEC's ongoing head-hunting for government officials who had useful connections. NUMEC's General Counsel Jack R. Newman was former counsel for the Joint Congressional Committee on Atomic Energy giving NUMEC unprecedented access to a key regulator of the AEC. In 1968 Irwin Becker had joined up with NUMEC after three years as a NASA planning engineer. NUMEC's most controversial hire was James Lovett, formerly Acting Chief of the Operations Branch, AEC Division of Materials Management before hiring on with NUMEC at the peak of the 1965 inventory crisis.

Shapiro considered canceling his plans to meet with the Israeli Government Investment Authority in New York City on February 19, 1969. On the 21st an Israeli official at the Embassy in Washington contacted Shapiro to say that he was sorry about "the cables back and forth regarding tests, and in the future he would come to the United States and have Shapiro's firm conduct the tests and send drawings and specifications for fuel elements needed in Israel."[173]

Shapiro was clearly worried about enhanced AEC regulatory oversight. On April 14, 1969 he discussed with another NUMEC executive an upcoming AEC nuclear safeguards conference scheduled in New York City. Both "expressed concern over the possibility that new safeguards and vigorous inspections by AEC might be adopted in addition to international inspections."[174]

Shapiro was—as always—closely following politics in Israel, advising a confidant that Golda Meir was a "shoe-in" for premier, while Defense Minister Moshe Dayan was "biding his time." Shapiro read with "trepidation" about Al Fayah bombings in Jerusalem and military actions along the Suez Canal.

Within a very short time, Atlantic Richfield ownership of NUMEC would mean that Shapiro and his Israeli visitors would no longer have completely unfettered access to HEU and classified documents at NUMEC. Israel was still awaiting further shipments of NUMEC irradiators, even as a new shipping problem arose. The AEC was no longer allowing the casual shipment of such equipment in wooden crates, and instead specified special metal shipping casks. Shapiro fretted this would raise the total cost of these items. On March 30 Shapiro admitted that he was having "difficulty getting an irradiator to Israel." Even with AEC and Department of Transportation permission, it would be impossible to get it to Israel within the next two months."[175]

Shapiro's elevated concern about getting the irradiator to Israel just as soon as possible was as suspicious as Israeli visitors suddenly touching down to briefly meet with Shapiro in motels. Shapiro claimed the irradiator would be used for "preservation of potatoes and onions, and the Israeli crop of those vegetables is due in May". In discussions with an unknown male, he asked for Avraham Hermoni's help on the matter, and proposed a meeting on April 1, 1969. Under observation by the FBI, Shapiro told the unknown male arranging the meet he had "some paper on the water supply and would bring that information with him."[176] But all of the talk about water may have been a cover story. Whatever was going on was truly urgent. On April 6, 1969 two Israelis from Sha'ar Negev suddenly arrived in Pittsburgh and arranged to meet at a Holiday Inn the very next day. Special Agents observed Shapiro leaving the hotel in the company of two stocky males carrying briefcases, bound for a conference in Shapiro's office. The FBI noted no further details such as their names, subject of the meeting, or why the presence of Hermoni was needed.

On April 12 Shapiro held another crash meeting at his Pittsburgh residence to discuss "fluoridation chemistry." Yet the principal guest was in charge of chemistry the Soreq research reactor, rather than an Israeli water treatment plant or related water security concern. The visitor did cancel a proposed visit to NUMEC's irradiator facilities claiming to be "more interested in chemistry" and wanting to discuss "nuclear techniques for analysis" and "ore analysis and scrap" with appropriate NUMEC personnel.[177] The FBI discovered this Shapiro visitor's work at the Soreq nuclear reactor involved "fission and heavy ion chemistry and analytical work using nuclear reactions either fluorescent or neutron activation...on uranium and plutonium fission..and things of that sort..."[178] On April 14 the FBI noted the visitor "was satisfied with the information he had received as a result of the day's activities at NUMEC."[179]

The FBI picked up other strange conversations revealing they weren't the only ones who were curious about Zalman Shapiro's HEU handling activities. On April 29, 1969 a nosy inquirer[xxviii] who actually may have been trying to issue a covert warning, brashly asked Shapiro's sister Zipporah Schefrin "whether she had seen the article concerning the atomic bomb in the May, 1969 issue of the *Esquire* magazine." The caller "added that the United States was apparently maintaining closer supervision of the ingredients utilized for making such a bomb." The cagey Schefrin said she was more interested in another subject "...than she was in NUMEC and then remarked that subject [Zalman Shapiro] did not need this headache, so if he [Bloom] did not like it, he could say good-bye."[180]

[xxviii] Likely the editor of the Pittsburgh Jewish Publication and Education Foundation, Albert W. Bloom, cross-referenced as "Supra" in the same FBI file page

Even as Zipporah Schefrin brushed off Bloom, only one establishment newspaper, *The New York Times*, had published a prominent piece that might have led large numbers of discerning readers to connect NUMEC with lost HEU. But another enterprising reporter at a supermarket advertising broadsheet long ago connected the dots, which promptly led to a visit by the FBI.

Frank K. Noll worked as assistant to the president of Jessop Steel Company of Washington, Pennsylvania. But Noll was also a public relations man who moonlighted. His Fourth Allegheny Corporation handled promotional activities for local shopping malls and other firms including Jessop and Sipes Paint Company. Noll produced a weekly publication called *The Advertiser* and another called *The Washington Mall*. These free, weekly publications were financed primarily through advertising. Noll generally wrote sensationalist feature articles with titillating headlines and draw readers in to pick up a copy. After receiving hostile phone calls over his provocative pieces about Israel and Arabs, Noll started writing under the pen name "Mack Truck." The February, 1969 issue of *The Advertiser* published an ominous, preachy article titled "Middle East 'Powder Keg' May be Fissionable Type if Israel Loses Friends." In it, "Mack Truck" sternly warned that Israel had unfairly been named the villain of the Middle East crisis, and that all its friends had abandoned Israel except for the U.S. and U.K. The essay pointed out that "the nuclear problem comes up as Israel has the know-how to produce the hydrogen bomb and the question is whether or not Israel has been pushed to the point it may be necessary to use the 'H' bomb."

It was the small inset box carrying another piece titled "Dr. X" that piqued the curiosity of the FBI. It read "Once or twice a month at Greater Pittsburgh airport a Pittsburgh physicist takes off for JFK International Airport where he then boards an El Al plane for Tel Aviv. Dr. 'X' is one of the nation's best-informed men on nuclear materials. He once worked for Westinghouse at Bettis until he and a few others left to start in business for themselves. Now he's a 'consultant' to several Israeli firms and the government on handling radioactive materials and related matters."

The FBI tracked the writer down. Under questioning, Noll quickly volunteered that the "Dr. X" in the article was in fact Dr. Zalman Shapiro. In 1962, Jessop Steel Company had sent Noll on a trip to Israel in order to develop a sales channel. On the plane, Noll met Shapiro, and discovered they shared experiences as residents of the same region (Washington is only 30 miles from Pittsburgh) and firsthand experience with Westinghouse. Shapiro mentioned he was a chemist who studied the uses of fissionable material. Noll told the FBI his pieces were based on sensationalizing a story in *Life Magazine* fused into his memorable airplane ride with Shapiro in 1962.[181] Yet Noll's intuitive leap was the first unvarnished account of what would only slowly become undeniable in the coming decades—that Shapiro was working for Israel's atomic bomb project and that NUMEC was the very latest generation smuggling front.

6. NUMEC and the reprocessing industry

NUMEC and its nurturing parent, Apollo Industries, were mere saplings in a Redwood forest. Despite its small capitalization, NUMEC managed to attract over 22 tons of government supplied highly enriched uranium.[182] NUMEC holds the official record for the highest pre-1968 "materials unaccounted for" losses of any reprocessing facility according to the Department of Energy. This latest official MUF estimate—issued in the year 2000—does not account for the possibility of undetected diversions through overseas straw buyers, particularly in France, since it appears that no investigator ever checked entities on NUMEC's export licenses to see whether they were bona fide buyers, or whether they had received all of the shipments claimed by NUMEC. While the U.S. reserves the right to oversee all licensed HEU exported overseas from reactor use to reprocessing to burial in the case of NUMEC this right has never been invoked. Also, the majority of NUMECs export records were flawed and incomplete.

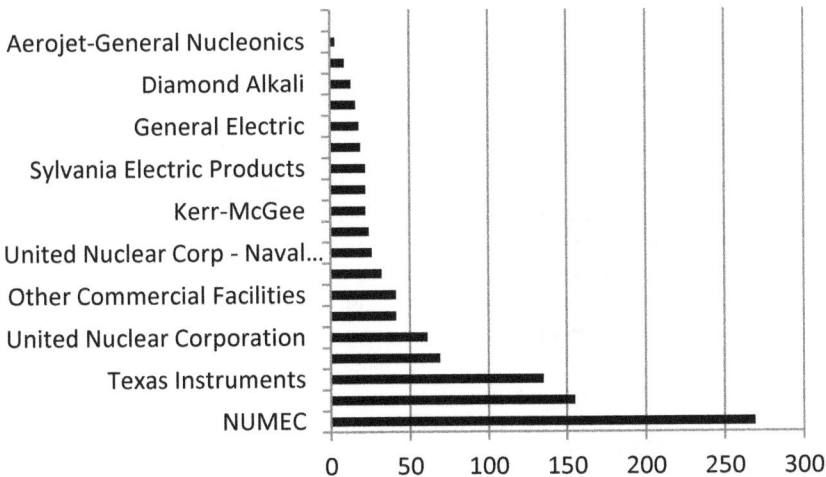

Figure 11 2001 DOE data NUMEC vs. cohort HEU losses before 1968 (KG)[183]

Comparing basic indicators of Apollo's cohort nuclear industry parent companies listed in the Department of Energy 2001 audit reveals a startling fact. NUMEC had almost twice the next competitor's MUF. Losses exceeded industry averages by several times.

	Revenue (Million)		Net Income (Million)	
	1967	1968	1967	1968
Aerojet-General Corp.	$443.80	$444.80	-$10.25	$7.45
Diamond Shamrock	$490.00	$514.83	$41.24	$34.20
General Electric	$7,740.00	$8,380.00	$361.40	$357.10
General Telephone and Electronics	$2,643.78	$2,927.06	$219.24	$222.69
Kerr-McGee	$423.83	$473.30	$34.71	$36.40
United Nuclear Corp.	$56.90	$61.70	$3.31	$3.41
Texas Instruments	$568.51	$671.23	$22.86	$26.33
Apollo Industries	$8.26	$6.24	-$0.85	$1.35

Figure 12 NUMEC vs. cohort comparative financials.

Comparative financial figures are also revealing. The average 1967 revenue of Apollo's cohort companies — $1.8 billion — was enormous compared to Apollo's *total* annual revenue of only $8.3 million. In 1968 Apollo's cohorts grossed on average $1.9 billion while Apollo brought in a paltry $6.24 million. Even United Nuclear Corporation, the next smallest industry participant with annual revenues of $61.7 million, was almost ten times the size of Apollo in 1968.

As NUMEC's "de facto" parent company and key backer in an increasingly regulated, government contract dependant, capital intensive field, Apollo had remarkably little to offer NUMEC in terms of a financial safety cushion. While diversified conglomerates such as GE or GTE could reallocate revenue from healthy consumer or industrial divisions into their nuclear divisions as needed (the entire business rationale of conglomerates in the volatile 1960s stock market) Apollo's financial architecture was so fragile that a single major problem with regulators or unforeseen costs would be enough to send NUMEC crashing down. Modern financial analysis reveals Apollo was not really a trapeze artist parent company with no safety net, it was a shell corporation for the Israeli government's nuclear weapons program.

NUMEC's loss of $2.6 million in 1966 put it into technical default of its loan agreements. By March of that year it had already begun to shop itself around to wealthy but inexperienced potential buyers looking to break into the nuclear industry, such as Peabody Energy.[184] On February 28, 1967 NUMEC had 323,390 shares outstanding, with 19,588 (6% of the total) directly held by Zalman Shapiro. By early March to align itself for acquisition, NUMEC filed amendments reorganizing the total number of outstanding voting shares to 321,465. In a March 20, 1967 letter to its thirty-odd shareholders, NUMEC recommended a "reorganization" within Atlantic Richfield Company. During an April 14 meeting in the ballroom of the Chatham Center Office Building in Pittsburgh, 68 percent of NUMEC shares voted for acquisition by Atlantic Richfield, with 4 percent against and 28 percent abstaining. When the deal settled each NUMEC shareholder, with the exception of a single holdout, received .255 shares of Atlantic Richfield stock for every share of NUMEC. In its 1967 annual report Atlantic Richfield recorded it had issued 84,500 shares to acquire NUMEC. With the price per share of common stock trading around $90 at the end of April this meant the company Shapiro and Lowenthal built was valued at $7.6 million.[185] Atlantic Richfield formally assumed all of NUMEC's liabilities and debts.

What did Atlantic Richfield, a company focused on fossil fuel energy development, actually have to gain in buying the struggling AEC contractor's nuclear business? Its own massive AEC contracts. Less than three months after agreeing to acquire the faltering NUMEC, the AEC awarded Atlantic Richfield a five-year contract to manage a $270 million chemical facility at the sprawling Hanford site near Richland, Washington. It formed a new subsidiary, Atlantic Richfield Hanford Company, and took over the operations in September of 1967. [186] By the end of 1967 Atlantic Richfield stock traded at nearly $104 per share.

The AEC granted the contract to Atlantic Richfield in exchange for its acquisition of NUMEC. This fact is buried deep in the contracting documents. The AEC issued a press release on February 27, 1967 "announcing that the AEC was inviting a number of firms to submit proposals." The AEC claimed it was looking for large, financially robust firms with competence in "management of complex chemical or allied operations."[187] Among the formally invited firms, only Atlantic Richfield stood out for its suspicious lack of nuclear material handling experience.[xxix] Only by acquiring NUMEC could the firm hope to receive the cost-plus AEC contract.[xxx] The NUMEC quid-pro-quo was implicit in the contract negotiation meeting minutes and even written into the contract.

By May 15, Dow Chemical, Aerojet and Isochem had submitted competing bids.[xxxi] Although all of the proposers were thought to be qualified to run "the 200 Areas" at Hanford, on June 29, 1967 "the [AEC] Commissioners authorized staff to proceed with discussions with Atlantic Richfield leading to the selection of that Company as operating contractor." [188] The agenda for the final July 5, 1967 contract negotiations with Atlantic Richfield was explicit about not allowing a repeat of the NUMEC fiasco. "[AEC] Commissioner Johnson stressed the need for proper emphasis on nuclear materials management. He pointed out the recent interest that the Joint Committee [JCAE] had shown, the Headquarters recent reorganization of nuclear materials and the troubles NUMEC had encountered."[189]

Rather than mention NUMEC's problems at Apollo and Parks Township in an official report justifying the selection, the AEC highlighted the experience of NUMEC's secretive Lewiston operations. The report gushed "The technical

---

xxix Other firms solicited by the AEC included Aerojet-General Corporation, Allied Chemical Corporation, Battelle Memorial Institute, Dow Chemical Company, W.R. Grace Company, Isochem Inc. (which was running the facility under a five year contract), J.A. Jones Construction Company, TRW Systems, and Vitro Corporation.
xxx In cost-plus contracts the contractor furnishes all material, construction equipment, and labor at "actual" cost, plus an agreed-upon fee for services.
xxxi The AEC originally set a deadline of May 1, but advanced it to May 15, 1967.

ability of Atlantic Richfield is considered outstanding and satisfies the requirements of the contract. Atlantic Richfield has not had a previous contract with AEC. However it recently entered the nuclear energy field by its acquisition of NUMEC. NUMEC's performance as operator of the Boron-10 Plant at Niagara Falls, NY, is considered sufficiently satisfactory to warrant requesting an extension of that contract." [190]In the first year the estimated payment to Atlantic Richfield for cost-plus was $30.4 million.[191]

In 1971, after gaining Hanford experience, Atlantic Richfield offloaded NUMEC to Babcock & Wilcox, while negotiating to renew the lucrative AEC Hanford management contract for another five years. By 1973 the U.S. government had committed $130 million to Atlantic Richfield's operations at Hanford.[192] By November 5, 1975 the AEC's successor advised Atlantic Richfield the total government obligation for running the plant "is increased to $335 million."[193] This was sweetened to $407 million by December.[194] An FBI interview of a former AEC official in 1979 characterized the NUMEC sale as "forced" because the AEC was insisting on a change of NUMEC's management—those running NUMEC were "unwilling or unable to go along with AEC demands."[195]

Atlantic Richfield, unlike NUMEC's first group of shareholders, assumed total financial responsibility for everything NUMEC did from 1958 until the moment Babcock & Wilcox took over the company. Shapiro, Lowenthal, and other executives responsible for bringing NUMEC to Apollo and running it during its highest period of SNM loss would be forever shielded from environmental liability. Although Atlantic Richfield characterized NUMEC's transfer to Babcock & Wilcox as a company sale, it provided no proceeds to the seller. As part of the B&W acquisition agreement, Atlantic Richfield retained NUMEC's liabilities and debts while agreeing to indemnify Babcock & Wilcox for any liabilities arising from NUMEC's operation of the facilities before the sale.[196] This was a major mistake. In its 1971 annual report, Atlantic Richfield reported no proceeds at all from the sale of NUMEC, but rather took "an extraordinary charge of $11,831,000 or 21 cents a share, from the sale of the Company's Nuclear Materials and Equipment Corporation operations in November 1971."

What caused the value of NUMEC to drop almost $20 million from a positive $7.4 million buyout of the original shareholders (including Lowenthal and Shapiro) to the negative $11.8 million sellout? And why did Babcock and Wilcox, which unlike Atlantic Richfield, had actual experience in the nuclear industry agree to acquire the company? The only plausible explanation is that B&W wanted to keep its massive U.S. Navy contracts.

In 1856 Stephen Wilcox patented an innovative water tube boiler. In 1902 B&W boilers powered the New York City subway, while a year later Commonwealth Edison Company of Chicago became the first utility station to use its steam turbines to generate electric power. To say that B&W was a major Navy contractor is an understatement. In 1916 revenues soared on sales of high explosive shell bodies as it operated plants "most of the year night and day." WWI hostilities generated sales that "exceeded those of any prior year in the company's history, both in stationary and marine boilers." The volume was "very widely distributed" causing "an increase in every industry which we do business" including public utilities. The U.S. Navy and Merchant Marine accounted for a huge share of its business.

In 1953-55 B&W designed and fabricated components for the USS Nautilus.[197] In 1955 Babcock and Wilcox earnings had grown to nearly a quarter billion dollars as it outfitted the "world's mightiest aircraft carrier" the USS Saratoga with "special B&W marine boilers." The company kept a watchful eye on the growth of atomic energy, promising shareholders it would fill "a role in the nuclear power business similar to that which we now fill in the conventional power business. We will make complete nuclear steam generators — beginning with the fuel elements on through to the marine turbine throttle."[198]

By 1963 B&W celebrated its "96th Year of Helping Industry Serve America" boasting revenues of almost $400 million. It achieved this as "a prime supplier to the nuclear Navy, thus extending B&W's traditional role as a major supplier of equipment for propulsion of naval ships."[199] By 1968 the firm's "backlog" had grown to such a significant extent it was ready for acquisitions. In 1971 Navy orders alone reached $251 million of the company's $959 million in shipments.

B&W's portrayal of its 1971 acquisition of NUMEC revealed an optimistic bid for future contracts tempered with informed foreboding. "The potential for our nuclear fuel business was broadened with the acquisition of the Nuclear Materials and Equipment Corporation. This enables us to supply uranium fuel and, most significantly, provides additional expertise in plutonium fabrication. This is of importance in recycling plutonium from light water reactors and in fabricating fuel for breeder reactors. The nuclear business continues to be adversely affected by lengthy proceedings and hearings concerned with both safety and environmental considerations." But B&W provided shareholders no details about the troubling terms of its NUMEC acquisition.

In 1974 Babcock and Wilcox stopped operating NUMEC as a subsidiary[200] and folded NUMEC into itself. This was a fatal mistake. [201] By the dawn of the 21st Century B&W was bankrupt and arguing in court with the remnants of Atlantic Richfield and the American Nuclear Insurers over whether $320 million in insurance coverage for NUMEC's Apollo and Parks Township facilities could cover the soaring costs of litigation and damage awards as more than 300 former employees sued for radiation-related cancer indemnities.[202] B&W completed the decommissioning of NUMEC's Parks Township plutonium facility in 2001.[203] After citizens groups complained that the NRC wasn't doing enough to remediate the shallow land disposal area next to the Parks facility, Congressman John Murtha mandated that U.S. taxpayers fund the cleanup by inserting funding for the U.S. Army Corps of Engineers into the 2002 Defense appropriations budget.[204]

In hindsight, NUMEC's shoestring Apollo and Parks Township operation only made sense if it's true purpose all along was tapping into the huge American stream of highly enriched uranium in order to siphon material into the Israeli nuclear weapons program. NUMEC was treated like a disposable asset by its founders, who nonetheless profited handsomely in their exit strategy. After Israel obtained what it could not yet mass-produce on its own, the entire NUMEC operation was spun off with the aid of the AEC, which would have been highly embarrassed by its collapse.

From a valuation perspective, the only assets NUMEC ever really developed was an uncanny ability to attract a steady stream of government contracts. A net-present value calculation of revenues from 1958 against the identified costs of cleaning up pollution in Apollo, PA in 2011 would reveal that NUMEC was financial toxic waste with a negative value in the hundreds of millions of dollars. Its externalized costs to the public vastly outstripped any meager income it delivered to owners after Lowenthal and Shapiro sold it to Atlantic Richfield.

NUMEC insiders privy to its true purpose managed to sell off the enterprise in a way that provided a shrewd disincentive to radiation sickness victims. Few would ever look too closely into why the plant was opened in Apollo or who really backed it. NUMEC's spinoff to Atlantic Richfield left the deep-pocketed energy company holding the bag for problems caused during NUMEC's dirtiest operational phase. This unfortunate handoff happened at the close of the 1960s, shortly before Israel's top spy swept in for one last radioactive asset-stripping run.

In hindsight, the FBI and U.S. Justice Department should have been on the lookout for something resembling NUMEC beginning in the late 1940's. J Edgar Hoover had already met with Jewish Agency representatives seeking special treatment when Israel's conventional arms smugglers were arrested. As someone who fought in Israel's War of Independence, David Lowenthal knew the importance of having the very best available weapons bought, stolen and smuggled to Palestine from the United States. This was accomplished only through the incorporation of a network of front companies in the United States.

7. David Luzer Lowenthal

The widened net of the FBI investigation logically closed in on Shapiro's trusted friend who had financed and given shelter to NUMEC in its startup phase—David Luzer Lowenthal. By January of 1969 informants confirmed to the FBI that Lowenthal's ties to the highest levels of Israeli intelligence had neither ended nor were they only taking place during his frequent trips to Israel. An informant "who has furnished reliable information in the past, advised on November 8, 1968, that General David Carmon, Military Attaché, Israeli Embassy, Washington, DC, is a key contact of David Lowenthal. [The informant] emphasized that General Carmon has been a leading personality in the Israeli Intelligence Service for several years..."[205] A source advised that "Lowenthal receives regular reports on Israeli activities, including military activities..."[206]

More than merely acting as NUMEC's venture capitalist and white knight, the FBI confirmed that Lowenthal was a "long-time personal friend of Dr. Zalman Mordecai Shapiro..." and that Lowenthal "is President of Raychord Corporation, immediately adjacent to the NUMEC plant at Apollo, PA. This source pointed out that Raychord Corporation, a steel fabricating company, occupied the very same building at Apollo, PA. with NUMEC prior to the time NUMEC moved to its new location on the outskirts of Apollo....Lowenthal and Shapiro lived near each other, both residing on Bartlett Street..."[207] The same source estimated that the intrepid Lowenthal traveled to Israel even more frequently than Shapiro "on average of once a month."

An FBI background check[xxxii] on Lowenthal's passport records revealed even more details about his 1956 visit to Israel. "On this [1956] application he indicated that he would travel by air from New York City to France and Israel and would remain in the latter named country as a delegate to the World Zionist Congress for one month." Earlier, Lowenthal had misled the U.S. State Department when he applied for a replacement passport at the U.S. embassy in France in 1948 in order to cover up his direct involvement in a complex, illegal, clandestine smuggling operation. By "losing" his previous passport, Lowenthal effectively wiped out all evidence (his entry and exit stamps) of visits to many ports of call during smuggling operations.

According to the FBI files first released in 2011, "Lowenthal filed an Application for Passport at Marseilles, France, on March 1, 1948, indicating that he had become stranded in that country in early 1947. He indicated that he was a seaman at the time and that he had lived in Brooklyn, New York, from 1934 to

---

[xxxii] The FBI received the author's FOIA for records on Lowenthal on April 12, 2010. Of 304 pages reviewed, only 49 were released on August 29, 2011 citing the CIA Act of 1949 and National Security Act of 1947 as reasons for retaining most of Lowenthal's file. The file is full of redactions at the request of OGA (other government agency, presumably the CIA). The CIA has refused direct requests for its Lowenthal equity content.

1947, and also in Pittsburgh.[xxxiii] He indicated that his father died in 1947, and his mother died in 1925, and that he was naturalized through his father's naturalization in the U.S. District Court at Brooklyn, New York, in July, 1929, while they were residing at 684 New Lots Avenue, Brooklyn, New York....He indicated that he served as a seaman aboard the 'SS President Warfield.'"[208]

Figure 13 Lowenthal's Brooklyn, New York residence.

Lowenthal's voyage actually took place on a ship formerly known as the *SS President Warfield*, but renamed the *1947 Exodus*. *Exodus* explains a great deal about Lowenthal's rise from a humble immigrant's apartment in Brooklyn and modest Pittsburgh duplex in the late 1930s and 1940s to a mansion in a leafy neighborhood in Pittsburgh helping to nurture NUMEC and a portfolio of other companies in the late 1950s and 1960s. The clandestine, largely cash-only funding for the provisioning of the *1947 Exodus* also explains Shapiro's unwillingness to fully divulge Lowenthal's structured financial arrangements with NUMEC to members of Congress in 1978. To Israeli nationalist smugglers, cash was always king. Checking account money trails had led to Al Schwimmer's criminal prosecution and almost took down Jewish Agency operative and Abraham

---

[xxxiii] 6223 Nicholson Street, a humble duplex unit.

Feinberg partner Nahum Bernstein. If the smugglers learned anything in the 1940s, it was that financial opacity was key to successful operations.

Lowenthal was a deck hand for one of the most elaborate and successful smuggling efforts conducted during Israel's formative period. In the summer of 1947 the Exodus carried 4,515 passengers in a bid to break British blockades and international policies on Jewish WWII refugees in displaced persons camps.[209] Ben-Gurion ordered Ze'ev Shind, an Aliyah Bet[xxxiv] operative, to obtain blockade running boats and establish landing sites and a resupply network in Europe. "I need ships, I need them now. And I need safe landing sites to avoid the Mandate [British] coastal patrols, I want them now!"[210] Shind, responsible for buying up diplomatic credentials from corrupt Latin American officials for use in covert operations, set up a front company called Weston Trading Company – Marine Surveyors in order to purchase a transport ship identified by the Haganah smuggling network, the *President Warfield*.[xxxv] Shind, conscious of possible U.S. counterintelligence measures, avoided using his office phone, preferring street and pay phones to avoid being tapped.

Like the War Assets Administration, the Maritime Commission was in charge of overseeing the sale and scrapping of a large surplus of U.S. WWII ships built up under programs beginning in 1939 through the end of the war. Shind contacted the Maritime Commission in Washington to make an offer on the *President Warfield*. Shind also formed a Panamanian corporation called Arias and Arias under the guidance of an expert ship broker. But ships could not be purchased for cash. He obtained a $50,000 check from a long time Jewish Agency operative for deposit into Chemical Bank to purchase the ship for $10,000, rename it, and prepare it for operations with laundered cash.[211]

Lowenthal was likely recruited through yet another front company set up to attract crewmembers for smuggling ships. Palestine Vocational Service operated on 76th Street in New York.[212] Palestine Vocational Service, like the Haganah smuggling network's use of established charitable organizations, tapped Zionist youth movement members in the U.S. for recruits. Candidates had to be cleared with the Haganah headquarters on Madison Avenue.

Fundraising for the smuggling voyages also counted on United Jewish Appeal donor lists. Abe Kay, a UJA leader, gathered 300 donors into his office, urging them to bring only gold and cash. The crewmembers of the *Exodus* were positioned alongside an open satchel as donors streamed by. UJA leaders told them "the British are trying to get our Treasury to stop any funds going to support what the British call 'illegal immigrants'...None of these you know. None will give you their names. But they are sailors...When they leave here tonight, your money will ensure that another ship will be brought to sail to

---

[xxxiv] The network facilitating the immigration of Jews to the British Mandate for Palestine, considered by the U.K. to be in violation of British White Paper of 1939 restrictions. Operative from 1934 to 1948.

[xxxv] Named after American shipping magnate Solomon Davies Warfield and launched in 1928. It served as a Chesapeake steamer until retrofitted with 20mm guns and 12 pound cannon in WWII and sailed by the Merchant Navy during the war.

Europe to collected maybe 800 Jews, maybe more, and bring them to Eretz Israel. Isn't that the best possible use you can make of your money?"[213] When the *Warfield* sailed from the U.S., it was seen off by Rudolf Sonneborn and other Zionist leaders who helped fund its purchase, refurbishment and provisioning.[214]

The New York Times ran a short story about how the ship was battered by a storm and forced to return to the U.S. The British government became "increasingly concerned that the Haganah was preparing to send 'a fleet of ships bought by wealthy Jews in the United States' and that this would trigger a full-scale war with Arabs that would ensnare the United States."[215] The *Warfield* was heavily dependent on Haganah operatives for refueling in such key locations as the Azores while it flew under a Honduran flag. The *Warfield* relied on Haganah agent Joe Baharlia to provide fuel for the blockade-running as well as sailing papers and fake ship manifests obtained through bribing French officials.[216] After retrofitting in Portovenere in Italy, the *Warfield* only narrowly escaped a gunboat blockade by forging a letter from the Italian admiralty.

Pursued and brutally attacked by the British, its refugees returned to DP camps, the *Exodus* provided a propaganda coup by reversing international sentiments about Jewish immigration to Palestine. Its desperate mission and self-righteousness became legendary. Branded an illegal immigrant ship by the British, the *Exodus* was for David Lowenthal a floating school of forgery, money laundering, smuggling, spy tradecraft and logistics. As Lowenthal pieced together Apollo Industries, the incubator for NUMEC, Zionism was the glue holding together the small group of mezzanine-level investors. Like the *Exodus*, NUMEC was pieced together within an old, defunct vessel on a desperate mission.

A.M. Oppenheimer who served as president of Apollo Steel Company beginning in 1918, sold the 69 year-old company in 1946 after peak WWII steel demand. Apollo Steel Company fetched a healthy $2.5 million, but then entered a steady period of decline.[217] By 1949, Apollo Steel Corporation was again looking for buyers. The local Chamber of Commerce gave townspeople until June 20 to raise cash for an employee buyout of the company in order to save 750 jobs. News reports were grave. "Company President Howard C. Keiser left Apollo for Chicago last night with a certified check for $125,000 — the townspeople's bid for economic salvation — in his pocket...The steel company is Apollo's only industry...Unless the company is saved, Apollo will become a ghost town."[218] The town and 600 jobs were preserved when N.M. Landley Company of Pittsburgh bought the mill and leased it out for operations using hand-rolling mills to convert sheet bars into galvanized roofing.[219] A temporary shortage of building materials for the postwar housing boom made the obsolete process economically feasible for a short time. But the plant again fell dormant by the mid-1950s. Pittsburgh buyers M.D. Wedner, Paul Sullivan and Audrey Sherman acquired the mill and incorporated Apollo Steel on March 1, 1955. In corporate filings, they claimed the "value of the property with which the corporation will begin business is $500."[220]

David Luzer Lowenthal subsequently acquired control of Apollo Steel and executed a complex merger in 1958 that created a holding company with two other major Zionists, Morton Chatkin and Zalman Shapiro, on the board of directors. The merger of Apollo Steel Co with San Toy Mining and American Nut and Bolt into Apollo Industries helped finance the creation of NUMEC.[221] It also shielded the obvious fact that that the primary interest of the core investors all along was NUMEC.

On paper, the three merged corporations were consolidated into San Toy, which was then renamed "Apollo Industries." The merger served to bring together some of the most ideologically committed (and later, outspoken) Zionists in the United States. The Apollo Industries corporate charter said nothing at all about nuclear reprocessing within a list of key business activities of the newly formed conglomerate.

San Toy was originally incorporated in December of 1901 to acquire and develop mines in Mexico and refine gold, silver, copper and other minerals. In turn-of-the century Chihuahua, such commercial exploitation also necessitated the construction and operation of all links in the production chain, from telegraph and phone lines, to railroad links, toll roads, worker dwellings and other infrastructure. San Toy's corporate charter authorized common and preferred stock of $2.5 million at a par value of one dollar. San Toy's original owners were F.L. Dutton, M.B. Ward and H.M. Heath, all of the state of Maine.[222]

In September of 1911, investors could buy a share of San Toy for fifty cents. By 1915, the struggling company reported income of $25,000, expenses of $95,000 with a deficit of $69,000.[223] A series of scandals about the true mining rights over the San Toy mines of Mexico thwarted the participation of larger investor syndicates looking to scoop up promising businesses in the pre-depression era stock boom. In 1949 San Toy President John W. Weibley and Treasurer Joseph H. Bialas registered the Maine company to do business in ore-hungry Pittsburgh.[224] But by 1951, San Toy's capital stock was slashed from $7 million to $570 thousand dollars. The par value of shares was reduced from $1 to ten cents a share.

By the 1958 merger into Apollo Industries, San Toy had been acquired by Ivan J. Novick and had 5.7 million in stock issued at a par value of ten cents.[225] Although it was used as the primary vehicle to form Apollo Industries, San Toy received only 23% of the newly issued Apollo's company stock. What did the floundering San Toy with its Chihuahua lead, zinc and copper properties really bring to the table in 1958? Ivan J. Novick.

San Toy Mining Company President Ivan J. Novick (1928-2009) was a committed Zionist who would become one of the Israel lobby's top American leaders. Between 1978 and 1983 Novick served as president of the Zionist Organization of America. He was also active in the World Jewish Congress, American Zionist Movement, Jewish National Fund, Israel Bonds, and American Friends of Tel Aviv University.[226] A *Washington Jewish Week* obituary noted his role as a "liaison" between the administration of President Ronald Reagan and Israeli Prime Minister Menachem Begin, who presented Novick with Israel's Jabotinsky Centennial medal. A review of the many issue ads Novick underwrote later in life reveal a world view commonly espoused by many of Israel's most ardent American supporters—that Israel was essentially above international law and that Americans breaking their own country's laws in service to Israel should never be punished.

A November 9, 1993 *New York Times* print ad urging the release of Israeli spy Jonathan Pollard placed by the American Zionist Movement listed Novick as a financial contributor. In 1981 as president of the ZOA, Novick supported an outright Israeli annexation of Syria's Golan Heights, urging the Reagan administration not to pressure Israel by suspending a strategic military cooperation accord.[227] The same year, as president of the ZOA, Novick affirmed the organization's coordination with Israel and support for its refusal to deal with the Palestine Liberation Organization "or to accept the concept of another Palestinian State in the Middle East, other than Jordan."[228]

Referring to the captured West Bank in the biblical terms "Judea and Samaria" Novick excoriated "the anti-Semitic taint" of an essay by Flora and Anthony Lewis who were cautiously advocating that "those who want peace and security more than territory deserve a hearing too." Novick thundered in retort "the Pan-Arabist vision, fundamentally opposed to American values and principles, simply does not allow for any non-Arab state to exist in the Middle East. It is for this reason that Israelis and their supporters around the world oppose a second Palestinian state, whether or not it is led by the P.L.O."[229]

Novick's outrage in 1979 over revelations that American Ambassador to the UN Andrew Young was meeting with members of the Palestine Liberation Organization can now be evaluated as the public relations component of a successful Mossad operation designed to destroy Young's career and derail U.S. rapprochement with the PLO. It was but one of many instances where Israeli intelligence or its U.S. lobbyists obtained classified or confidential information and tactically leaked it in order to destroy unwelcome diplomatic initiatives.

On September 1, 1979, the *Baltimore Sun* published Novick's diatribe. "After weeks of speculation, which resulted in considerable tension in Israel and pronounced anxiety within the American-Jewish community, President Carter broke his silence when he stated August 12 'I am against any creation of a separate Palestinian State.' Within 48 hours of the President's assurances, a sensational news story revealed that Andrew Young, America's ambassador to the United Nations, had already conducted a face-to-face meeting with the PLO representative at the UN."[230] The true source of the "news story" that destroyed Young's diplomatic career-by one credible first-hand account-was an Israeli covert intelligence operation.

According to retired Mossad operative Victor Ostrovsky, destroying Young and the "story" used by the ZOA was the result of a successful intelligence operation. Ostrovsky's fellow Mossad operative had been working in New York since 1978 to infiltrate Arab peace initiatives influencing President Jimmy Carter's Middle East peace talks. Secretary of State Henry Kissinger had famously pledged in 1975 that the U.S. would not recognize or negotiate with the PLO, and Carter had later promised to honor that pledge.

But in November 1978 Illinois Congressman Paul Findley carried a Carter message to Yassir Arafat in Damascus. In a meeting, Arafat indicated that the PLO would be nonviolent if an independent state were created in the West Bank and Gaza. Carter was already under suspicion by Israel and its U.S. lobby for having publicly called for a homeland in 1977. Compounding the threat, U.S. Ambassador to Austria Milton Wolf met with a PLO representative with the approval of the Carter administration. UN Ambassador Andrew Young, a black southern leader, opposed Israeli settlements in the West Bank, but wanted to delay an Arab PLO recognition drive in the UN. Young met with PLO representatives in the residence of the Kuwaiti ambassador. The Mossad notified its headquarters of the meeting on July 25, 1979. On July 26, the Mossad covertly recorded Young's meeting with Arab diplomats and PLO representatives. A complete transcript was flown to Israel on an El Al flight. Menachem Begin insisted that it be publicized, even at the risk of burning Israeli intelligence sources and methods. *Newsweek* magazine was given the juicy scoop that Young had met with the PLO in a cascading public relations campaign that led to his forced resignation and the subversion of U.S. diplomatic tenders to the PLO.[231] The same year *Newsweek* would publish another curious story alleging Shapiro's close ties with the CIA and prosecutorial immunity at the height of the FBI's final investigation of NUMEC diversion.

While it is no longer a secret that Novick was a very committed Zionist who abided an entirely different set of rules and principles for all things Israeli—this was not obvious during his time as a board member of NUMEC's parent corporation. Novick and others quietly put their shoulders into the campaign launched by the Mossad to help defeat U.S. rapprochement with the PLO in the late 1970s. But what was Novick up to in 1958? How did Novick really win Israel's Jabotinsky award? Today it's much clearer. Novick positioned San Toy's rusty mining cars at the railhead of one of the most precious and at that time difficult-to-mine metals in the world: U.S. government owned highly enriched, weapons-grade uranium. In forming Apollo, appearing to revitalize plant operations with a vertically integrated steel operation, Apollo Industries gave cover and legitimacy to NUMEC.

On paper, San Toy's access to conventional raw minerals clearly seemed to justify its integration into Apollo Steel Company. On the surface it also seemed to make business sense in the 1960s era of conglomerates to link up San Toy and Apollo Steel to downstream output, in this case the American Nut and Bolt Fastener Company of Pennsylvania. During the WWII fastener heyday almost 3% of American steel output was destined for industrial fasteners. Daily consumption reached half a billion new bolts, nuts and rivets churned out by thirty thousand workers in 250 companies. Industry output doubled from 1929 to 1943, reaching $250 million through sales of 400,000 unique standard items—from joiners of rough timber to precision aircraft rivets.[232] The industry was still growing strong at the dawn of the 1960s.

Prior to its 1958 merger into Apollo Industries, American Nut and Bolt had 1,775 issued shares with a par value[xxxvi] of $100 each, a total value of $177,500.

---

xxxvi The par value of a stock has no relation to its market value. The values stated in corporate charters are presented here as a relative measure of how the merged entities were valued and distributed among the owner/investors. The market value of the Apollo steel works had probably fallen from $500 closer to zero when it was finally acquired by Lowenthal. Similarly both San Toy and American Nut and Bolt Fastener were essentially bankrupt entities when merged

Yet it received 119,525 shares of new Apollo Industry shares with a par value of almost $600,000. As a manufacturer of industrial flat and lock wasters, Morton Chatkin, president of American Nut and Bolt Fastener Company and a long time Lowenthal business partner, was worth the $423,000 paper premium for his business acumen. Apollo Industries also acquired a unique insight into how federal law enforcement officials operated in Pittsburgh.

Another original Apollo Industry shareholder, Elliot W. Finkel, was a former Justice Department lawyer who was forced to resign over conflicts of interest. Finkel served in the Army Air Force during WWII and was invited by U.S. Supreme Court Justice Robert H. Jackson to clerk for him at the Nuremberg Trials to prosecute Nazi war criminals. Finkel then served as special assistant to the U.S. Attorney in Pittsburgh, PA. In 1950 he was ordered to stop simultaneously serving as counsel to the Sterling Steel Company Foundry which was in bankruptcy proceedings in the Western PA district court. Finkel's moonlighting violated conflict of interest rules codified in the U.S. Attorney's Manual.[233] An FBI investigation found Finkel was also providing legal services to another bankruptcy client in Federal court. "Finkel not only failed to carry out his promise made to the Department that he would discontinue his connection with a bankruptcy matter concerning the Sterling Steel Foundry Company, but shortly thereafter had himself appointed by the court as attorney or trustee in another bankruptcy matter." [234]

Finkel resigned before the Justice Department could fire him. He entered private law practice in his father-in-law's firm Kaplan, Finkel and Roth and later served as Chairman of the Anti-Defamation League which made him an honorary life member of its national commission.[235] Finkel's awareness of bankrupt assets available for pennies on the dollar in the failing Pittsburgh steel industry was likely vital for bringing NUMEC into existence. On paper, Apollo Industries appeared to be a holding company for vertically integrated mining, steel and fastener production which would restore the economic health of Apollo. Yet a review of the officers and directors of the company after the 1958 merger reveals that something entirely different was underway.[236]

---

into Apollo Industries, but provided some highly usable assets and a cover story for the NUMEC operation.

President	Morton Chatkin
Executive VP	Ivan J. Novick
Secretary Treasurer	David Lowenthal
Clerk	Joseph B. Campbell
Director	J. Farrell Bash
Director	Bernard Kaplan
Director	David S. Livingston
Director	Isadore Glosser
Director	Elliot W. Finkel
Director	D.T. Horviz
Director	Elwood M. Jepsen
Director	John M. Joyce
Director	Ivan J. Novick
Director	Frank J. Pohl
Director	Louis J. Reizenstein
Director	Alvin Rogal
Director	Herbert B. Sachs
Director	Zalman M. Shapiro
Director	M.E. Solomon

Figure 14 Apollo Industries board of directors May 31, 1958.

Shapiro and Chatkin both held leadership roles in the Zionist Organization of America. As an American membership organization founded in 1896 dedicated to the creation of a Jewish state in Palestine, the ZOA played an instrumental role in mobilizing U.S. political and economic support. But the organization also secretly clashed with the FBI and Criminal Division of the U.S. Department of Justice. The FBI received complaints that the ZOA was attempting to stir up popular U.S. opinion against Arab states, and that it should be treated as a foreign agent of the Israeli government. After the FBI investigation, on February 25, 1948 the ZOA was asked to openly register as a foreign agent of its parent organization, the World Zionist Organization. But in what was soon to be standard procedure, the Justice Department refused to take a proactive enforcement stance while the ZOA reconstituted itself.

On March 21, 1949, the FBI forwarded another frustrated memo to the Justice Department about ZOA's nation-wide solicitations of sensitive American industrial information through its "Economic Affairs Council." It wasn't until November 8, 1949 that the Criminal Division finally responded to the FBI. "As you know, the obligations of the subject organization under the terms of the Foreign Agents Registration Act have been a subject of study and consideration by the Foreign Agents Registration Section for the past several years. In April of 1947 a comprehensive memorandum was prepared by that Section outlining in considerable detail the activities and relationships of the subject in connection with the World Zionist Organization. As a result of this memorandum, a letter was transmitted to the subject again soliciting the registration under the Act. This letter resulted in a series of conferences being held between representatives of the subject. As a result of these conferences the subject organization materially changed its constitution and at the same time effected a change in the constitution of the World Zionist Organization in an effort to remove itself from its agency status. As a result all attempts to procure the registration of the subject organization were dropped."[237]

ZOA's leveraging its membership into technology transfer activities clearly warranted official foreign agency designation. However, simply by changing its bylaws, ZOA avoided further FBI or DOJ interest in the activities that triggered the Foreign Agents Registration Act investigation: secret and sensitive extra-governmental technology transfers to Israel and orchestrated public relations campaigns. For its part, the Justice Department unit responsible for enforcing the 1938 Foreign Agents Registration Act continued an almost unbroken string of enforcement failures over Israel lobbying groups engaged in similar activities. Whenever evidence of foreign agency through illicit activities was detected, Israel lobbying organizations either reconstituted into new shells or made immaterial changes in corporate documents, seemingly reversing corporate hierarchies.[xxxvii]

Unlike the domestic surveillance conducted on Zalman Shapiro, the FBI's efforts to understand the peripatetic Lowenthal were heavily dependent on the CIA. In February, 1969 special agents ordered reviews of long distance phone records, but could not locate any vehicles registered to Lowenthal or his family members. "Bureau of Motor Vehicles records, Harrisburg, PA checked in name of subject and members of his family, and under names of companies which he is affiliated have not been productive."[238] The FBI had trouble mapping and tapping Lowenthal's many foreign communications. On May 22, 1969 the FBI sent a memo "regarding Israel phone calls for dissemination to CIA in an effort to learn more about subject's contacts in that country" with "other phone numbers contacted from Lowenthal's Pittsburgh phone...being sent out in a report being prepared."[239]

[xxxvii] In 1951, the FARA section ordered AIPAC's founder to continue registering as a Foreign Agent after he left his position as an employee and lobbyist for the Ministry of Foreign Affairs. He ignored the order. In 1962 the Section ordered the American Zionist Council, an umbrella group of which the ZOA was a member, to register as a foreign agent of the Jewish Agency. The AZC shut down and transferred employees and activities into a group incorporated six weeks later, the American Israel Public Affairs Committee. On June 30, 1971, the Jewish Agency's American Section, found by the Justice Department to be a quasi-entity of the Israeli government, reconstituted as the foreign agent of the nonprofit World Zionist Organization on September 21, 1971, without changing staff, leadership or office space.

The increase in chatter was perhaps a sign that Tel Aviv was ordering a roll-up of NUMEC. But like Shapiro, Lowenthal was not one to be caught having incriminating phone calls with Israelis that could be picked up by wiretap. Phone calls were strictly for setting up meetings in secure environs. On April 15, 1969 an informant "advised that on December 3, 1968, a collected telephone call was made to Lowenthal's home at Pittsburgh, PA from Tel Aviv, Israel, from a person named Zieney (phonetic) at Tel Aviv phone number 771-545. According to the same source on January 17, 1969, an unnamed person made a collect telephone call to Lowenthal's home in Pittsburgh from Tel Aviv, Israel, phone number 241-111....PG T-3...advised on January 13, 1969 that Lowenthal went to New York City on January 12, 1969, en route to Israel, where he was expected to remain for several days. The purpose of this trip by Lowenthal was not known.... Lowenthal has been to Israel several times in the last 18 months."[240]

Solid tradecraft for work on Israel's nuclear facility involved summons to Israel for sensitive discussions taking place in secure facilities far beyond the reach of the CIA or FBI. Today, only the CIA knows who Lowenthal visited in Israel at this critical time and perhaps why. But the CIA won't let the FBI publicly release its equity content which makes up the largest portion of David Lowenthal's file.

8. AEC Commissioner Seaborg defends Zalman Shapiro

AEC commissioner Glenn Seaborg kept copious office diary notes during his term as chairman of the Atomic Energy Commission from 1961 to 1971. The record consists of a diary written at home each evening, correspondence, announcements and meeting notes. Seaborg's amazing career from his discovery of plutonium and work on the Manhattan Project, rise from instructor to Chancellor of the University of California at Berkeley to head of the AEC is stunningly well-documented. Few details of Seaborg's many speeches, meetings, dinners, family life and travels were too mundane for his daily personal journal (which he started in 1927) scrap books and more official office journals. International Atomic Energy Commission seating charts, a 1966 contract signed with family members reserving Seaborg's broadcast news viewing times in exchange for purchasing a color television, and official AEC meeting minutes were all compiled by the Department of Energy into bound volumes.xxxviii

Seaborg's vast donated collection of correspondence, trip logs, contacts with the White House and photographs is held in over a thousand boxes at the Library of Congress Manuscript Division. Two recurring themes impacting Seaborg's treatment of the NUMEC affair run through his files. Seaborg was highly sensitive about the treatment and concerns of the many Jewish scientists working in the U.S. atomic energy program. Seaborg also viewed Israel's dedication to scientific development as a positive function of Zionism.

Seaborg always publicly maintained his steadfast commitment to the theory that NUMEC and Zalman Shapiro had not diverted SNM to Israel. Privately, however, Seaborg wondered whether he became too much Shapiro's advocate. Seaborg knew that using the full powers of the AEC to investigate or sanction NUMEC would have led to more than its bankruptcy and AEC embarrassment. Atomic Energy Act violations carried the death penalty for In the most severe violations. Exposure of the AEC's mishandling of NUMEC could have led to the agency's own early abolishment. Seaborg wasn't about to let that happen on his watch.

On June 19, 1953 Julius and Ethel Rosenberg were executed after passing atomic bomb information to the Soviet Union. David Greenglass served 10 years of a 15 year prison sentence for his role in supplying documents to Julius Rosenberg which were taken from the Los Alamos National Laboratory. Harry Gold, having given up Greenglass to law enforcement, also served 15 years. Morton Sobell served almost 18 years of a 30-year sentence after being tried with Julius and Ethel. Sobell denied any personal role until 2008 when he finally admitted he was "in a conspiracy that delivered to the Soviets classified military

---

xxxviii A significant amount of material considered sensitive or classified by various agencies, including the Department of Energy, has been removed from the Glenn T. Seaborg papers.

and industrial information and what the American government described as the secret to the atomic bomb."[241]

Sobell's name appears several times in Seaborg's AEC notes as he struggled over how to unwind House Committee un-American Activities charges against Julius Robert Oppenheimer. Seaborg's fellow physicist at the University of California at Berkeley was the true "father of the atomic bomb" and provided indispensible leadership to the Manhattan Project. After becoming a chief advisor to the AEC after WWII, Oppenheimer lobbied heavily for international control of nuclear energy in order to defuse an arms race with the Soviet Union and stem nuclear proliferation. This put him in the cross-hairs of hardliner American cold warriors.

After the public humiliation of losing his security clearance in 1954, Oppenheimer continued to teach and lecture while working in physics. Richard G. Hewlett and Jack M. Holl, the authors of *Atoms for peace and war, 1953-1961: Eisenhower and the Atomic Energy Commission,* interviewed many AEC officials who characterized Oppenheimer's security clearance loss as "perhaps the single most controversial act of the Commission during the 1950s." Seaborg responded positively to Federation of American Scientist clemency campaigns and worked diligently behind the scenes with other AEC commissioners to rehabilitate Oppenheimer. These efforts paid off when President John F. Kennedy finally awarded Oppenheimer the Enrico Fermi Award.[xxxix] President Lyndon B. Johnson presented it on December 2, 1963, somewhat restoring Oppenheimer's professional standing.

Glenn Seaborg claimed he was not a lifelong secret friend of Zalman Shapiro. He first met Shapiro after moving to Washington to assume the chairmanship of the AEC in 1961. But Seaborg was steadfast in arguing his processing loss theory and that Shapiro was "innocent until proven guilty." He strongly resisted lifting Shapiro's security clearances without the due process of a hearing even as evidence of Shapiro's ongoing contacts with Israeli intelligence operatives and serial deceptions piled up in the late 1960s.

Few of the folders containing Seaborg's papers have been as thoroughly "sanitized" by the Library of Congress for classified information as those about Israel. But Seaborg's papers, including notes to his biographer Benjamin Loeb, reveal his generally favorable attitude about Zionism even as Israel lobbyists worked tirelessly to bring him into their sphere of influence and use his fame to advance Israel's prestige while warding off unwelcome scrutiny.

Seaborg won his 1951 Nobel Prize in Chemistry for "discoveries in the chemistry of transuranium elements,"[xl] the subject of his many lectures to students, scientists and laymen. Seaborg responded to a solicitation by Democratic Party fundraiser (and later Israeli nuclear weapons funding coordinator) Abraham Feinberg on November 24, 1952. It was apparently his

---

[xxxix] Enrico Fermi was an Italian-born naturalized American physicist most recognized for his work developing the first nuclear reactor.

[xl] In chemistry, transuranium elements (also called transuranic elements) have atomic numbers greater than 92 (uranium's atomic number). None of these elements are stable and radioactively decay into other elements.

first, but far from last, contact with Feinberg. Feinberg was then serving as president of the American Committee for the Weizmann Institute of Science[xli] and asked Seaborg for a message to include in a booklet "saluting" the Weizmann memory to be presented in a December 1952 fundraising event at the Waldorf Astoria hotel in New York. Seaborg obligingly wrote "The life of Chaim Weizmann was, as he put it, a 'tug of war' between Science and Zionism. Yet in another sense the forces which struggled for fulfillment at his hands were not opposed, rather they were motivated by the same burning faith in the dignity of the human mind and soul. And there have been few whose lives could offer as much testimony to the victory of such faith in so many of the issues and enterprises of humanity. In the passing of Dr. Weizmann, science has lost a brilliant investigator, Israel a faithful son, and the world a truly great man."[242] [xlii]

Seaborg's papers reveal no direct involvement in John F. Kennedy's final high stakes nuclear negotiation with the Israelis. JFK's final demands were that there be internationally credible inspections and oversight of the Dimona complex. Practically alone, Kennedy was the greatest single threat to Dimona, NUMEC, future sales of nuclear-capable jet aircraft, and the unregulated operations of Israel's lobbyists in the United States. Kennedy's assassination on November 22, 1963 came as both a shock and challenge to Seaborg. Seaborg asked that he personally be allowed to conduct a neutron analysis of assassin's bullets in order to help resolve the crime.[243] The AEC was hosting an official entourage of Soviet nuclear scientists touring non-sensitive laboratories across the country at the time. After briefly stashing the Russians in Yosemite National Park, Seaborg and LBJ decided that allowing the tour to proceed was the best course of action as the administration scrambled to deal with the assassination. The dour and concerned countenances of Soviet scientists in posed photographs in front of civil nuclear power plants contrasts sharply with images of much happier foreign and "Atoms for Peace" visits.

Under LBJ, U.S.-Israeli nuclear activities quickly accelerated. On February 10, 1964 LBJ highlighted nuclear projects underway at a keynote speech to the Weizmann Institute.[244] The launch of a nuclear desalinization agreement with Israel was celebrated on October 14, 1964. The influence of Israel's nuclear funding coordinator on the LBJ administration may not have been entirely apparent to Seaborg. Although he noted longtime LBJ aide Walter Jenkins "moral" arrest on October 7, 1964 (for "disorderly conduct" with another man in a YMCA bathroom) and his own interview on Jenkin's character with the FBI, Seaborg had no idea what a crisis this caused in the administration. It forced White House spokesperson Bill Moyers and Meyer Feldman to move a quarter million dollar stash of cash Feinberg raised for LBJ from Jenkin's safe to a safer location during the investigation. Jenkins subsequently resigned.[245]

[xli] Later involved in nuclear weapons design.

[xlii] Weizmann revolutionized the production of acetone, vital for the efficient production of gun powder. Weizmann was a Zionist leader, the President of the Zionist Organization and became the first President of the State of Israel. He was elected in February of 1949 and served until his death in 1952.

Although protective of the AEC and sensitive to alienating top Jewish scientists working on U.S. nuclear programs, Seaborg demonstrated no obvious tilt toward Israel until his own address to the Weizmann Institute in 1971. During a visit to the LBJ ranch on December 10, 1965 McGeorge Bundy had to shoot down Seaborg's proposal that the AEC balance the Israeli desalinization project with a similar one for Eqypt.[246] LBJ's determination to advance U.S.-Israeli nuclear ties on a more exclusive basis slowly became clear to the AEC. In March the AEC loaned a cutting-edge food irradiator to Israel to boost agricultural exports. [247] Little did Seaborg know that Zalman Shapiro could use AEC-approved irradiator shipments as an opportunity to stuff the equipment with hundreds of pounds of U-235. But employee eyewitness accounts of such illicit shipments would not emerge until 1981.

Seaborg would soon be visiting Israel, aware of the U.S. failure to obtain inspection rights over Dimona. It was by then an open secret in Washington that the French-built Dimona plant was a nuclear reactor rather than a factory, the ruse publicly advanced by Israel. Seaborg referred to Dimona in coded language, writing that German diplomats promised him that they were not helping with "that plant so much discussed in Washington."[248] Seaborg also viewed with great curiosity that Israeli Prime Minister Eshkol assumed chairmanship of the Israeli Atomic Energy Agency from Ernst Bergmann on April 26, 1966. Israel's nuclear program had necessitated other such strange moves. Former Prime Minister David Ben-Gurion resigned rather than accept JFK's final demand for inspections of Dimona. As Eshkol helped to gradually roll back U.S. demands for transparency with LBJ's acquiescence, Eshkol himself was heading the agency in charge of Israeli nuclear weapons development. Seaborg skeptically noted "The Prime Minister took over the chairmanship, we understand, at the suggestion of personnel within that agency." The Israeli seemed defensive to Seaborg "He is reported to have stated that such a move by a head of government was not without precedent." Returning to his usual optimism Seaborg penned "Newspaper reports from Israel indicate that although Prime Minister Eshkol also holds the portfolio of Defense, he is not expected to give a military orientation to the Israeli nuclear energy program but, rather, to emphasize peaceful uses of atomic energy. This does not seem an unreasonable assumption as the Prime Minister was already overseeing all matters concerning nuclear desalting before assuming the atomic energy post."[249]

During his 1966 visit to Israel Seaborg parried reporter questions, and confirmed that he would not be "visiting the Dimona complex." But he refused to allow his visit to officially take place under the auspices of the Israeli Atomic Energy Agency, instead making arrangements with the U.S. State Department (which couldn't even provide a car and driver due to the presence of U.S. Congressional delegations on separate missions). Although he gave a lecture at the Weizmann Institute, Seaborg first visited Arab East Jerusalem. He noticed the Jordanian machine gun emplacements ominously overlooking the tense crossing point as he was whisked through the Old City and historic sites that would fall under Israeli control by mid-1967. [250]

After Israel attacked, Seaborg closely tracked the course of the 1967 Six-Day War writing on June 6 that "Israel had practically completed its conquest."[251] Two days later, Israel attacked and attempted to sink the USS Liberty, an intelligence gathering ship monitoring the hostilities. Secretary of Defense Robert McNamara and LBJ recalled two separate fighter aircraft rescue waves dispatched from a nearby aircraft carrier. A subsequent official cover-up claiming that the Israeli attack was a mere case of mistaken identity was debunked in 2002 when retired Capt. Ward Boston, a member of the inquiry board conducted 10 days after the attack, came forward. The board's conclusion was intended to let Israel off the hook, claimed Boston. No serious, credible formal inquiry to determine if the attack was deliberate was ever conducted.

Naïve and out of the loop, Seaborg again proposed dual Israel-Egypt nuclear powered desalinization plants as a "lever" for getting Israel into the Nuclear Non-Proliferation Treaty. "It seems to me that the recent events probably increase rather than decrease the danger that one of the Middle Eastern countries will feel, however mistakenly, that its best interest in the future would be served by the acquisition of nuclear weapons." Senator Henry "Scoop" Jackson, whose office was a nesting ground for many staffers now known as neoconservatives, quickly caught wind of the proposal and phoned Seaborg at home to pump him for information. Perhaps as a result, Seaborg's June 6, 1967 proposal to remove water as a cause for conflict on the Middle East chess board was again rejected as promptly has it had been at LBJ's ranch.

In early 1968 as LBJ and Israeli Prime Minister Levi Eshkol again publicly celebrated bilateral U.S.-Israeli agreements on nuclear desalinization, Seaborg noted an undersecretary of defense expressing the Pentagon view that Israel had embarked on a nuclear weapons program.[252] On February 18, 1969 Seaborg wrote "...Israel and West Germany might eventually sign the NPT. Following the hearing I proceeded directly to the Roger Smith Hotel and had a quick lunch with [AEC staffers] Kratzer, Labowitz and Rubin. Later in the afternoon I discussed with Howard Brown and Julie [Julius] Rubin the growing concern in the CIA and State Department that an officer of a certain industrial nuclear facility [Zalman Shapiro] may have diverted appreciable amounts of enriched uranium-235 to Israel over the last several years. This possibility has apparently been brought to the attention of the President."[253] On Thursday, February 20, 1969 Seaborg met "in Executive Session...to discuss the sensitive matter concerning alleged diversion to Israel of enriched uranium-235 by a senior officer of a fuel fabrication concern. J. Edgar Hoover has written to [William T.] Riley suggesting rather stringent Commission action. However, in our opinion, the evidence is far from proving that the alleged diversion took place." [254]

Seaborg's institutionally protective inclination to quash the diversion theory received support from the Defense Department. The NUMEC problem sprang up again on February 25, 1969 during a top-level meeting about nuclear materials cut-off discussions. "[Defense] Secretary Laird raised a question about a story he heard up on the Hill about our losses of nuclear materials. I briefly reviewed the matter of the alleged diversion of enriched uranium to Israel by the executive of a nuclear fuel fabrication plan. Mr. Laird's reaction was that we should make every attempt to squelch this story and not have it result in any Congressional Report."[255]

But Seaborg did not follow Laird's advice regarding Congress, and insisted on involving a Congressional oversight committee during a meeting the very same day with Attorney General John Mitchell. [xliii] "Howard Brown accompanied me to my meeting with the Attorney General.... I said that there was no proof that the material had been diverted to Israel and that it would be a mistake to prosecute on the assumption that an adequate case could be made. I also asked permission, which must be granted by the Attorney General, to pass on the latest information (that is, the Presidential and FBI interest in the case) to the Joint Committee. Mr. Mitchell said that he would let us know whether the Department of Justice planned to prosecute and whether we might inform the Joint Committee of the turn of events."

On April 3, 1969 Seaborg moved to advise the Joint Committee on Atomic Energy. "I called Attorney General Mitchell to tell him I am sending a letter to Chet Holifield, Chairman of the JCAE essentially along the lines that he (Mitchell) suggested in his letter to me of March 21. I said I thought we should be ready to expect that Holifield will ask to see the FBI file on this and this would be a request we would have to accommodate sooner or later. I told him I planned to try to deliver the letter myself in order to explain to him the sensitivity of the investigation. I told him I would be sending a letter to him today more or less recapitulating where we stand and putting forth our views as to what the best method of procedure is. I said I am referring to the person involved as 'the subject' rather than identifying him."

JCAE Chairman Chet Holifield was already well aware of the case and promised to get back to Seaborg after some official travel. Seaborg nervously fretted that it could "blow up into a big thing."[256] He was right. Seaborg's single page April 3, 1969 letter to Holifield cautioned that the FBI was investigating Zalman Shapiro's relationships with the Israeli government. It did not mention the possibility of HEU diversion. Another Seaborg letter to Attorney General Mitchell on the same day outlines the political concerns of handling Shapiro more directly. After reviewing some of the FBI's reports, Seaborg declared that no evidence of diversion existed sufficient to trigger review Shapiro's security clearances or termination of NUMEC's classified contracts. Seaborg argued that while no formal process would be needed to terminate contracting with NUMEC, Shapiro would likely insist on a detailed explanation for why business was being cut off. He did not reveal to Mitchell how any such cutoff would likely backfire on the AEC, which had lured Atlantic Richfield to buy NUMEC in order to obtain lucrative Hanford contracts.

Seaborg argued that to even interview Shapiro, AEC protocol would require advising him of his right to involve legal counsel and that his testimony could be used against him. But interviewing Shapiro could also ward off the swarm of Israeli intelligence operatives that surrounded Shapiro. "An interview by the AEC might serve to clarify information on the subject's associations with Israel and hopefully might provide additional information on his contacts with

---

[xliii] Mitchell served as Nixon's campaign manager in 1968 and 1972. During the Watergate scandal, he became the only attorney general ever to be convicted of illegal activities.

members of the Israeli government and Israeli embassy staff. An interview could also serve to diminish Israeli interest in the subject as a possible source of assistance and possibly inhibit his activities on behalf of Israel." Seaborg thereby tacitly acknowledged that Shapiro's relationships were indeed highly questionable and that the suspicions raised were largely self-inflicted.

Seaborg did not want to reveal the full extent of FBI surveillance or have Shapiro unleash a public relations campaign charging government agencies of discrimination or anti-Semitism, a highly likely outcome. "On the other hand, a substantive interview would almost certainly establish that an investigation had taken place and he might deduce the nature of it. We could not be sure that an interview would not evoke public charges by the subject that he was being victimized by the AEC and FBI because of his support of the Israeli cause and had been subjected to an unlawful invasion of his privacy." This, of course, would be precisely the case Shapiro and his legal counsel would make during later testimony and via allies in the news media, all the while avoiding direct questions that would reveal his position within Israel's nuclear program.

Seaborg seemed to struggle over the relevancy of Shapiro's loyalty to the United States. At his direction the AEC prepared a 181 page draft report about SNM safeguards in 1967. The report stressed how import verifying operator loyalty to the U.S. through the security clearance process was to safeguarding SNM against diversion. But the draft report was never finalized or presented to the JCAE. Seaborg simply penciled "not used" over the draft report cover page and filed it away.[257]

Seaborg was nevertheless very interested in a November 3, 1968 meeting at Shapiro's home staked out by the FBI, and any evidence about whether classified information stored at NUMEC had exchanged hands.[258] If Seaborg had fully understood exactly who Shapiro was meeting with, it might have fundamentally changed his opinion of NUMEC. But the relevance of Shapiro's contacts became available much too late, even as Seaborg appeared to remain almost willfully ignorant of other key facts he should have known about as they emerged.

The reality for Seaborg was that AEC's congressional appropriations were at risk if the NUMEC matter were not handled carefully. Seaborg worried about the coming media frenzy as a congressional staffer confided to him on April 4, 1969 "Holifield is still sweating it out how to get this information out without hurting programs, individuals, etc. because this type of loss cannot stay hidden..."[259] But in the end, the lengthy, exhaustive FBI investigation of Shapiro and NUMEC brought no Justice Department actions. In a letter to the FBI dated April 15, 1969, the Department of Justice "advised that based on the merits of the FBI investigation conducted to date, the facts of the matter were not such as to warrant action against Shapiro."[260] But the matter was far from over.

On May 2, 1969, Attorney General Mitchell declined approval for the JCAE to review the FBI file on NUMEC and Shapiro. Seaborg fretted that "It seems certain that this will lead to a demand by Holifield that this file be made available. I shall probably call Attorney General Mitchell to discuss with him the implications of his refusal and the possible need for reconsideration, in view of the fact that the outcome can probably only be eventual assent to this request."[261] But the Justice Department held firm in its refusal, alerting the AEC on May 7. Seaborg suggested AG Mitchell call Holifield and explain that his request for review was being carefully considered.

The Shapiro problem seemed to be cloning itself. On May 13, 1969, Seaborg panicked. "I called Herman Pollack [State Department's international scientific and technological affairs bureau director] to alert him that [Edward] Teller is considering accepting a part-time position (which will involve four weeks a year) as chairman of the board of the Israel Research and Development Corporation, Ltd. I said this will be a source of concern to some people because of Teller's background dealing with the H-bomb. Teller is adamant that he will not divulge anything to the Israelis; in fact, he has told them that if he learns anything in Israel, he will certainly pass it on to the U.S. Pollack said he would pass this on to the proper people at State; I asked that, if he receives any adverse reaction, he let me know because Teller says he would need to know within a week." [262]

On June 22, 1969 Seaborg proposed bringing in Peter Flanigan, the Nixon administration's "Mr. Fixit" for any crisis involving business interests and public policy, about the problems that might arise from interviewing Shapiro. Mitchell agreed to help and suggesting wiring in John Ehrlichman as well. [263]

At a June 25, 1969 summit between Flanigan, Seaborg, and AEC security officials, it was concluded that "the benefits of an interview outweighed the risks and that the AEC should proceed." Howard Brown sensed the AEC's cold feet in a June 30 file memo, that despite the approval "...another consideration should be the possible, however remote, embarrassment with respect to our relationships with the country involved; and that, therefore, we intended to advise the Department of State of our intentions..."[264] But the State Department did not object and on August 14, 1969 the AEC interviewed Shapiro at its H Street offices in Washington.

The AEC's official account of its August 14 interview of Shapiro, titled "Informal Interview of Dr. Zalman Mordecai Shapiro," was written up by the AEC Office of the General Counsel and William T. Riley (Division of Security). The AEC positioned the interview to Shapiro as an action in "connection with the recent FBI investigation to determine the nature and extent of his relationship with the Government of Israel and certain Israeli nationals."[265] Unbeknownst to Shapiro, the FBI investigation had escalated to an "intelligence basis" in the spring of 1969 "in view of the subject's pronounced Israeli sympathies." The FBI had even received permission from the attorney general in March to continue electronic surveillance[xliv] of Shapiro for another three months and distribute the information to the AEC, State Department and the Department of Justice. [266] Based on what the FBI found, action was to be taken on the "subject's security clearance and classified contracts held by this company."

The FBI found plenty. Summaries of wiretap transcripts reveal extensive contacts with a cagey Avraham Eylonie, who tried desperately to calm down Shapiro about Israel's military situation during a phone call on March 19, 1969,

---

xliv The FBI carefully recorded the cost of that surveillance involved coverage of Shapiro's home, mobile surveillance, three telephone trunk lines tapping three private lines for $69.90, 72 hour per week monitoring by a GS-9 and GS-13 special agents and employees.

telling Shapiro he would "disclose his reasons during their meeting."[267] The next day the FBI wire-tapped Shapiro complaining about statements made by the Egyptian minister of information on a local radio station, suggesting that Albert Bloom, editor of the Pittsburgh Jewish Public and Education Foundation be allowed to broadcast a rebuttal. [268]

By March of 1969, the FBI observed Shapiro was clearly not in good health. His sister Zipporah Schefrin described him as "near physical or mental breakdown" and "visiting a psychiatrist on a regular basis." [269] There appeared to be good reason for Shapiro's near breakdown. On March 29, 1969 an employee at NUMEC phoned Shapiro to inform him of a major plutonium shortage. Shapiro quickly calculated it would cost the company $600,000. Then "they discussed the possibility of an AEC inspector learning of the shortage." Later, two and a half kilos were located "leaving 13.5 kilos unaccounted for." The same day, Shapiro contacted another NUMEC employee to discuss shipping "difficulties" relating to an irradiator manufactured to preserve potatoes and onions for Israel.[270]

The FBI surveilled and photographed Shapiro's meeting with Israeli embassy Scientific Attaché Jeruham Kafkafi on April 1, 1969. The FBI later came to believe Kafkafi was an Israeli intelligence agent working under Avraham Hermoni's LAKAM espionage team.[271] On April 14, Shapiro commiserated with a NUMEC employee that a forthcoming AEC Nuclear Safeguard conference might impose new inspections and safeguards "in addition to international inspections." [272]

By May Shapiro was chafing under Atlantic Richfield's ownership of NUMEC and executive prerogative. In discussions with the Israeli Government's "Israel Investment Services" representative on May 2, 1969 over a new venture to manufacture sterile hypodermic syringes for Jewish hospitals, Shapiro revealed how he had almost told Atlantic Richfield to "go to hell." He had requested company permission to go to Israel in June but "his employers wanted him to visit Europe" on company business. Shapiro could no longer direct NUMEC's operations to benefit Israel. But a more rational Israeli Government official "discouraged the subject from taking this cause of action by informing subject that he was too valuable in his present position with the Nuclear Materials and Equipment Corporation (NUMEC), Apollo, PA." [273]

On May 5 FBI wire tappers caught wind of a dangerous nuclear spill at NUMEC (which were only fully declassified late in the year 2011) due to knowingly illegal storage practices. A day later, Shapiro met with AEC Commissioner Glenn Seaborg. The subject of their meeting may have been persuading the Justice Department not to make Shapiro's FBI file available to the JCAE. Seaborg subsequently reversed himself and contacted the Attorney General in May to extract his promise that the file would in fact not be made fully available to the AEC and NUMEC's congressional overseers.[274]

On May 8, Shapiro fretted over the loss of a $250,000 Navy fuel contract that "leaves NUMEC with no classified material." [275] NUMEC was a repository for thousands of classified documents relevant to its work for the government. Lamenting the loss of access to government classified documents (or possibly the SNM) rather than the revenue from a large contract raised yet another red flag about whether NUMEC was a nuclear reprocessing business or a covert front operation. Other conversations revealed new information gathering tasked by

Avraham Hermoni to the attendees of the previous 1968 meeting in Shapiro's home.

On August 14, the AEC carefully explained to Shapiro that he was being interviewed because the AEC routinely reinvestigated key officials to "assure their continued eligibility for access to classified information." The AEC was blunt. Despite a recent reinvestigation it still had questions about Shapiro's Israeli relationships. Shapiro claimed he could not recall the names of all of the individuals who attended a 1968 meeting in his home under the guidance of Avraham Hermoni. He claimed the meeting was organized by a University of Pittsburgh professor and focused on detecting contamination of Israel's water supply, technical methods to thwart "saboteur type attacks" particularly infrared detection. Shapiro said he merely contributed his views about how to resolve long-life power source challenges.

Given that much of the discussion guided by Hermoni was related to military affairs, Shapiro could not credibly explain to AEC interrogator Riley why Hermoni had not instead formally "requested information on these topics from the U.S. military." [276] However it is clear from the AEC's interview report that the Israelis were interested in establishing direct relationships with their elite American science and business contacts in ways that avoided official U.S. government involvement, knowledge, regulation and possible refusal. The singular importance of this relationship — whether the ideology driving such direct assistance was deep enough for Israel to obtain nuclear materials directly from its civilian "helpers" in the U.S. — was left largely unexplored by Riley. Shapiro claimed the University of Cornell scientist who called together this "Committee on Science Based Industry" (a group to boost Israeli technology exports) midway between Cornell and Washington because of the "lack of coordination between several committees which had been established to assist Israel." Another group disclosed by Shapiro, the "Prime Minister's Economic Committee," sought investment funds in the U.S. for Israel. According to Shapiro "people who were trying to help [but] were going in all directions and that as a result very little was being accomplished in the U.S." [277] Given his prior run-in with the Foreign Agents Registration Act, it would have been worthwhile to ask Shapiro how a committee organized under the auspices of a foreign prime minister could operate legally without registering under the Act.

Shapiro displayed a studious and suspicious lack of interest about Hermoni's background and was vague about exactly when he first met him, thinking it was probably when he was a professor at a university.[xlv] He claimed he did not know whether Hermoni was associated with the Ministry of Defense, saying "many Israelis have dual occupations." Yet Shapiro had been handed off by Colonel Hillel Aldag like a prized covert agent at an Israeli embassy event held so Shapiro could renew his acquaintance with Hermoni, Aldag's successor.

Riley then entrapped Shapiro. Privy to FBI surveillance complete with photographs of a suspicious meeting between Shapiro and Jeruham Kafkafi on June 20, 1969, Riley asked Shapiro when he had last met with the Scientific

---

[xlv] Likely Technion, Israel's "MIT" and a key player in nuclear weapons development.

Counselor to the Israeli Embassy. Shapiro claimed he had seen Kafkafi about half a dozen times, with the last occurring in May. Shapiro did not admit to meeting with Kafkafi in the Pittsburgh airport on June 20 until pressed by Riley. Surprised and off-balance, he claimed the meeting was about payment of an outstanding bill and that Kafkafi had suggested they meet in Washington, but traveled to Pittsburgh at the suggestion of Shapiro.[xlvi] He further claimed the airport meeting was an opportunity to talk about whether a power supply unit for an "incursion detection device" could be adapted to Israel's environment and also discuss difficulties in potato irradiation.[278] This deft maneuver put Riley in a position of interrogating an innocent defender of Israel, eager to improve the country's security. This was a place many investigators of NUMEC frequently found themselves.

Shapiro told Riley, who was in a position to know, that the only classified work being done by NUMEC was producing Navy fuel, and that he had never been asked to or passed classified documents to persons not entitled to receive them.[279] To have admitted such activities could have been prosecuted under both the Espionage and Atomic Energy Acts. Shapiro then claimed NUMEC had been "blackballed" over the processing losses of 1965 and that it had been recently denied a contract he expected to receive, though he did not mention his real disappointment over losing access to classified material. Shapiro thought the entire Riley interview was taking place because "the JCAE had indicated he had probably carried the unaccounted-for material to Israel, presumably for use in the Israeli weapons program." He characterized that theory as "preposterous" and purposely leaked to the *New York Times* and *Wall Street Journal* to write "asinine" articles about the "whereabouts of the material" which injured the reputation of NUMEC.[280]

True to AEC form, Riley rushed to reassure Shapiro that "insofar as the losses of material at the NUMEC plant were concerned, the Commission had conducted a complete investigation and that this matter was closed. We pointed out that we had not brought up the issue of the losses of material and that this interview was not in fact caused by the previous material losses at NUMEC."[281]

Shapiro then wondered out loud what the interview was about, if not the 1965 losses. On August 26, 1969 he found out. The AEC again contacted Shapiro to flesh out his portrayal of his June 20, 1969 meeting with Kafkafi. The AEC stated flatly that discussing an unpaid bill, problems with irradiators and power supplies did not seem to warrant a rushed meeting at the airport. Would Shapiro like to add anything to his characterization of the meeting?

Shapiro said there was some chit-chat about Israel's political and economic situation as well as speculation over who would be the next prime minister. But Israel's bill was not only delinquent, it was in the neighborhood of $32,000. Shapiro claimed he had escalated the payment matter to Hermoni as well. He also claimed victory because "the account was settled shortly thereafter." Beyond

---

[xlvi] When interviewed by members of Congress in 1978, Shapiro would claim it was another Israeli who had met with him, and that the Israeli was visiting his daughter in Dayton, Ohio and had simply stopped along the way to meet.

that "Dr. Shapiro stated he had nothing further to add in respect to the meeting at the Pittsburgh airport." [282]

Seaborg eagerly seized on the most positive aspects of the interviews. "While the subject was not entirely candid in the interview...he did state that throughout his associations with Israeli officials he has never been asked to furnish classified information, has never furnished and would not, if asked to, furnish classified information to Israeli officials or to other unauthorized personnel. We are sending letters with the gist of this information to Mr. J. Edgar Hoover, Mr. J. Walter Yeagley (Assistant Attorney General), Congressman Chet Holifield and will also inform Mr. Flanigan at the White House and a representative of the State." [283] But Shapiro's lack of credibility was already well-established. Had Seaborg really expected some kind of confession or inadvertent disclosure?

An AEC letter to the JCAE on August 27, 1969 candidly documented Shapiro's evasiveness, particularly over the 1968 meeting. "The Commission staff who interviewed Dr. Shapiro report that in his discussion of the subjects covered at a meeting with an Israeli official that was of specific interest to the AEC, Dr. Shapiro's recollection was consistent with other reports that the AEC had received of the meeting. However they reported he appeared to be less than completely candid in the discussion of his relationship with some officials of the Government of Israel to the extent that he was vague and uncertain as to details that he should have been able to recall of several meetings he had had with these Israeli officials." However, based on Shapiro's assertions about classified information, the AEC told the FBI and Justice Department it 'does not contemplate further action in this matter at this time.'" [284]

FBI Director J. Edgar Hoover may have been livid as he fired off a September 3, 1969 memo to the AEC. "We have conducted a thorough and extended investigation of Shapiro for more than a year, including substantial physical surveillance coverage. We have developed information clearly pointing to Shapiro's close contacts with Israeli officials, including several Israeli intelligence officers. It is believed most unlikely that further investigation will develop any stronger facts in connection with the subject's association with Israeli officials."

The FBI director pleaded "The basis of the security risk posed by the subject lies in his continuing access to sensitive information and material and it is believed the only effective way to counter this risk would be to preclude Shapiro from such access, specifically by terminating his classified contracts and lifting his security clearances. However, after careful consideration, including an interview with Shapiro, you have advised that your agency plans no further action in this matter at this time." But the AEC wasn't about to unwind its dubious processing loss theory or huge success stashing NUMEC within the hapless Atlantic Richfield.

The FBI director caved in. "Under these circumstances, we are discontinuing our active investigation of the subject. We will, of course, continue to keep interested agencies advised of any pertinent information concerning the subject which may be received from our sources." [285] For Hoover the missing HEU, Shapiro and his ongoing Israeli intelligence associations were a toxic mix that presented an obvious risk to U.S. national security. All of the warning signs that

Shapiro was a key player in an Israeli nuclear espionage network are present in the FBI reports circulated to the AEC, State and Justice Departments. Shapiro's high "value" to Israel derived of his role running a nuclear reprocessing facility, his participation in an economic intelligence gathering of the type that previously put the ZOA under a FARA investigation. This exacted heavy psychological stress requiring psychiatric treatment. The troubling new allegations of significant amounts of missing plutonium that had to be kept hidden from the AEC were dropped by the FBI. Shapiro's burning worries of ever more stringent AEC inspections were left unexplored, along with his concerns that NUMEC was being cut off from the flow of government classified documents and nuclear material. None would probe Shapiro's care-free attitude about dangerous toxic spills in the environs of Apollo. But why?

Private memos from Henry Kissinger to President Richard Nixon written in July of 1969[xlvii] reveal that Israeli demands for advanced jet fighters, the secret nature of its nuclear weapons programs, and Nixon's fear of the Israel lobby rendered the administration incapable of dealing with nuclear theft from NUMEC. In a July 19, memo, national security advisor Kissinger told Nixon that "The Israelis, who are one of the few peoples whose survival is genuinely threatened, are probably more likely than almost any other country to actually use their nuclear weapons." Nixon was at the time preparing for a visit by Prime Minister Golda Meir during which nuclear weapons issues would be discussed. Kissinger was blunt in another lengthier memo about NUMEC and Shapiro. "There is circumstantial evidence that some fissionable material available for Israel's nuclear weapons development was illegally obtained from the United States about 1965."[286]

Kissinger's July 19 top secret memo to the president bears the subject line "Israeli Nuclear Program" and recalls to Nixon that two top-level meetings attended by CIA Director Richard Helms and other top cabinet members considered analysis from "a small working group." Kissinger recalled that Israel already had 12 nuclear capable surface-to-surface missiles bought from France and would take delivery of more to achieve a force of 30 by 1970. Although they had promised "not to be the first to introduce nuclear weapons into the Near East" in exchange for U.S. Phantom aircraft, Kissinger noted this was just an Israeli word game. By Israel's contorted definition "introduction" only meant making public, testing, or being detected deploying nuclear weapons. The issue was assuming new urgency because the Phantoms Israel wanted were capable of carrying nuclear weapons.

[xlvii] The papers were released by the National Archives and Records Administration on November 28, 2007 as a result of Mandatory Declassification Review filings. Declassification authorization was given by equity holders on June 4, 2007. But the Kissinger memo (included in the appendix) about theft of nuclear material was never posted online by NARA with other contemporary documents and is only available to researchers who either travel to the Nixon Presidential Library document room in Yorba Linda, California or file a special request. Many NARA facilities hold such important public, yet hard to access, files.

The U.S. had already signed a contract with Israeli ambassador Yitzak Rabin, clearly specifying that "possession" constituted "introduction" and that the U.S. could cancel planned August and September 1969 aircraft deliveries if Israel was found to be in possession of nuclear weapons. In a section titled "What We Want" Kissinger was clear that "Israel's secret possession of nuclear weapons would increase the potential danger in the Middle East, and we do not desire complicity in it." But then Kissinger justified a mass deception campaign that continues to this day. "Public knowledge is almost as dangerous as possession itself. This is what might spark a Soviet nuclear guarantee for the Arabs, tighten the Soviet hold on the Arabs and increase the danger of our involvement. Indeed, the Soviets might have an incentive not to know."

Kissinger proposed that "while we might ideally like to halt actual Israeli possession, what we really want at a minimum may be just to keep Israeli possession from becoming an established international fact." While most of the U.S. working group wanted to actually prevent Israel from going nuclear by taking firm action, Kissinger felt that the Israel lobby in the U.S. would strongly react to any withholding of arms. "Our problem is that Israel will not take us seriously on the nuclear issue unless they believe we are prepared to withhold something they very much need—the Phantoms or, even more, their whole military supply relationship with us. On the other hand, if we withhold the Phantoms and they make this fact public in the United States, enormous political pressure will be mounted on us. We will be in an indefensible position if we cannot state why we are withholding the planes. Yet if we explain our position publicly, we will be the ones to make Israel's possession of nuclear weapons public with all the international consequences this entails."[287]

There is historical consensus that Nixon agreed to an "ambiguity" policy during his meeting with Prime Minister Meier on September 26, 1969. Israel would conduct no nuclear tests, declarations and would hide deployment while the U.S. would "stand down" pressure and pretend that by failing to acknowledge or discuss the obvious, Israel's nuclear arsenal would never become an "established international fact."

In hindsight, Kissinger's presentation of such a limited set of options demonstrated a dangerous lack of imagination. What if Nixon had instead publicly insisted the Israelis subject themselves to inspections and sign the Nuclear Non-Proliferation Treaty in exchange for Phantoms and their future U.S. military relationship? What if he rejected outright the notion of Israel orchestrating yet another covert U.S. public relations campaign and instead rolled up all known members of the LAKAM network by prosecuting Shapiro, Kafkani, Hermoni, Lowenthal and any other operatives captured within the U.S. such as Rafael Eitan? Nixon could have declared loudly to the world that his commitment to non-proliferation applied to both rivals and friends—and taken away leverage Israel would soon use to blackmail America and delay bona fide negotiations on land and reparations for the Palestinians.

But such thinking was beyond the capability, ideology or private political commitments of Nixon's advisor Kissinger. The documentary record reveals that "standing down" resulted in calling off the FBI and Justice Department investigation of Shapiro and NUMEC.[288] While Nixon's capitulation rewarded

Israel's deceitful treatment of the U.S. during Israel's build-out and launch of Dimona, it had a longer-lasting domestic impact. The accord taught parastatal operatives functioning within the Israel lobby and Israeli covert agents that even the most outrageous clandestine operations targeting America would not be punished if they were justified on the basis of securing Israel. They could count on the U.S. government secrecy to avoid disturbing the public, which might become alarmed at the decrepit state of U.S. governance if they are ever revealed. Journalist John Fialka dubbed NUMEC a "self-concealing problem"[289] inside the U.S. government, which continues to devote considerable resources toward keeping official information about Israeli nuclear weapons out of the reach of concerned citizens.

Israel, which had benefited enormously from such crooked arrangements since its founding, immediately took advantage of the Nixon-Meir accord. Shapiro quickly maneuvered himself into place within another private corporation. Continued access to classified material would soon no longer be an issue for Zalman Shapiro and his handlers, or so they hoped.

9. Shapiro joins Kawecki Berylco Industries

Shapiro resigned from NUMEC in June of 1970, a little more than three years after it was acquired by Atlantic Richfield.[290] Discouraged by NUMEC's prospects for winning government contracts, experiencing heavy psychological stress, and chafing under ARCO's leadership, Shapiro began working his network of contacts in the summer of 1969 for another job. On July 10, 1969 Shapiro discussed leaving NUMEC with Jack Goldman, chief scientist of Xerox Corporation. Goldman went on to found the Palo Alto Research Center in 1970. Shapiro confided in Goldman his disappointment that "he made inquiries regarding job opportunities in Israel but found nothing to his complete satisfaction."[291]

Shapiro had many attractive options that would allow him to unleash his expertise in civilian or naval nuclear power systems. Given his career trajectory and financial resources, he did not need to take just any job. Under the terms of the AEC engineered buyout, NUMEC shareholders received .255 shares of Atlantic Richfield for every share they held. Shapiro held 19,588 shares of NUMEC stock.[292] At the time of the 1967 buyout, his 4,995 converted shares of Atlantic Richfield would have been worth $449,545. Shapiro was also a director of Apollo Industries, which also benefitted from the NUMEC sale. On July 16, 1968 Atlantic Richfield implemented a two-for-one share split. If Shapiro retained his shares, which is normal for key executives during buyouts, Shapiro would have held 9,990 shares valued at $492,002 ($2.8 million in today's dollars) by the time he finally left NUMEC.

As if seeking to test the outer limits of the AEC and Nixon administration's patience, Shapiro quickly signed on at Kawecki Berylco Industries as a corporate vice president in July of 1970. It was a new position specially created for him at the company. In the first week of August, Shapiro requested a classified briefing which prompted a company check of his level of security clearance.[xlviii] Upon confirmation that Shapiro did not have AEC clearance for advanced nuclear weapons work, the meeting was promptly canceled.[293]

The Shapiro controversy was once again on AEC Commissioner Seaborg's desk. On October 14, 1970, Seaborg despairingly wrote "in executive session we discussed the case of the individual whose clearance has been in question and who is now requesting clearance which includes access to weapons information in connection with the requirements of his new company affiliation. This is a very difficult situation, and we will have Security staff discuss it with him with the view of his possibly having a Personnel Security Board review, a method of

---

xlviii Q and L are non-military security clearances. Most cleared NUMEC employees had L level clearance, a lower level access to confidential and secret classified information, which permitted unescorted access to protected areas. Q clearance gives access to information classified as "Top Secret" such as nuclear weapons design information.

procedure which I inherently deplore. We will also check with the White House to get their views." [294]

On October 29, Acting Assistant Attorney General John F. Doherty filed a proposed Security Statement noting that "Subject is now employed by Kawecki Berylco Industries, Inc. which has available all current and advanced weapons concepts and techniques related to beryllium[xlix] weapons components...subject has knowingly established an association with individuals reliably reported as suspected of espionage...and engaged in conduct which tends to show that subject may be subject to influence or pressure which may cause him to act contrary to the best interests of national security."[295] Shapiro was placed into the FBI's "security index," a paper based system to track U.S. citizens considered to be dangerous to national security, which was distributed to the AEC and other government agencies and relevant FBI field offices. Derived from FBI reports dating back to 1963, Shapiro's security index listing was unusually blunt. (See the appendix for more a declassified version.)

One entry about Shapiro's voluntary association with Avraham Hermoni notes that Hermoni was "known to be engaged in the establishment of a technical intelligence network in the United States; and that prior to his assignment as Scientific Counselor of the Israeli embassy, Washington, D.C. he was intimately connected with Israeli efforts to develop nuclear weapons in his capacity as Technical Director, Armament Development Authority, Ministry of Defense."[296]

The question over public disclosure of Shapiro's lengthy FBI file arose again for Seaborg. "We have sent to the Justice Department a draft of a proposed letter of notification that would be sent to the individual, provided Justice would be willing to release the information contained in the FBI reports; without the FBI information, a meaningful hearing could not be developed. Of course, release of this data would involve the names of people in the Israeli embassy. I said I am not at all sure that we should hold a hearing; alternatively, we could grant him clearance, but keep close surveillance over his activities involving weapons aspects. "[297]

The Justice Department once again refused to release the FBI file on Shapiro as it had earlier done with the JCAE's request. The AEC held a meeting on November 23, 1970, scrambling to implement an ad-hoc honor system solution, custom-built for Shapiro. "We decided, on the basis of the Department of Justice's decision that they couldn't make available in a hearing the information they have, that we will not conduct a hearing in connection with granting clearance for weapons information to the individual whose case I had been discussing with Attorney General Mitchell; instead, we will ask that the individual sign an affidavit under oath, that he has not passed on sensitive information to any unauthorized person, that he will not do so in the future, and that he doesn't intend to move to another country..."[298]

This ad hoc approach did not satisfy Justice Department concerns. Assistant Attorney General Mardian huffed "it is the considered opinion of the Department of Justice, including John Mitchell and the White House and perhaps involving

---

[xlix] Beryllium has been used since 1956 in fusion boosted fission to both increase yield and to lighten and miniaturize weapon size.

Henry Kissinger and even the President, that clearance should be denied without offering a hearing." Seaborg was despondent. "This causes some consternation among [AEC] Commissioners because this will be the first instance in which such an action has been taken in the history of the Atomic Energy Commission. It was agreed that I would talk to Mitchell about this, apprise him of our views, and seek to determine whether some way of handling the situation through informal contacts with the man or his company might be tried."

Shapiro, annoyed that his higher security clearance was inexplicably being held up, retained the services of top trial lawyer Edward Bennett Williams of the firm Williams & Connolly and began pressuring the AEC. During the month, he again tried to get a classified briefing from his employer, who reminded him that "he could not discuss the project since KBI had not yet received an AEC authorization to grant Dr. Shapiro access to this information." Shapiro, still demanding access, instead received a low-level briefing on project staffing requirements and production schedules. KBI's President sternly informed management that Shapiro did not "have appropriate AEC clearance and should not be given access to project information."[299] Shapiro once again found that he was no longer his own boss and free to shape his activities to Israel's benefit. But the chance to work on state-of-the-art weapons design at KBI was worth fighting for.

On January 21, 1971 Seaborg met with Attorney General Mitchell as a Shapiro advocate, explaining "the charges were essentially without substance and that I strongly opposed denying a clearance without going through the hearing process. I said this would be the first time that this had been done in the history of the AEC and also that it had never been done before by any other government agency." Mitchell was unmovable. The Justice Department view was that "charges were serious enough so that the man should not have access to sensitive weapons information." Mitchell thought that given Shapiro's retention of legal counsel, the matter should now be resolved by the courts. Seaborg felt the newspapers, not the courtroom, would be the primary field of battle and that the AEC would be the ultimate loser. "I emphasized that this could lead to a sensational public relations problem since the man is being defended by Edward Bennett Williams and [Harold] Unger and they intend to put up a public fight in order to vindicate the man's honor." Once again, Seaborg insisted on first consulting other executive departments. Mitchell agreed to this, suggesting Seaborg contact former Attorney General Bill Rogers, who was now Nixon's secretary of state and National Security Advisor Henry Kissinger, to be followed up by a meeting between the four.

In a January 22, 1971 briefing, Seaborg convinced Secretary of State Rogers that there was an imminent danger of orchestrated publicity. "I explained the background for the case, including the early allegations that the individual had diverted enriched uranium. I emphasized that a denial of a hearing would have very severe public relations aspects and that I thought the person ought to be cleared, but with somewhat limited access to information." Rogers saw the dangers of denying a hearing. [300]

On January 25 Seaborg and a small delegation sat down with Henry Kissinger at the White House. Seaborg became an even more unabashed Shapiro advocate, apparently convincing himself that all diversion suspicions were "false" and that denial of full access to advanced weapons secrets was unwarranted. "I described the case involving security clearance, including his [Zalman Shapiro] early involvement with a nuclear fuel processing plant and the false allegations of diversion of material, the present request to upgrade his clearance in order to receive weapons information, and the Department of Justice's recommendation that he be denied clearance without affording the opportunity for a hearing. I said that the AEC and the Administration would be very vulnerable should this procedure be followed because this would be the first time it would ever have been used, and scientists all over the country would be up in arms over it. I indicated that the Commissioners and staff were unanimous in their opposition to this course of action. Kissinger expressed great doubts about such a denial of clearance and said he would get in touch with Mitchell and I meeting to discuss the matter." [301]

Seaborg also lobbied Nixon's Science Advisor Ed David, downplaying the risk of giving Shapiro access to advanced weapons designs. "I had lunch with Ed David in the special dining room of the White House mess. I described the case involving security clearance in some detail, as I had with Kissinger earlier. I emphasized the concern that scientists would have if such clearance was denied without a hearing, and the injustice of such a procedure, and the fact that the information for which he would be cleared is not so critical and that the Commissioners and staff are unanimous in opposing the proposed procedure. David was very sympathetic to our point of view and indicated that he would be ready to talk to Mitchell about it." [302]

On January 27, Kawecki Berylco Industries Vice President Edmund Velten met with AEC Commissioner James T. Ramey to ascertain why there was still a hold on Shapiro's security clearance upgrade. Ramey explained "in a general way the problem involved and this led to the suggestion that the individual be permitted to carry on limited activities in connection with his weapons responsibilities without upgrading his clearance." [303] But rather than calm KBI executives, the news set off liability alarm bells.

According to a January 28 Seaborg meeting memo, KBI was becoming very concerned about the overall negative business impact of the Shapiro controversy and also began piling on pressure. "Ramey reported on his discussion with Ed Velten of Kawecki about the present personnel security case active at the Commission level. The company's senior management was reported to be increasingly concerned about the delay in availability of the individual. Mr. Abeles, Chairman of the Board of Kawecki, or Mr. Lowry, President may be contacting me shortly to try and resolve the matter. Ramey reported he was not encouraging about near-term solution of the problem but urged that no action be taken that would affect the individual's employment. Hollingsworth was instructed to take a new look at the significance of access to weapons information under this contract and separately to investigate whether alternate contract work might be substituted for present activities of the company. [304]

On February 3, Shapiro's attorney Edward Bennett Williams personally called Seaborg demanding action. "He said that his client's employer, Kawecki Berylco Industries, is getting impatient and won't wait very much longer to find out whether this individual has a clearance." Seaborg dialed up Ed Davis about Shapiro's security clearance the very next day. "He said that he and Flanigan are working on it with the attorney general. He said he took the line that he hasn't gone into the merits of the case, but we want to see everybody treated fairly, and if there is a decision to do anything other than grant the clearance, then it has to be done in accordance with tradition and accepted procedures; otherwise he would not go along. I said I had a call from his [Shapiro's] lawyer yesterday." [305]

The Justice Department did not budge and firmly insisted on denial of security clearance without a hearing.[306] Henry Kissinger lobbied Attorney General Mitchell, but, as Seaborg nervously penned on February 8 "Mitchell was still of the view that the individual in question should be denied clearance without a hearing." [307]

On February 8, 1971 Seaborg presided over yet another AEC meeting overshadowed by the Shapiro controversy. "In an executive session at the end of the meeting we discussed further the security case and decided that we would have Vinciguerra...interview the subject with the view of trying to convince him to withdraw his request for clearance—Vinciguerra will offer to help find him another position." John C. Vinciguerra was a director of the AEC division of contracts, giving him a high degree of influence over the huge corporations active in the nuclear industry. Seaborg realized that the potential for sensitive technology leaks to Israel was much broader than just Shapiro. "We also discussed the case of another scientist who, following a trip to Israel, reported to the CIA that he thought officials of KMS have given information on laser-induced fusion reactions to officials in Israel. When Riley and Hennessey investigated this with the KMS officials, these officials denied it in a convincing manner." [308]

On February 9, 1971 Seaborg and Ed David strategized over how to resolve the Shapiro security clearance issue. "Ed David called to tell me he plans to see the attorney general on our security clearance case. David said he has had some reaction now from the White House. He said that Mitchell needs to be convinced that if the request for clearance were handled in the way Justice is advocating, there would be a tremendous outcry. He asked how we could convince him of this. I said we have no precedents; however the reason I feel there will be an outcry is that this individual has hired these particular lawyers and he had indicated that he will try to vindicate his honor."

Why was Shapiro willing to fight so hard for the KBI position? It clearly wasn't about the salary or the possible economic consequences of an employment interruption. The FBI drolly noted "Father Abraham Shapiro, Mother Minnie Shapiro, deceased. Subject mother-in-law, who died in approximately September, 1968, left estate valued at $150,000. "[309] Why was Shapiro willing to expend a portion of his NUMEC buyout windfall fighting the federal government with high-priced lawyers? Seaborg continued to fret. "I said we are considering approaching this gentleman to try to get him to withdraw his request on the basis that it would hurt everybody, and then we would try to help him find a job elsewhere. [Ed] David asked what the chances are of denying this clearance on the basis of previous falsification of records. I asked whether he was referring to the matter of the loss of fissionable material and said I would check, although I don't think the case against him is very strong. He asked me to try to provide him with any information where the academic community—or any other group—has indicated its interest in due process in such matters, and to come up with arguments to impress on Justice the importance of going through the established procedure in order to avoid raising a ruckus." Perhaps in desperation, Seaborg began to think more creatively about pursuing the falsification of records angle. What if the MUF situation at NUMEC could be used as the reason for denying clearance on the basis of fraudulent reporting?

Seaborg checked in with his contracting chief. "I called David back a little later and told him I asked Vinciguerra regarding falsification of records, and he said there definitely was none. What did happen was that the individual did not report the loss of material on schedule; we called him in and got him to admit that he had delayed reporting in the hope that it could be found. I said there are others who are not very prompt in making such reports also...Vinciguerra's comment was that David would be walking into a lion's den if he were to go to Justice alone. He then asked that I and some of our staff come over to brief him on procedure. Vinciguerra's views are the same as mine: we have such a tradition of having a hearing that nobody ever assumed we wouldn't have one, and that's why we feel there would be such an impact. I suggested the possibility of sounding out some other scientific opinion, like PSAC,[1] for instance. David said he didn't think he'd want to get involved in that. I told him that Vinciguerra has a meeting with Mardian this morning."

John Vinciguerra, Joe Hennessey and Bill Riley met with Robert Mardian at the Justice Department, briefing him on their plan to pressure Shapiro to withdraw his request for an upgraded security clearance. Mardian would not agree to the gambit unless Seaborg made a firm commitment that if it failed, he would drop his opposition to denial of clearance without hearing. Mardian dismissed Vinciguerra, Hennessey and Riley "in a rather summary fashion." Seaborg then complained to Ed David who calmly smoothed his feathers, commenting that "We can't agree to that. Let's not have any of that. I think it's just a power play." [310]

The next day the AEC chief realized that the unsubstantiated public stance that no NUMEC diversion had taken place derailed the "falsification of records" option for dealing with Shapiro. This position now limited Seaborg's options. He had to continue defending Shapiro to the hilt. Seaborg lamented in a meeting with Ed David that "the MUF was a closed case and that it would be completely untenable to proceed along those lines—in fact, the Commission is on public record as indicating that they believe evidence of illegal diversion of such material is lacking. We discussed the aspects of the case in detail and emphasized that we felt strongly that the man should not be denied a clearance without a hearing. Toward the end of the meeting Pagnotta offered the opinion that the best solution was to grant clearance under conditions that would include surveillance and as much limitation of access as was consistent with the man

---

[1] President's Science Advisory Committee, a body established in the Truman administration bandy later disbanded by the Nixon administration.

doing his job. David indicated, as he closed the meeting, that it was clear to him what needed to be done, but he didn't reveal what was on his mind." [311]

The Justice Department could have chosen to reveal its wiretaps of a 1969 toxic spill at NUMEC caused by Shapiro's out-of-compliance waste handling. This would have provided a rationale for revoking security clearances by showing Shapiro could not to be trusted to abide by the rules for handling sensitive — in this case toxic — material. The files also documented NUMEC's bid to acquire government-owned equipment under the false pretext that it had been contaminated.[li] But the FBI wiretap transcripts, interviews and recordings were never used as a basis for revocation of security clearances or warranted criminal charges.

Ed David and Glenn Seaborg spoke again on February 12. Seaborg perceived David's growing nervousness. "I said he must have run into the feeling around town that this man actually did divert some material. This is absolutely wrong — no one in the AEC believes that. The amount of material lost is the same as other processing outfits lose.[lii] He said there are a lot of people involved in the case and he has been advised that he shouldn't get into a head-on collision with [Attorney General] Mitchell. He said he is going to ask for a meeting of everyone involved (State, Justice, White House, AEC) to see if we can get this matter resolved in a way that will do the least damage to the country."[312]

Seaborg walked back to David's office to discuss the security clearance. David described a harrowing hour and a half meeting with Mardian "in which all aspects of this case had been explored in detail." David naively reasoned to Mardian that Shapiro "probably already has all the information necessary in order to make a simple nuclear weapon, and thus it would be a mistake to possibly drive him out of the country as a result of unfair treatment in the clearance process with respect to his present job." It evidently did not occur to David that Shapiro might have been fighting hard for the job in order to be able to relay the latest boosted and miniaturized weapons designs to Israel. Lacking deep expertise on the latest hydrogen bomb designs, David emphasized to Mardian "the additional information he [Shapiro] would gain from further clearance would be of very little additional value to him in view of the vital information he already has." Mardian agreed to another meeting in order to arrive at a "satisfactory conclusion." [313]

Later that day, Seaborg received another call from Shapiro's lawyer, Edward Bennett Williams. "He said he keeps getting 'bugged' by his client and his client's prospective employer as to when there will be an answer on his client's clearance. Williams said that if his client loses the job on the basis of an adverse decision, that is one thing; but if it happens because of no decision being made, then that's something else. He asked whether his partner, Harold Unger, and possibly he himself, could come to see me. An appointment was made..." When

---

[li] A likely violation of the False Claims Act, an American federal law that imposes liability on persons and companies (typically federal contractors) who defraud governmental programs.

[lii] Seaborg had no factual grounds for making this argument, and the 2001 Energy Department study found the losses far in excess of industry norms.

Mardian indicated he would like to be present at the meeting. Seaborg was relieved. "...I agreed that this suits us fine because it would bring the Justice Department into the act, which is what we have been hoping to achieve." [314]

Impatient, Shapiro again pressed ahead in his new position. Word quickly made its way back to Seaborg. "In executive session we heard the report from Riley and Vinciguerra that our clearance subject has been pressing for information that he is not entitled to from officials of his company — information that he can only receive after he has been cleared." [315]

On February 18 Seaborg visited the CIA to discuss NUMEC over lunch. Did they know something he did not? "Howard Brown and I went to CIA headquarters in Virginia and had lunch with [CIA Director] Dick Helms in his private dining room on the 7th floor. We described to him the present status of our security case and the impasse that has developed between Robert Mardian and the AEC. We particularly emphasized our view that the subject did not divert enriched uranium in the episode of a few years ago. Dick assured us that he didn't have any information concerning this individual that had not already been given to us."

Perhaps lowering Helm's confidence in his good judgment, Seaborg also lobbied the CIA director not to be overly concerned about the alleged leakage of classified information from KMS to Israel. "We also described the situation in this country, and throughout the world, with respect to laser-induced controlled thermonuclear fusion, with the aim of giving him the background so that he wouldn't become too concerned about cases involving alleged leakage of this type of information. We described, in particular, the recent situation involving KMS personnel and a University of Rochester scientist's allegations concerning them as a result of the latter's visit to Israel." [316] It is unlikely this testimony set the CIA director's mind at ease.

On February 24, Mardian reneged on his promise to attend Seaborg's meeting with Shapiro's lawyer. "...Mardian wants to cancel the meeting. (We learned later that he wants this done without indicating Justice Department involvement)." Mardian did agree to make it perfectly clear to Shapiro's legal counsel that the Justice Department opposed Shapiro receiving a security clearance. The AEC would use the following careful language provided by the DOJ. "I have been advised by the Department of Justice that the granting of Sigma Access to Dr. Shapiro would be inconsistent with the Presidential Executive Order and the applicable AEC regulations issued pursuant thereto. I have been further advised to deny this access pursuant to Section 9 of Executive Order 10865 and the implement regulations..." Mardian again emphasized that the DOJ was absolutely adamant that Shapiro's clearance be denied without a hearing. [317]

After another meeting of Commissioners the following day, Seaborg got cold feet about delivering this legalese. "We are going to prepare a letter to Mitchell which will indicate that we would prefer to seek a compromise solution by talking with the subject, and failing that, we would prefer that the subject be given a hearing." Seaborg also considered using the AEC's sway to persuade KBI to change Shapiro's responsibilities. "The second step we decided to take was to have Ramey and Larson meet with officials of Kawecki-Berylco Industries to try to convince them to, while retaining the subject as an employee, transfer him to work that wouldn't require this additional step-up of clearance."

Ed David at the White House also began to get cold feet. When told that the Justice Department wanted to issue a formal denial of clearance statement for delivery to Shapiro's lawyers, David—sensing an ensuing PR fiasco—dialed down his own involvement by telling Seaborg "he [David] does have to be very careful so that his involvement doesn't then involve the President. He then alluded to his philosophy that when he's in a tight spot where there is everything to lose and nothing to gain, the best thing is to delay, and delay, and delay..."[318]

In March 5 talks with KBI executives, the potential for shifting Shapiro to non-classified work proved utterly impossible. However, the AEC's preferred solution of shunting Shapiro off to another company was shared with KBI executives as they sought to protect their own company's interests. Alarmed over the entire matter, KBI executives insisted that the AEC at least reaffirm Shapiro's lower "Q" security clearance in writing. KBI also expressed hopes that the security upgrade matter would be resolved in one to two months and that the Commission speak directly to Shapiro rather than through intermediaries. By March 11, KBI had received a comforting AEC telegram affirming the continuation of Shapiro's "Q" clearance, which Seaborg hoped "might be helpful in relieving some of the tension" as the AEC worked furiously to find another Shapiro another job within the vast industry it so heavily influenced. [319]

An inbound March 15 phone call indicated Shapiro's legal counsel had no idea about the magnitude of forces now arrayed against a higher security clearance. Seaborg scrawled "Unger asked that that meeting be set up again. He indicated that the last meeting was postponed because he understood we would resolve this case in a week or so. He said that time was dragged on and they don't know any more now than they did at the time, and he indicated that he felt the situation was 'becoming virtually intolerable.'" Seaborg told him the matter was complex because other agencies were involved. Unger said he thought the State Department would be involved, and later added the Defense Department as a possibility. Seaborg confided that there were still others, and Unger said he hadn't any idea what other agency might be involved. Seaborg put him off by saying he would "look into this." [320]

By March 18, AEC officials had extracted a promise from Westinghouse to offer Shapiro a job. [321] As it had during Atlantic Richfield's acquisition of NUMEC, the AEC was finally seeing light at the end of the tunnel. "...Ramey took some calls regarding our clearance case. He talked to John Simpson of Westinghouse who is willing to offer the subject a salary of $60,000 a year. He also talked to the subject himself, indicating to him that he should be patient—in this conversation no reference was made to the fact that we were helping to locate another position for him..."[322]

Westinghouse offered Shapiro a job in a "senior advisory technical capacity." While he briefly considered it, another AEC official discretely met with Shapiro "to describe some of the problems in the clearance upgrading." The AEC official was tasked to, in Seaborg's words "make it clear that the JCAE has not been involved and that the decision of the clearance upgrading is beyond our control and the prospects do not look very good." [323]

On April 2, 1971 Seaborg received good news. Shapiro had finally given in. "The subject in our security case had agreed to accept the job offer from Westinghouse." Unger called insisting that his client's "Q" clearance be maintained for a while, even if the new position did not require it. [324] But Shapiro was also informed that the key agency blocking the security clearance had been the Department of Justice. The AEC's emissary to Shapiro, according to Seaborg "had met with the subject last Thursday afternoon, and on the basis of his explanation, which included revealing the role of the Justice Department, the subject decided to accept the job with Westinghouse and phoned Simpson immediately. Ramey explained the situation to Bauser[liii] at a function at which they met on Thursday night. Bauser had learned about it because the subject had come to him on the mistaken assumption that the problem arose as a result of intervention by the JCAE. Actually, the JCAE had not even been informed. Ramey had lunch with Bauser on Friday and explained to him the sensitivity of the situation without having to reveal the role of the Justice Department. He is going to meet with the subject on Thursday of this week and give him a further explanation..."[325]

Seaborg's victory lap included a trot over to Mr. Fixit. "I told Flanigan that we had found a solution to the touchy problem concerning the clearance of our subject along the lines that the subject is being offered a position by another company, which he is accepting. Flanigan was delighted and agreed that Mitchells' course could have led to extreme embarrassment to the White House because of the nature of the publicity that might have accompanied such a course of action."[326] On April 19 Seaborg was notified that Shapiro had formally asked that his request "for upgrading of clearance be withdrawn." For Seaborg "this very difficult case is now closed."[327]

Although the NUMEC diversion case was far from over, Shapiro's lawyer-powered locomotive barreling from HEU access to advanced bomb making designs was derailed. At the beginning of January, 1972 Shapiro signed on as assistant in charge of the Breeder Reactor Division, AEC, Westinghouse Electric Corporation in Monroeville, PA. He later held no security clearances and gained no further access to classified national defense materials.[328]

In Seaborg's self-congratulatory final journal entry on the security clearance matter, he records his final interactions with Shapiro's lawyers. "I talked to Ed Williams and he expressed satisfaction with the way I handled the case involving his client; I told him there was more to the case than he knew and I will reveal more details to him someday." On the very same office diary page, Seaborg notes with appreciation the help he was receiving for his elder son Pete. "[Arthur] Goldberg offered to help Pete in connection with possible acceptance at American University Law School and Mrs. Goldberg said she would think in terms of helping him find a job. Arthur said I should look up David Feller of Boalt Hall Law School."[329]

Goldberg, an illustrious lawyer, served in the OSS during WWII. He was persuaded by LBJ to step down as a Supreme Court Justice in order to replace

---

[liii] Edward J. Bauser was the Executive Director of the Joint Committee on Atomic Energy of the U.S. Congress.

Adlai Stevenson as U.S. Ambassador to the UN. Following the 1967 Six-Day War, Goldberg drafted resolution 242, passed unanimously by the General Assembly affirming "the inadmissibility of the acquisition of territory by war and the need to work out a just and lasting peace in the Middle East in which every state in the area can live in security." In seeming contradiction, Goldberg later argued that Israel did not have to withdraw from occupied territories. He later assumed the presidency of the American Jewish Committee.

Seaborg was surrounded and constantly courted by such influential Israel supporters ever since gaining fame for the discovery of plutonium. He would soon cash in on attractive speaking fees at the Weizmann Institute and offers piling in for lucrative board memberships when he left the AEC. Seaborg was interacting with many highly talented Zionists, and not just in the ranks of Jewish nuclear physicists in AEC regulated operations, but across the entire federal government. Between the lines, his heavy concern about "bad publicity" and "involving Congress" can only be seen as a political assessment that there was nothing to be professionally gained from challenging Israel's extensive lobby if he leaned too hard on Shapiro, which may be why he turned the AEC into Shapiro's job placement agency before resigning.

Years later Seaborg would return to his office diary entries about Zalman Shapiro and NUMEC along with his coauthor Ben Loeb to write the book *The Atomic Energy Commission Under Nixon*. Published in 1993, Seaborg devoted an entire chapter to NUMEC non-diversion, even after an avalanche of facts to the contrary continued to pile up.[liv] In his memoir *Adventures in the Atomic Age: From Watts to Washington* posthumously published in 2001 Seaborg suggested that Mitchell was entirely unreasonable. "We had no power to obtain the [FBI] files...we could not even reveal to Shapiro the whole story of the reason for the delay. Mitchell simply told us to deny the security clearance without a hearing. Never in history had any government agency done that, and we AEC commissioners were not about to be the first...Mitchell was unmoved and it would be fruitless to butt heads with him since he was one of Nixon's closest associates. Mitchell was satisfied simply to let Shapiro fight it out with the government in the courts...When it came to Mitchell, the watchword 'tough' was better spelled 'unpleasant.'"

By the time Seaborg started work on his memoirs, it was already public knowledge that Rafael Eitan, the handler of Israel's most notorious spy Jonathan Pollard, had visited NUMEC in 1968 on false pretexts facilitated by Shapiro's invitation and permission granted by the AEC's own security office. The LAKAM network was finally exposed and struck a severe blow by Pollard's 1985 arrest, yet Seaborg and Loeb chose not to honestly deal with the meaning of the Eitan visit in their words about NUMEC and Shapiro. In 1986 Middle East analyst Anthony Cordesman told the press bluntly "There is no conceivable reason for Eitan to have gone [to the Apollo plant but for the nuclear

---

liv Shapiro wrote a congratulatory letter after Seaborg sent a copy of the book claiming he had little idea of the magnitude of the attorney general's level of opposition to his bid for a higher security clearance. It is duplicated in the appendix.

materials."[330] Rather than tackle such damning assertions, Seaborg sprinkled his chapter "A matter of Justice" with subtle charges of anti-Semitism penning "Contributing to the suspicion of Shapiro was his personal background. Son of an orthodox rabbi from Lithuania, he never made a secret of his sympathy for the Zionist causes and, once it had been established, for the state of Israel. He had active and open relationships with the Israeli government, which he served as a technical consultant and for which he provided a training and procurement agency in the United States. He started a subsidiary of NUMEC of Israel, in partnership with the Israeli Atomic Energy Commission, to develop machinery for the preservation of fruits and vegetables by irradiation. He employed at least one Israeli, a metallurgist, in the Apollo plant, and he regularly received visits from Israeli officials, including the Israeli embassy's scientific attaché."[331]

Yet in the end it was Seaborg, not Shapiro's inquisitors, who had been duped. What would the Nobel laureate who discovered plutonium have thought if he were made aware that the entire Israeli Atomic Energy Commission was (and still is) the primary front organization in charge of Israeli's nuclear weapons development? What if Seaborg had been advised, in a timely fashion, that scientific attaché Avraham Hermoni was actually technical director of Israel's nuclear bomb project at RAFAEL and also had a role in running the highly decentralized Dimona project? Would it have led to warranted and penetrating questions about the real roles of other AEC-approved Israeli visitors to NUMEC in 1968 that also infiltrated under false pretenses? Would the fact that none of Shapiro's four Israeli visitors to NUMEC in 1968 had any expertise in "thermo electric devices" which was the stated purpose of their visit have shaken Seaborg's theories about processing losses?

Seaborg recognized that some of Shapiro's explanations were not all that credible. After the initial AEC audit detected materials unaccounted for, Shapiro claimed they could be recovered from waste burial pits, yet they never were. Seaborg's book also records Shapiro made unfounded public claims about the levels of NUMEC's losses. In a 1991 book, Seaborg admitted that "the AEC did not have much of a basis for comparison—there were only a handful of firms engaged in comparable activity—the staff tended to believe that the losses at NUMEC had indeed been excessive."[332] This contradicted Seaborg's February 12, 1971 office diary blanket assertion that "This is absolutely wrong—no one in the AEC believes that. The amount of material lost is the same as other processing outfits lose." In 2001 the DOE finally publicly released accurate figures that obviated the need for "belief" based industry estimates. Through 1968, the time when Shapiro was in charge of NUMEC, the plant had almost twice the MUF as other U.S. operators.[lv]

In his 1993 book, Seaborg claimed AEC commissioners were convinced that Shapiro and NUMEC had not diverted any nuclear material "either to Israel, or to any other country for transshipments to Israel." Yet this was also largely a belief-based theory now revealed as the most politically convenient path toward congressional appropriations rather than the result of carefully tracing shipments

---

[lv] And ten times the rate of loss NUMEC experienced after he left, as explored in Chapter 14.

or a conclusive search for overseas front companies. Seaborg wrote a lengthy account of how AEC safeguards during the period of NUMEC were largely based on an honor system that counted on fuel reprocessors to minimize loss of materials since it was valued at "four times the price of gold" and had to be repaid if lost. But even Seaborg candidly admitted that this minimal pretense of oversight ended at the U.S. shoreline.

Despite the lack of any credible monitoring system, Seaborg was steadfast that absence of evidence was evidence of absence. "We reached this conclusion in large part because there was no evidence to support a charge of diversion. It was hard to believe that such an effort could have occurred without leaving some trace of evidence for AEC, Joint Committee, and FBI investigators. We reached our conclusion also in consideration of the difficulties any would-be diverter would have had to face. We did not believe it possible that diversion could have occurred without the knowledge of many of the employees of the plant. It seemed inconceivable that every one of these employees would have clandestinely agreed to suppress knowledge of a traitors act. As to Shapiro himself, it was not reasonable to suppose that he would have undertaken such a hazardous course of action, whose penalty could have been a sentence of death. Nor was it consistent with his character, as we came to know it, to believe that he would have been motivated to commit an act of disloyalty to the United States..."[333] But Shapiro's ideology considered acts that advanced Israel were by definition an advancement of U.S. interests. By the time Seaborg penned his theories of NUMEC employee "suppression" several had already stepped forward detailing to the FBI precisely how the HEU had been smuggled out of the U.S.

It is both ironic and suspicious that Seaborg, who had applied the full powers of his agency to compel government contractor Westinghouse to rehire Shapiro, seemed to overlook the single most obvious explanation for why NUMEC employees did not talk to AEC inspectors swarming the plant in the 1960s. They did not have to support "a traitor's act" or proactively engage in a plant-wide ruse. In the end, the singular motivation of keeping up a steady paycheck in a one-company town—and avoiding armed thugs at the loading dock—were reasons enough.

Seaborg left the AEC in 1971 in the midst of growing questions about NUMEC and Shapiro. The AEC's inability to manage materials accountability would become one of the primary rationales for splitting up the AEC. "...am I trying too hard on these pages to argue NUMEC's case?" Seaborg privately queried his coauthor Benjamin Loeb.[334] History reveals he was indeed.

The Weizmann Institute continued courting Seaborg, mailing its annual "Report on Scientific Activities" and inviting him to events held across the United States. On July 21, 1970 Seaborg accepted Joseph Kaplan's invitation to lend his name as a sponsor for a dinner honoring Nobel Laureates, expressing regrets that he could not personally attend.[335] Kaplan was a professor of physics at the University of California and Chairman of the California Committee for the Weizmann Institute. On October 22, 1970 Kaplan lamented that the planned November 11 event in Los Angeles had to be canceled since "in light of Israel's very serious economic situation, all Israeli academic and scientific institutions, including the Weizmann Institute, have decided to forego all major activities in the United States in order to focus the energies of its supporters on aiding Israel directly." [336]

Just nineteen days after President Richard Nixon accepted Seaborg's July 21, 1971 resignation as chair of the AEC, the Weizmann Institute's American Committee in New York[lvi] circled back to Seaborg with an offer he couldn't refuse. Weizmann's President, Albert B. Sabin invited Seaborg to be the guest of honor for the annual dinner at the Waldorf Astoria. Seaborg's predecessors, gushed Sabin, included President Harry S Truman (1953), JFK (scheduled for December 3, 1963), and LBJ (1964). Seaborg, who typically accepted honorariums totaling no more than $500 on the understanding that they would be given as grants to the University of California, was offered $2,500 plus expenses to give the keynote. Seaborg's AEC salary at the time was only $30,000 per year.

It was an immense propaganda coup and show of force for the Weizmann Institute. By that time President Nixon had already acquiesced to accepting a policy of "ambiguity" over Israeli nuclear weapons and backing off U.S. pressure during a summit with Israeli Prime Minister Golda Meier on September 26, 1969[337] But what to do about Israel stealing know-how and material from the U.S. was still an outstanding issue for Nixon and subsequent administrations. In 1970 a popular tell-all book titled *The Pledge* by Leonard Slater provided a devastating account of 1940s Haganah arms smuggling operations across the United States. *The Pledge* touched off angry pleas for overdue Justice Department prosecutions by such prominent Americans as Norman F. Dacey. By gathering 1,500 guests, including "many world leaders of the scientific community," Feinberg could stave off warranted calls for bona fide investigations of nuclear smuggling through a huge show of force even though his prime political benefactor, LBJ, was now out of office.

The Institute wanted a talk of no more than 30 minutes covering "the future of atomic energy...desalting, etc." in "that part of the world." The Weizmann Institute flattered Seaborg, telling him that President Nixon himself was the only other candidate under consideration for the keynote. Seaborg accepted on September 2, agreeing to deliver a talk about Israel in a speech dubbed "State of Knowledge." But the Weizmann Institute was leaving nothing to chance. On September 17, Lillie Shultz asked for "two copies of the text of the address" no later than October 20th. Seaborg's speech, by subtly affirming Nixon's adopted

---

[lvi] The Weizmann Institute's American Committee was located in the same offices, 515 Park Avenue, as the Jewish Agency's American Section. After WWII, the Jewish Agency established arms smuggling front companies in the United States while lobbying for the creation of the state of Israel. The founder of the American Israel Public Affairs Committee worked at the Jewish Agency until resigning to work as an agent of the Israeli Ministry of Foreign Affairs. The Jewish Agency later funded the American Zionist Council and the founder of AIPAC's newsletter, the Near East Report. During 1971 the Jewish Agency American Section was fighting an order by the U.S. Department of Justice to register as an Israeli foreign agent after its 1953 Covenant with the Israeli government revealed access to government revenue and power in the Knesset. It reconstituted itself into the World Zionist Organization – American Section, at the same address on September 21, 1971 to dodge regulation under the 1938 Foreign Agents Registration Act.

policy of "strategic ambiguity" could have been written by the Israeli Atomic Energy Commission.

The event was a "bait and switch" for Abraham Feinberg. After 27 years, Dewey Stone finally resigned as Chairman of the Board of Governors of the Weizmann Institute in September of 1971. Only after Seaborg accepted the keynote did Feinberg suddenly publicly emerge as the newly elected Chairman of the Board of Governors of the Weizmann Institute in both Israel and over its American Committee. Democratic Party fundraiser Abraham Feinberg had until recently, as Seymour Hersh put it, "enjoyed the greatest presidential access and influence in his twenty years as a Jewish fund-raiser and lobbyist with Lyndon Johnson. Documents at the Johnson Library show that even the most senior members of the National Security Council understood that any issue raised by Feinberg had to be answered."[338]

There is no indication that Seaborg had any idea who his new benefactor really was. Feinberg long operated on a different plane than fellow Americans, who were generally subject to rule of law. Feinberg organized the "whistle stop" campaign and cash raising bonanza that saved Harry S. Truman's 1948 presidential campaign from certain defeat. But Feinberg was also a WWII draft dodger who had an influential friend secure him a deferment. K. Bertram Friedman, a sympathetic U.S. attorney in the Southern District of New York, let it slide and declined to prosecute Feinberg.[339]

Like Zalman Shapiro, Feinberg was in constant contact with Israeli intelligence operatives such as Rueven Shiloa, the founder of the Mossad. Like David Lowenthal, he also played a key role in the Haganah arms smuggling network, organizing with other leaders such as Teddy Kolleck while under observation of the FBI. When 70 individuals were identified for possible indictments for arms smuggling in Los Angeles, the Israeli Ambassador to Israel in the U.S. Eliahu Elath was told by the Israeli government to "have Abe take it up with his friends, to 'squash it once and forever.'"[340] In the end, thanks to the efforts of Feinberg and others, only a handful of Haganah arms smugglers were ever convicted. All later received presidential pardons after heavy Israel lobbying.[lvii]

The FBI investigated Feinberg as an Israeli foreign agent in 1952 over his role as publisher of *Israel Speaks*, the successor to a periodical called *Haganah Speaks*, a propaganda trumpet of the Israeli government. Feinberg wriggled out of registering by simply shutting down the publication, after a failed bid to sell it to the *Jerusalem Post*. [341] On October 31, 1958 Israeli Prime Minister David Ben-Gurion recorded in his diary a conversation "he had with Abraham Feinberg, a wealthy Jewish businessman and major Democratic fund-raiser" to raise funds for Israel's nuclear weapons program among "*benedictors*" in the United States. Feinberg not only fulfilled this role, but laid out smoke screens to undermine warranted U.S. press coverage.

President John F. Kennedy fought for biannual international inspections of Dimona and verifiable Israeli assurances that it was not a nuclear weapons production facility. In 1960 Feinberg made a series of payments to the American

---

lvii Charles Winters (2008), Herman "Hank" Greenspun (1961), Adolph "Al" Schwimmer (2001)

Israel Public Affairs Committee (AIPAC) whose director then mounted a U.S. disinformation campaign through the *Near East Report* that Israel's nuclear program would be incapable of ever producing weapons. AIPAC continued to carefully monitor how Dimona was being covered in the U.S. press, though JFK's assassination in 1963 marked the gradual end to inspections and the U.S. nonproliferation drive as it applied to Israel.

The Kennedy administration immediately requested Feinberg's lengthy FBI file when it took office. [lviii] LBJ—upon hearing reports of Israeli nuclear weapons development—presumably had no problems connecting the dots back to his most important campaign contributor. The implications of Israel's nuclear weapons funding coordinator having such influence over a U.S. President have never been adequately explored by historians. But Feinberg's legacy casts new light on the oft-reported reaction of LBJ when CIA director Helms informed him of the Israeli nuclear weapons program. "Don't' tell anyone else, not even Dean Rusk and Robert McNamara." [342]

Figure 15 Abraham Feinberg and Glenn Seaborg, November 3, 1971.

Feinberg spared no expense on the lavish 1971 Waldorf Astoria event. For his part, Seaborg delivered exactly the rhetoric Feinberg needed to stave off any

---

[lviii] Released to the author under FOIA in March of 2011.

consequences for Israel's parastatal organizations and supporters who had once again stolen what they needed from America.

Seaborg began his address referring fondly to his 1966 trip to Israel. "To be in the company of men and women who are steeped in the legacy and love of learning always gives me a warm feeling and one of hope in this troubled world. I have had this feeling before—and noticeably so during my visit to the Weizmann Institute of Science during a trip to Israel."

In the speech Seaborg deplored the misspent sixties decade of contemporary youth who rejected straight careers in the sciences or liberal arts for altered states of consciousness. He lauded Israel's shining example to the world as a state with few resources creating a "thriving human society of free men out of the sands of the desert and the rocks of the highlands." Not until the very end did Seaborg finally tackle the matter Feinberg had really paid him for.

"During my tenure as Chairman of the AEC I was asked on numerous occasions whether I thought Israel was a nuclear power—or less euphemistically, did she have the bomb? My usual reply was that she was among those nations that had the knowledge to build one if she wanted to, and I speculated no further. Now in retrospect I often wished I had said, 'Yes, she is a nuclear power, the kind that knows of, and makes use of, the atom's power for peace. She is a member of the International Atomic Energy Agency, her nuclear laboratories do advanced research in most nuclear fields—nuclear physics, solid state physics, reactor physics, radiation chemistry, inorganic chemistry, nuclear chemistry. She uses radioisotopes in medicine, in agriculture and food preservation, in industry and in other areas of research. Her professors in nuclear sciences and engineering are excellent and their students enthusiastic and enlightened. Of course Israel is a nuclear power—the kind that more nations should strive to be.' I recommend those or similar words now should such a question be asked."

Seaborg's full-on public embrace of "strategic ambiguity"—dodging warranted questions about Israel's nuclear weapons arsenal, delivery vehicles and use policy—was as lucrative as it was eloquent. But Seaborg's ability to officially douse the flame of any regulatory oversight that threatened to rekindle traumas sparked by the Rosenbergs and Oppenheimer ended when he left the U.S. nuclear regulatory bureaucracy. Shapiro's more skeptical inquisitors would continue to dog Seaborg until his death in 1999. When investigative reporter John Fialka interviewed Seaborg about the Shapiro's security clearance crisis, Seaborg denied even having a role in finding a new job for Shapiro.[343]

What financial connections Abraham Feinberg ever may have had to NUMEC remain unclear. If any lesson had been learned by Feinberg and his partner Nahum Bernstein during their Haganah smuggling days, it was to leave no traceable financial trails. LBJ aide for Jewish Affairs Myer Feldman confirmed to investigative reporter Seymour Hersh that "Abe only raised cash—where it went only he knows." Feinberg readily admitted that the entire purpose of cash fundraising was to limit public disclosure and accountability. "A lot of people were afraid publicly to give as much as they could, so they arranged sub rosa cash payments. It had to be done laboriously— man-to-man. Raising money is a very humiliating process, people you don't respect piss all over you."[344]

NUMEC had not yet entered the broader public consciousness. Not until late 1974 would the other newspaper of public record, the *Washington Post*, finally publish a significant MUF story actually naming NUMEC. While the buried lede of the article "Possibility of Attempted Nuclear Thefts Causing Deep Concern" was somewhat accurate, the account also contained major errors. "The most celebrated MUF took place back in the 60's in the Apollo, PA plant of Numec. The factory had just taken a big order to process and fabricate 2,200 pounds of fully enriched uranium for Westinghouse Astro-Nuclear, which was making the fuel for the nuclear-powered rocket. In the fall of 1965 Numec was told to make an inventory of its uranium. It came up short by 207 pounds, worth at that time more than $1 million. It was also enough to make several large bombs. For a while, China and Israel were both under suspicion as the possible thieves." The story mentioned "The AEC closed down the plant and began to look for the missing uranium."

The *Post* continued on that "Some was found in the air filters, about 12 pounds in the 730 filters that kept uranium from blowing out the smokestacks. Another 14 pounds was found in a burial pit on a mountaintop eight miles away. The factory spent $100,000 digging up the burial pit looking for the missing metal. At the' end of the search, 148 pounds of uranium was still missing. Numec was forced to pay the AEC $834,000 for the missing metal. Diversion was still suspected, so the AEC interviewed every employee in the plant and every one of its past employees. The AEC concluded there was 'no evidence' of diversion, but there are still a few people there who suspect China and Israel."[345]

In reality, China was never the primary suspect for diverting HEU from NUMEC. It had none of the obvious cultural, social, philanthropic and business connections to the plant that Israel did. AEC internal concerns about materials accountability long preceded the date of the cited 1965 inventory, which were in turn prefaced by Rickover's insistence that Shapiro's Israeli employees be promptly ejected from the plant. Any careful review of AEC meetings transcripts, correspondence and its flawed interview of NUMEC employees reveals the AEC was determined from the very onset of the crisis to present its "theory" about processing losses as fact in the NUMEC MUF crisis rather than seriously pursue a diversion investigation.

Still, the *Washington Post* story moved NUMEC into the crosshairs. Investigative reporters such as John Fialka and chatty CIA officials soon blew the lid off NUMEC, causing headaches and a new round of FBI investigations that finally yielded credible eyewitness accounts of diversion directly from NUMEC's loading dock to Israel.

10. Ford Justice Department reopens the NUMEC investigation

The AEC's impossible mandate to both promote and regulate the nuclear industry finally ended in early 1975. On January 19, AEC was split into the Nuclear Regulatory Commission (NRC) and the Energy Research and Development Administration (ERDA). The NRC assumed responsibility for nuclear regulation while ERDA took over nuclear development and industry promotion. While NRC continues to operate, the short-lived ERDA was folded into the Department of Energy by late 1977. The NRC's more serious dedication to materials accountability and attempt to rectify the past failures of the AEC put it on a collision course with the NUMEC cover-up.

NRC Division of Safeguards employee James Conran was frustrated. Tasked with reviewing safeguards over commercial U.S. nuclear facilities, the engineer was particularly interested in reviewing information developed about any covert Israeli nuclear theft in the U.S. Conran was denied access to an NRC intelligence report he thought would be invaluable for future efforts to reduce potential HEU theft. The secret NUMEC report was so sensitive that the ERDA at first refused to even acknowledge the report existed. Conran began pestering top officials at the NRC for a classified briefing so he could write an accurate report about the history of safeguards.[346] NRC Chairman William Anders, formerly an Apollo 8 astronaut, sought to allay Conran's concerns. He made a decision he would later come to regret. Anders invited CIA Deputy Director for Science and Technology Carl Duckett to brief NRC officials.

Duckett served in the U.S. Army in Europe and the Pacific during WWII. He stayed on to oversee missile intelligence at the White Sands range after the war and chaired a standing committee on missiles and astronautics on the U.S. Intelligence Board. He joined the Central Intelligence Agency in 1963 and served thirteen years in various senior positions. He became a senior advisor on strategic arms limitations negotiations and helped develop programs for aerial reconnaissance and surveillance. He received the CIA's Distinguished Intelligence Medal twice for innovative operations in electronic and submarine intelligence.

Brandishing a thick folder stamped "Top Secret"[lix] Duckett told a stunned NRC executive audience not only that CIA believed Israel had illegally obtained HEU from the Nuclear Materials and Equipment Corporation, but that the stolen

---

lix The CIA has not released Duckett's briefing folder despite many requests. A July 15, 2011 letter to the author in response to a Freedom of Information Act request for Duckett's briefing material stated "We have previously conducted searches on behalf of earlier requesters for records concerning the subject of your request. No responsive records were located. Although our searches were through and diligent, and it is highly unlikely that repeating those searches would change the result..." A few weeks later, the agency quashed a bid for public release of FBI files bearing hundreds of pages of CIA content.

material was used to produce Israel's first atomic bombs.[347] Reporter John Fialka wrote "In fact, the eleven NRC officials at the briefing had been so awed by the secrecy surrounding the file that they did not even ask to see it."[348]

Duckett confirmed that the CIA believed Israel had already assembled nuclear weapons by the mid-1960s. Israel began to practice A-4 jet bombing run maneuvers that were only warranted if the explosives being delivered were atomic rather than conventional. Such practice runs to guarantee aircraft and pilot survival "would not have made sense unless it was to deliver a nuclear bomb."[349] Direct evidence of diversion from NUMEC was also gathered up by the CIA in Israel. On June 21, 1978 Bill Knauf and Jim Anderson from the Department of Energy's Division of Inspection visited Glenn Seaborg, informing him exactly what that evidence was: traces of HEU of the type provided to NUMEC ended up in Israel.

Seaborg recorded this unwelcome news in his journal. "Some enriched Uranium-235 which can be identified as coming from the Portsmouth, Ohio Plant has been picked up in Israel..."[350] (See Appendix) Knauf and Anderson interrogated Seaborg from a list of questions crafted for secret report being drafted by the Energy Department.[351] Despite the stunning news, Seaborg went on the defensive after the investigators probed the AEC's handling of NUMEC. "I asked them if any responsible persons feel that Shapiro actually diverted material to Israel. They replied that nobody with a scientific background believes this but that it is difficult to convince some members of Congress." Seaborg wrote "such enriched material has been sold on an official basis to Israel and this could be the source of the clandestine sample." But Seaborg did not request details of the sampling or further ponder the significance of why Portsmouth U-235, the source of much of NUMEC's HEU, could be found in Israel. [352] His immediate appeal to authority (whether "responsible persons" suspected Shapiro) is all the more ominous given his subsequent avoidance of FBI investigators. When FBI special agents attempted to question Seaborg in March of 1979, he simply refused to speak with them.

According to a leaked page from a classified NRC report about the Duckett briefing (see Appendix), NRC Chairman Anders later apologized to Duckett for the large number of NRC employees who attended the CIA briefing. Anders announced that "in light of the sensitive nature of the information, he was going to the White House." Anders "did not realize how sensitive the information was" and said that if he had "he would have restricted the attendance even more."[353] The NUMEC genie was again out of the bottle. Duckett was at a unique stage of his career. He was more than ready to talk to more of those he felt had a "need to know" causing self-inflicted headaches the CIA continues to suffer to this day.

The CIA staged a state-of-the-world cocktail buffet for 150 members of the American Institute of Aeronautics and Astronautics. The briefing was part of a new public relations program. When an executive asked during questions and answer sessions whether Israel had an atomic arsenal Duckett was unequivocal. Israel "has ten to 20 nuclear weapons ready and available for use" he quickly responded. [354] Duckett's remarks appeared in the *Washington Post* four days later, angering CIA Director George H. W. Bush and Senator Frank Church, Chairman of the Senate Select Committee on Intelligence. Church called Duckett's revelation "the biggest goof in the history of leaks that I have ever seen" but there was no formal denial other than one meekly emitted by Israel. What kind of goof was it, and who did it really damage? LBJ had little faith in Church, once conspiratorially telling Glenn Seaborg on December 27, 1966 that Frank Church and Senator Joseph Clark often unwittingly met with foreign "secret agents" without specifying the country.[355] In June of 1976 Duckett resigned from the CIA stating privately that his reason for leaving was that he became aware George H.W. Bush would never promote him to Deputy DCI despite his many qualifications.

Anders quickly passed along Duckett's revelations to James Connor, a former AEC director of planning and analysis now working as assistant to President Gerald Ford. This new "set of circumstances" prompted Connor to start checking into the 1960's NUMEC investigation. Connor was appalled at the superficiality of the NUMEC investigation. The AEC interview of Shapiro was simply "not of a penetrating nature." There was no bona fide effort to explore the true nature and meaning of his many Israeli contacts. There was no documented justification for why the investigative effort had just suddenly "came to a stop." Connor possessed additional inside information from an associate who was the White House science advisor's secretary in the late 1960s.[356] Connor was both candid and blunt about the fact that some of "the stuff came from Pennsylvania" to build Israel's nuclear weapons.[357] He motivated Attorney General Edward Levi to formally look into the matter.

Levi duly briefed President Ford that in the 1960s the FBI had been essentially talked out of investigating NUMEC losses as a diversion issue by the AEC which promoted its highly dubious processing loss theory. Levi formally ordered the FBI to reopen the NUMEC investigation in order to determine whether criminal statutes covering unauthorized release of classified information and uranium diversion had been violated. But Levi went much further. He also authorized investigating whether a government cover-up had taken place."[358]

The National Security Council and CIA were provided scientific analysis about "the number of nuclear weapons that could have been made if the material had been diverted." [359] The FBI needed much more scientific methods to pursue its long-delayed Atomic Energy Act investigation. As a preliminary step, on May 20 an FBI special agent met with Samuel C. T. McDowell of the ERDA to learn what specialized techniques could trace uranium samples picked up in Israel to materials processed by NUMEC in 1963-1965. McDowell told the FBI that there were "two possibilities" for tracing the material.[360] McDowell had led the AEC survey team of NUMEC at the height of the uranium loss crisis in the mid-1960s.

Levi had the FBI prepare a report dated April 22, 1976 on the NUMEC affair for congressional testimony about government safeguards over nuclear materials. Levi sent a memorandum to President Gerald Ford the very same day with the FBI report attached. Titled "Possible Violation of Criminal Statutes" Levi outlined the potential criminal violations which included collaboration with officials inside the U.S. government. Levi felt that since no serious diversion investigation was ever undertaken, the Department of Justice could not definitively state that there was no evidence which would support a criminal charge. Levi went on to write that "the facts available with respect to this matter indicate that the following criminal statutes may be involved" totaling a stunning ten possible violations of the Atomic Energy Act and various other criminal statutes.

Two of the Atomic Energy Act violations covered unauthorized dealing in special nuclear material and transportation of dangerous articles, presumably over the unlawful removal of nuclear material from the NUMEC facility. The last three statutes targeted persons operating from within the federal government as accessory after the fact, misprision of felony[ix] and conspiracy. But the statute of limitations clock was ticking, necessitating urgency according to Levi, who wrote "Because the statute of limitations may not have run with respect to any [of the three government official related] offenses that may be involved and because of the responsibility to consider whether any dismissal or other disciplinary proceedings may be appropriate with respect to any persons presently employed as federal officials who may have participated in or concealed any offense, I believe it necessary to conduct an investigation. Section 2271 of the Atomic Energy Act provides that 'the Federal Bureau of Investigation of the Department of Justice shall investigate all alleged or suspected criminal violations' of the Act." [361] Levi may have been signaling President Ford there were persons still employed by the federal government suspected of having "participated in or concealed" the diversion of nuclear material from NUMEC. [362]

On April 29, 1976 the criminal investigation of Atomic Energy Act violations got underway, with "no limitations on scope." Although a 1969 amendment to the Atomic Energy Act of 1954 had substituted "imprisonment for life" in lieu of "death or imprisonment for life" imposed by a jury, it only applied to "offenses committed on or after December 24, 1969."[363] Any hard evidence developed against Shapiro could theoretically have led to a death sentence. Depending on what the FBI uncovered, prosecution under the Espionage Act was also a possibility since NUMEC had been a repository for up to 2,400[364] classified documents necessary for its government work.

The FBI was not entirely eager to reopen this can of worms. Its final advice a decade earlier—terminate Zalman Shapiro's security clearances and all HEU contracts to NUMEC—while not completely ignored, had been carried out in ways that avoided criminal prosecutions and future deterrence. The FBI began its work reluctantly, giving the renewed investigation a new administrative code word, DIVERT.[365]

On May 3, FBI special agents interviewed Former Deputy Director of AEC Safeguards and Security Robert E. Tharp. He assured them that no diversion could have taken place because "Shapiro could not have done it alone." Tharp pointed out that two GAO investigations had found no evidence of wrongdoing, and that while Shapiro had never admitted "juggling" material between contracts, the AEC was sure NUMEC suffered processing losses "in line with that experienced by other manufacturers."[366] On May 11, Deputy Director of Safeguards at the NRC Ralph G. Page quantified that the AEC found the loss at NUMEC to have been "about 60 kilograms" between 1965 and 1969.[367]

[ix] Concealing knowledge, usually by a government official, of a felony committed by another person.

Deputy Attorney General John C. Keeny ordered the FBI director to be exhaustive in its work interviewing former NUMEC employees, including President's Assistant Bertran Schwartz who worked under Shapiro, as well as Manager of Security Bruce Rice, Robert Oliger, Floyd Joyner and Sylvester Weber, who managed NUMEC engineering and accountability. Keeny ordered that "summaries of investigative results should continue to be furnished directly to the office of the Attorney General."[368]

The statute of limitations clock slowly ran down along with the Ford administration's hold on the presidency. The FBI Washington Field Office special agent in charge doubted that additional interviews would any yield new insights, recommending more direct approaches to Shapiro. This request was denied and the WFO was ordered to conduct the additional interviews requested by the Justice Department with confidence that only then would FBI headquarters "request departmental approval to administer a polygraph to Shapiro."[369]

In 1977 the counselor to the outgoing Ford administration passed its file of classified NUMEC documents on to the head of President Jimmy Carter's transition team. When Carter took office January 20, he had to make fast decisions about how to proceed with the NUMEC investigation. The documents inherited by the Carter administration included a classified 1968 memorandum from CIA Director Richard Helms to Attorney General Ramsey Clark, a 1969 letter from FBI Director J. Edgar Hoover to Helms, a 1969 memorandum from Helms to President Richard Nixon, and a 1976 memorandum from CIA Deputy Director for Science and Technology Carl Duckett to Helms. [370] [lxi]

Tasked to complete a long list of interviews,[lxii] from the irascible Admiral Hyman Rickover to former AEC high officials and NUMEC employees previously interviewed during the AEC's flawed investigation, the FBI again suggested cutting to the chase. On May 4, 1977, the FBI director drew a line in the sand with the assistant attorney general of the Criminal Division. "It is the opinion of the Federal Bureau of Investigation that in order to fully resolve this matter, a polygraph examination of Shapiro is necessary. Consequently, please advise this bureau of the Department of Justice interposes any objection to such action."[371] But once again the FBI's sensible solution was stymied. Attorney General Griffin Bell succeeded Edward Levy. The FBI would once again interview the cagey and evasive Shapiro, but its formal request to polygraph Shapiro was denied.

[lxi] The CIA has repeatedly refused all requests to fully declassify and publicly release these Carter Library documents.

11. Carter administration grapples with NUMEC

Zalman Shapiro agreed to be interviewed by FBI agents on June 27, 1977 only after consulting his lawyers. Shapiro had moved 2 miles northwest of his Bartlett Street home neighboring David Lowenthal's residence to another part of Pittsburgh. After reading a "Waiver of Rights" form pushed toward him by the FBI agents, Shapiro "voiced certain reservations relative to its purpose and was concerned that display of the form implied certain unlawful acts on his part." He stated that signing the form would be agreeing that he was complicit in illegal activity. 372

After a conference call to his lawyer, Shapiro agreed to an interview limited only to his tenure as president of NUMEC. This meant that any line of questioning by the FBI aimed at fully fleshing out Shapiro's previous or subsequent trips to Israel, how he came to be a member of Apollo Industries board of directors, work for the Zionist Organization of America, or his established contacts in Israel was off the table. The still unanswered question of exactly how and when he had met David Lowenthal was also a taboo subject. After an hour setting up the restrictive ground rules, the FBI interview began at 11:11 AM.

Shapiro began the interview by insisting that "there was never any diversion of enriched uranium from the NUMEC facility and he termed the chances of any individual or group of individuals successfully diverting such material as 'miniscule.'" Though he emphatically denied he was in any way connected with or responsible for diversion, by the year 2009, Shapiro would no longer so broadly defend NUMEC, and instead only ask that the NRC clear him personally of diversion.

In his 1977 interview, Shapiro claimed the complex manufacturing processes, particularly for the NERVA fuel, involved such fantastic temperatures, solid formation, pounding and crushing, which necessitated constant recovery of microscopic materials, that losses were inevitable. It was lost in pipes, pumps and valves, and "clung even to the clothing and shoes of workers." It was both impossible and economically unfeasible to recover everything.

Shapiro claimed that NUMEC had adopted the very same continuous production processes developed by Oak Ridge National Laboratory. This made it "impractical and uneconomic to attempt recovery of material from waste and scrap on a batch by batch basis" since that would require dismantling all equipment. Instead, "it was the practice of NUMEC to blend its scrap and waste" and recover uranium at the conclusion of each contract period, which NUMEC called "campaigns."

Figure 16 Zalman Shapiro's Pittsburgh residence in 1977.

Shapiro emphasized to the agents that "significant diversion of SNM at NUMEC was out of the question and added that Atomic Energy Commission (AEC) officials had, in fact, assured him personally that NUMEC MUF compared favorably with MUF experienced throughout the rest of the industry."

Shapiro also claimed that during the 1965 MUF crisis, the AEC had recommended NUMEC hire away Lovett, rather than mentioning him as a way to demonstrate the existence of a wide pool of available candidates. "At the time it was the understanding of Shapiro and the other officials at NUMEC that the AEC inspection as well as all paperwork attendant to the inspection, including a final report, had been completed." Shapiro claimed that NUMEC was sure Lovett had already given his full input for the AEC final report before he hired the AEC official, and that it was not "an effort to influence...findings...nor an effort to mute the AEC report of the inspection." He did not explain how NUMEC managed to accomplish this review in such a short time frame.

Shapiro also claimed he had never released any classified information to persons not entitled to receive it. He was not aware of any contact with "foreign intelligence officers or organizations." Tellingly, Shapiro now claimed NUMEC's involvement with ISORAD was a way of skirting Food and Drug Administration opposition to radiation-based food preservation and medical sterilization, or as he put it, ISORAD was merely an avenue to "proceed with the venture without interference from the FDA." This was the first time Shapiro ever made such a claim to criminal investigators about ISORAD, who later documented employee accounts that the venture provided an ongoing pretext for rushed shipments of large containers bearing radioactive warning labels off to Israel.

ISORAD also necessitated "meetings between Shapiro and the scientific attaché at the Israeli embassy" although when pressed for the identity of the highest level contact, Avraham Hermoni, Shapiro claimed he "could not recall this individual's name." Shapiro did claim to recall his first meeting occurred while the attaché was working as a teacher at a university. Atlantic Richfield Company, which was heavily involved in Algerian oil exploration "became nervous over the ISORAD connection with Israel" and simply gave its stock back to the Israelis. Atlantic Richfield's low valuation of ISORAD, along with its quick divestment of NUMEC, adds to evidence that it only acquired the company in order to obtain lucrative Hanford management contracts dangled by the AEC.

Shapiro complained that although regulated and inspected by both the AEC and U.S. Navy, they were both "quick to criticize." NUMEC's requests for guidance or standards that could actually be translated into operations were rarely answered. Shapiro wondered out loud at the conclusion of the interview when the "matter of NUMEC would finally be laid to rest" seemingly oblivious that his own evasiveness and evolving story line were not helping to reach that conclusion. The AEC and GAO had already made in-depth inquiries into the situation, pled Shapiro, claiming a failing memory. He "personally had been interviewed at great length by AEC officials. This all occurred at a time when records and memories were fresh. Shapiro questioned first the reason for the additional inquiry into the matter of NUMEC and the ability of anybody to reconstruct, in detail, the situation at NUMEC ten years ago."[373]

FBI counter-intelligence was well aware that Israel's LAKAM technical intelligence operation was run by the scientific attaches of Israeli consulates and embassies. But the FBI Field Office, anxious to wrap up the investigation, did not compare Shapiro's new testimony about mixing scrap between contracts to his earlier assurances to the AEC that NUMEC maintained strict separation by contract. Within a year, Shapiro would proudly disclose to Congressional investigators what NUMEC had stridently denied in the 1960s, that it was actually commingling HEU between contracts. The FBI also did not think creatively about Shapiro's evasion of regulatory bans on irradiation equipment. If Shapiro felt empowered enough to thwart FDA regulations based on his own judgments and financing from a foreign government, what else was he capable of?

In October of 1977 President Jimmy Carter's Press Secretary Jody Powell, a former Georgia farm boy and trusted aide, publicly announced that "four years of continuing investigation" by the AEC, FBI, and General Accounting Office had "failed to reveal" a diversion of uranium to Israel. But the Carter administration soon discovered that inheriting the NUMEC problem meant it had to do more than make such unsubstantiated public announcements. It opened a White House file marked "Proliferation: Apollo."[374] In the House of Representatives, Chairman of the Subcommittee on Energy and Power Congressman John Dingell waged a bitter battle to put all official government findings about NUMEC into the public domain in a third GAO report. This congressional action created serious rifts with Carter's administration.

On November 6 the *New York Times* openly speculated Powell's official denials about NUMEC were related to the Carter administration's concerns with "its relations with Israel." Harold Unger, Shapiro's Washington lawyer, broadcast his client's assertion that "he never diverted a single microgram of nuclear material to Israel or anyone else and does not believe that anyone else did so at the plant." Predictably, Unger dealt out anti-Semitism cards. "If they are pursuing Dr. Shapiro because he is Jewish and a Zionist, for which he offers no apologies, it's a hell of a basis for an investigation."[375]

Gerald G. Oplinger worked as National Security Advisor Zbigniew Brzezinski's aide for global issues. Oplinger held a position at the Nuclear Regulatory Commission before joining Brzezinski's staff. Oplinger's Global Issues group raised the NUMEC affair with Brzezinski on December 7, 1977 while under sustained pressure from John Dingell's congressional investigation. Oplinger "...reviewed the 30 lengthy responses drafted by DOE to Congressman Dingell's questions on the Apollo-Numec-Israel issue. Dingell's pursuit of the question of whether the CIA withheld information from other agencies strikes me as potentially dangerous, so I will ask Hoskinson[lxiii] to review these also..."[376]

Senator John Glenn sponsored legislation calling for sanctions against rogue states trafficking in nuclear materials and technologies outside the NPT. [lxiv] Glenn also began pressing the White House for comprehensive information about NUMEC and Apollo. On September 22, 1979 a U.S. VELA satellite witnessed an Israeli nuclear test over the Indian Ocean. The clandestine weapons test—observed from orbit only because of an unpredictable break in cloud cover—was under official observation by at least two Israeli Navy ships and other Israeli personnel working with the apartheid South African Navy.[377]

Senator Glenn hunted down more information, grilling the Energy Department and putting Oplinger on the spot. Oplinger wrote his boss that "I have a Top Secret memorandum from you to the President written by Jessica Mathews in August, 1977, concerning the above subject, i.e. missing nuclear material from the NUMEC plant in Apollo, PA. It reports everything Jessica was able to learning about this in briefings by ERDA, FBI and the CIA. Senator Glenn has for some time been pressing John Deutch[lxv] for his views on this matter. Since John will be speaking for the administration, he and I believe it is important that he should know the contents of the memorandum in order to avoid stepping into

---

[lxiii] Samuel L. Hoskinson was the National Security Council's chief intelligence aid, responsible for legislative and budget issues affecting the intelligence community.

[lxiv] The Glenn Amendment (Section 670) was later adopted in 1977, and provided the same sanctions against countries that acquire or transfer nuclear reprocessing technology or explode or transfer a nuclear device as a predecessor law called the Symington Amendment, which limited aid under the Foreign Assistance Act of 1961. The Glenn provision, as amended, is now contained in Section 102 of the Arms Export Control Act (AECA).

[lxv] Then at the Energy Department, later DCI of the Central Intelligence Agency

unknown pitfalls in this sensitive matter..." Oplinger went on to recommend that Deutch be allowed to review the classified documents in Oplinger's office. Brzezinski disapproved, instead ordering Oplinger to take a more deniable route. "Brief him orally."[378] Into what NUMEC pitfalls could a Carter administration DOE official fall? What GAO investigation so upset National Security Advisor Brzezinski with its irreverent suspicions of CIA involvement in NUMEC?

On August 12, 1977 Congressman John D. Dingell had requested the GAO investigate all official information that had been developed on the diversion of material from NUMEC and whether investigations performed by the Federal Government had been effective. The bombshell December 18, 1978 report *Nuclear Diversion in the U.S.? 13 Years of Contradiction and Confusion* was classified "secret" at the insistence of the CIA and FBI.[lxvi] The GAO had intended to make the report public in order to respond to growing citizen concerns about NUMEC. Rep. Dingell, who had chartered the inquiry, had been assured as little as six months before it was issued that only the most sensitive sections of the report would be classified. Only later did the CIA and FBI finally insist that the entire report be classified at the "secret" level over the strident objections of Dingell, who was quoted expressing his disappointed to the press at the time. "I think it is time that the public be informed about the facts surrounding the ... affair and the possible diversion of bomb-grade uranium to Israel." The GAO was apologetic about the secrecy in its report transmittal letter to Dingell. "We made every attempt to issue an unclassified report on this matter. However, neither the Federal Bureau of Investigation nor the Central Intelligence Agency was able to provide us with a declassified version of the report."[379]

Nuclear Diversion in the U.S.? 13 Years of Contradiction and Confusion covers the period between 1957 and 1967 when NUMEC received over 22 tons of uranium-235. The GAO was specifically chartered by Congress to investigate four allegations about what happened to the uranium. The first was that "the material was illegally diverted to Israel by NUMEC management for use in nuclear weapons." This was a result of early AEC and FBI investigations into the activities of Zalman Shapiro. The second theory was "the material was diverted to Israel by NUMEC management with the assistance of the Central Intelligence Agency (CIA)." The final theories explored by GAO were more general, that "the material was diverted to Israel with the acquiescence of the United States Government" or "there has been a cover-up of the NUMEC incident by the United States Government."[380]

Although GAO solicited all data and analysis developed by the CIA, FBI, Department of Energy and AEC, it documented in the report being "continually denied necessary reports and documentation...by the CIA and FBI." GAO attempted to fill in gaps or outright refusals by key officials to cooperate through

---

lxvi Although GAO is not subject to the Freedom of Information Act, the report was released to the author on May 6, 2010. While FBI FOIA officials claimed cooperation with the GAO security officer releasing their equity content, the CIA did not. The CIA later refused direct author requests for release of its equity content in the GAO report.

directly interviewing former government employees, including FBI special agents.

The GAO report lambastes the FBI's refusal to investigate potential diversion. "The FBI, which had the responsibility and authority to investigate the alleged incident, did not focus on the question of a possible nuclear diversion until May 1976–nearly 11 years later. Initially, the FBI declined DOE's request to conduct an investigation of the diversion possibility even though they are required to conduct such investigations under the Atomic Energy Act...." The FBI's Foreign Agents Registration Act investigation of Shapiro during the 1960s did not lead to a Justice Department registration order. When the FBI zeroed in on NUMEC management and recommended openly barring contracts and security clearances, its recommendations were largely ignored. The GAO report recognized this. "The FBI became so concerned about the security risks posed by NUMEC's president that they asked DOE whether it planned to terminate his security clearance or stop the flow of materials to NUMEC. According to the FBI's liaison with GAO, the FBI recommended that NUMEC's operating license be taken away...." While the GAO report correctly noted that the FBI dropped the entire investigation between 1969 and 1976, it doesn't speculate what would have happened if NUMEC contracts had been promptly cut along with Shapiro's security clearance after a formal order to register as a foreign agent.

The GAO seemed puzzled that it took a direct order from President Gerald Ford in 1976 for the FBI and Department of Justice to finally "address the diversion aspect." But the GAO did not have access to Nixon and Kissinger top secret deliberations and commitments to "ambiguity" that led to dropping NUMEC the matter.

The public reversals of official, but fundamentally unsubstantiated, U.S. government positions on NUMEC were among the few positive developments of the renewed NUMEC investigation. "Until the summer of 1977, the only publicized Government view on the NUMEC incident was that there was no evidence to indicate that a diversion of nuclear material had occurred." By February 1978, the Nuclear Regulatory Commission (NRC) announced it had "reconsidered" its previous position that there had been "no evidence" to support diversion.

But GAO accurately concluded that the 11-year gap "obviously hampered" the effort to understand what really happened. The DOE's nuclear materials safeguards, which before 1967 tracked the monetary value rather than the precise mass of the uranium, were seriously flawed according to GAO. NUMEC's claims that key records covering a period of heavy uranium loss were destroyed during a "labor dispute" in 1964 were clearly suspect. NUMEC's fine of $1.1 million for 206 pounds of missing uranium in 1966 — enough for "at least four or five nuclear weapons" — which ended the DOE case was also highly suspicious to the GAO. An FBI agent in charge told the GAO it did not even investigate the source of funds used to pay NUMEC's fine anticipating "legal difficulties." So the GAO investigated the matter, placing its own telephone calls to Mellon Bank to verify the existence of sufficient credit lines and other financial wherewithal used to pay the fines.[lxvii]

NUMEC's hiring away one of the AEC's chief on-site investigators[lxviii] to enhance the appearance of serious materials control and accountability also struck the GAO as odd. The GAO found that even by 1978 the FBI had not contacted key individuals in the affair as required by the newly reopened investigation. But the GAO report levels most of its criticism toward the CIA: "From interviews with a former CIA official and with former and current officials and staff of DOE and the FBI we concluded that the CIA did not fully cooperate with DOE or the FBI in attempting to resolve the NUMEC matter." The report was inconclusive about exactly what happened at NUMEC, but not about the performance of agencies involved in the investigation through 1978. While the FBI provided an excuse that too much disclosure could "jeopardize an ongoing investigation of the alleged diversion incident" the CIA provided no justification at all for its total denial of GAO requests for access.[lxix][381]

The GAO, as the investigative arm of Congress, was bitter in its final conclusion. "We believe a timely, concerted effort on the part of these three agencies would have greatly aided and possibly solved the NUMEC diversion questions, if they desired to do so....Based on the totality of GAO's inquiry, we believe that the allegations have not been fully or adequately answered....Based on its review of available documents...GAO cannot say whether or not there was a diversion of material from the NUMEC facility." The GAO recognized that the DOE and FBI had "begun new investigations of the incident" but expressed no optimism whatsoever about the cooperation of the only agency with robust overseas investigative capabilities.

The deputy attorney general, after reviewing recommendations of the GAO report *Nuclear Diversion in the U.S.?* accepted its recommendation that because there was a "failure of DOE, the FBI and the CIA to coordinate their efforts on the suspected diversion when it occurred...." that the attorney general should "take the lead in establishing an interagency plan" under its authority granted by the Atomic Energy Act of 1954.

In a redacted February 28, 1979 letter to the president, Deputy Attorney General Benjamin R. Civiletti laid out the recommended policy. "A recent GAO report...recommended that DOE, NRC, CIA, and Justice establish a plan for coordinated interagency action to detect and investigate the theft or diversion of nuclear material in the future. [paragraph redacted] We believe, therefore, that as regards diversions from the United States to foreign nations, the CIA's participation is essential, because it is the only agency of the government with the necessary capability to obtain information abroad. [paragraph redacted] We have accepted GAO's recommendation that this Department take the lead in establishing the interagency plan. As the initial step, the Attorney General is

---

[lxvii] The GAO did not investigate AEC's role in engineering NUMEC's sale to Atlantic Richfield.

[lxviii] James E. Lovett

[lxix] For a current day expert assessment about the CIA response, see the epilogue.

directing the Interdepartmental Committee on Internal Security (ICIS)[lxx] to study the problem and develop a comprehensive plan, with the CIA and other responsible agencies, to deal with nuclear threat situations."[382]

The main outcome — to establish better control over bureaucracies tasked with averting diversion, rather than prioritizing warranted prosecutions for past diversions — was a relief for the Carter administration. The year 1979 saw the Soviet invasion of Afghanistan, the Islamic Revolution and overthrow of the Shah of Iran and brewing crisis in Central America. The final FBI files developed on NUMEC in the early 1980s, some with tantalizing new leads — likely fell near the bottom of the administration's list of priorities.

Nothing was ever passed from the Carter team to the Reagan administration, which never opened any of its own NUMEC proliferation files.[lxxi] Although the Carter administration did not distinguish itself by settling any outstanding public questions over NUMEC, Carter later became the first former president to publicly admit the glaringly obvious. In May of 2008, while attending the Hay-on-Wye current affairs book festival, Carter confirmed that Israel had 150 weapons in its secret arsenal.[383] Carter was the first former president to break with "ambiguity" in order to make the argument that it would be impossible for Iran, a signatory to the NPT, to secretly develop nuclear weapons and delivery vehicles given the history of Israel, the U.S., Russia, China, Britain and France.

[lxx] The Interdepartmental Committee on Internal Security was established as an independent interdepartmental committee of the National Security Council in March of 1949. It was transferred to the Department of Justice by President John F. Kennedy on June 9, 1962. The Committee was responsible for the coordination of all phases of internal security, except those pertaining to investigative agencies. The Committee consisted of representatives of the Departments of State, the Treasury, Justice, and Defense.

[lxxi] Based on queries and discussions of finding aids with Reagan Library archivists based in Simi Valley

12. Congress interviews Zalman Shapiro

On Thursday, December 21, 1978 the Committee on Interior and Insular Affairs held a small hearing with Zalman Shapiro in the Longworth House Office Building on Capitol Hill. Committee Chairman Morris K. Udall peppered Zalman Shapiro and his legal counsel with questions about NUMEC. Udall took great pains to emphasize that the Committee was mainly concerned about adequate nuclear safeguards and was only investigating NUMEC because "in early 1977, our Committee was informed that government files contained documents indicating the possibility that a substantial quantity of missing bomb grade uranium had been diverted from a processing plant operated by the Nuclear Materials and Equipment Corporation in Apollo, Pennsylvania."[384]

Although the CIA had largely denied GAO access to its files, selected staff members from John Dingell's Subcommittee were allowed limited access under controlled conditions.[385] Recalling a meeting with Carl Duckett on January 25, 1978, Udall staffer Henry Myers recalled that "Duckett believes that the totality of circumstantial evidence supports the conclusion that there is a significant likelihood that Apollo uranium went to Israel."

But Duckett had resigned two years earlier and no longer spoke officially or semi-officially for the CIA. When Myers solicited even more sensitive and authoritative information from the CIA he was informed that director Stansfield Turner would have to decide whether or not to release anything. Turner objected to revealing the agency's espionage operations targeting Israel in a stern February 21, 1978 letter to the NRC. "CIA's findings or conclusions relating to the alleged diversion of nuclear material from the NUMEC facility should be classified SECRET. To do otherwise in an open hearing would be a public official acknowledgement of CIA's intelligence role as it relates to Israel, which could have serious repercussions."[386]

Myers later ruefully relayed to Udall that "We have not been getting a straight story from either the FBI or the CIA." By May of 1978 Myers's grim assessment was that "the further we get into the NUMEC matter, the more we see that either the FBI has not leveled with us or that their investigations have been inexplicably truncated." [387] Assistant Attorney General Patricia Wald only added to Udall's frustration, writing him in June of 1978 that "You will understand the difficulty involved in disseminating information concerning a pending and active investigation." [388]

Udall was about to get his own taste of just how difficult it was to investigate NUMEC by interviewing the principals as he tried to get an evasive and slippery Shapiro to respond to direct questions. Although Shapiro was not placed under oath, he was informed that a transcript of the proceeding would be turned over to the Justice Department.[lxxii] That remarkable transcript follows.

[lxxii] The dialogue that follows is taken from the preliminary congressional transcript.

Chairman Udall: "While the losses in question occurred in the early or mid-1960's, I believe it is important that we explore whether there are lessons that might be derived from the events and circumstances of that period. To the extent possible, the facts should be placed on the table in order that we be able to reach the sound decisions necessary to protect against the hazards and risks of using nuclear power.

Our inquiry has several specific objectives. First, we want to understand any defects in safeguards as they existed in the 1960's so that we can be sure corrective measures have been taken. Our second objective is to know whether the Atomic Energy Commission's responsibility to encourage nuclear development forestalled its conducting with appropriate vigor an adequate inquiry into the NUMEC situation. We need to understand this to help us judge whether the old conflict between the government's promotional and regulatory activities have in fact been resolved by dissolution of the AEC and establishment of the NRC. We have, in addition, a third purpose. It amounts to this. NUMEC has been the subject of speculation and rumor for more than a decade. All of us who have responsibilities in this area have an obligation to work toward venting what is known. The America people, who paid the bills and incurred the risks, should be able to make their own assessment as to whether their government exercised adequate caution in protecting nuclear materials, and whether all reasonable steps have been taken in search for an explanation of NUMEC's uranium losses. Underlying this and previous inquiries in the matter is the fact that the AEC determined in late 1965 that some tens of kilograms of bomb quality uranium had been lost by NUMEC since the plant had begun operations in 1957. After some months of investigation, the AEC accepted NUMEC's explanation that the material had been lost in the normal course of plant operations. The AEC was, however, unable to pinpoint the time at which the losses occurred or the mechanism that led to them. In accepting NUMEC's explanation, the AEC also stated that it could not unequivocally rule out the possibility of a diversion.

Our inquiry to date has found that safeguards requirements were minimal in the early and mid-1960's. They were based on the assumption that those who possessed nuclear explosive materials would not engage in complex and illegal diversions, and I might add that the financial responsibility of the companies involved would be a further deterrent. I am reading this almost verbatim but I have changed a word or two.

We have also found that NUMEC's material control procedures were repeatedly criticized by those who conducted material surveys of the Atomic Energy Commission. In communicating with the Committee on a matter related to this inquiry, NRC Chairman Hendrie stated in December 1977 that he " ... conclude(s) that for regulatory purposes we must assume the circumstances were such that a diversion could have occurred, and must construct our safeguards requirements accordingly."

Our inquiry has shown that the Atomic Energy Commission was particularly concerned about the public reaction were it publicize the uranium losses at NUMEC. In addition, the AEC, as a consequence of its concern for the wellbeing of the nuclear industry, acted with restraint both with regard to enforcing its own regulations upon NUMEC and imposing generally more stringent safeguards requirements.

I believe it was also out of concern for the adverse publicity and the impact of a crackdown upon the industry that led the AEC and its successor agencies to play down for more than a decade the possibility that a diversion did occur. While we were being assured by both ERDA and NRC officials as recently as the summer of 1977 that there was "no evidence" of a diversion, there were responsible officials in the government who had believed for years that there was circumstantial evidence indicating that materials had, in fact, been diverted.

We know now that as far back as 1966, at least one staff member of the Joint Committee on Atomic Energy questioned the propriety of then AEC Chairman Seaborg having testified that he thought that "there has not been any material diverted from peaceful to military uses." We know also that in December 1971, William Riley, former director of AEC security, wrote in a memorandum that Dr. Shapiro reported to Riley that he (Dr. Shapiro) had been told by officials of the government of Israel that an unnamed AEC Commissioner had suggested to representatives of a foreign government that Dr. Shapiro had been involved in diversion of materials to the Israelis.

We know now that some high level officials of the CIA believed there was evidence of a diversion. Former NRC General Counsel Peter Strauss is reported to have said following a CIA briefing at the NRC that he "...got the impression that the CIA had a fairly strong belief that the inventory discrepancy represented material taken to Israel. He said that if the CIA's information was accurate, there was a strong circumstantial case—missing material, motive and opportunity." At a meeting of NRC Commissioners on November 12 of this year, Commissioner [Victor] Gilinsky stated that he had told Mr. Gossick [Executive Director for Operations Lee Gossick], with respect to the CIA briefing at the NRC, that the one thing that I felt was significant that should not be set aside easily, was the fact that whatever the fact, the CIA seemed to be convinced of the fact something had occurred there, that there was a diversion."

I would also like to note for the record that earlier this month we received from NRC Chairman Hendrie, the Commission's view that "...here are many people familiar with the information on this subject (i.e. Apollo/NUMEC) who seriously suggest a diversion occurred, and they have arguments that do have substance. We have seen no hard proof, one way or the other, but there are various circumstantial items that keep the question unanswered."

Given that allegations about a diversion necessarily affect the reputations of those who were in charge of NUMEC, we have believed it important to ask Dr. Shapiro, as NUMEC's former president, whether he might wish to provide information on this matter which would help resolve, either by stilling the suspicions that exist, or shedding light otherwise.

We are aware that the investigations of this Committee could intrude upon the privacy of Dr. Shapiro. We believe, however, that the seriousness of the allegations has required us to pursue this matter as far as we can, making every effort to be fair and open.

I would say, in addition, that if there had been more openness over the years both by NUMEC with regard to conceding its losses, and the AEC in admitting the true nature of the situation, much of the current suspicion might have been put to rest long ago.

We have agreed that this hearing would be held informally, with the attendance of myself, Representative Baumen, our staffs, Dr. Shapiro and his attorneys. Because of the informality of the hearing, Dr. Shapiro will not be put under oath; we will have a transcript and a recording of the hearing which will be provided to Dr. Shapiro as soon as they are available. At that time Dr. Shapiro will be permitted to insert into the record whatever supplemental or clarifying material he feels is necessary to present the facts as he sees them. We plan, at the conclusion our inquiry, to turn over all our documents to the Department of Justice.

It is important to me that Dr. Shapiro not feel coerced or intimidated in any way. If he chooses not to address any question put to him this afternoon, he will not be pressed further on that issue. Bob Terrell, does Congressman Bauman have any preliminary remarks to make?

Mr. Terrell. No, he doesn't, Mr. Chairman.

Chairman Udall. Do you wish to respond, Abe, to this?

Mr. Krash. Not at all, Congressman.

Chairman Udall. All right. Let me get involved in the questions, then. Dr. Shapiro, NUMEC was founded in 1956, 1957, somewhere in that period.

Dr. Shapiro. 1957.[lxxiii]

Chairman Udall. What was your role in getting the company started?

Dr. Shapiro. I was responsible for organizing it.

Chairman Udall. You were one of the original organizers of the corporation.

Dr. Shapiro. Yes.

Chairman Udall. You had been employed just previous to that at Westinghouse?

Dr. Shapiro. Yes, I was.

Chairman Udall. Doing what? What were your duties?

Dr. Shapiro. I was at Bettis, the Bettis laboratory operated by Westinghouse.[lxxiv]

Chairman Udall. Just basically and generally, what was the motive and purpose in establishing NUMEC?

Dr. Shapiro. I wanted to own a business for myself.

Chairman Udall. Okay. And can you tell us something about the stockholders and the incorporators, who the main investors were, and so on?

Dr. Shapiro. Initially the investors were private investors. I believe there were about 25 of them, and the funds were raised by—through the Apollo Steel

---

lxxiii NUMEC was incorporated on December 31, 1956 according to articles of incorporation filed with the Commonwealth of Pennsylvania, Department of State, Corporation Bureau

lxxiv In 2009 Shapiro divulged a wealth of information about his history at Bettis and role working to overcome engineering problems on the Nautilus submarine, as well as his close association with Hyman Rickover, in an application for a National Medal for Technology and Innovation. In 1978 Shapiro avoided discussing these details even when Udall asked direct questions.

Company[lxxv], and I believe they owned somewhere in the neighborhood of 30 percent of the stock initially.[lxxvi]

Chairman Udall. Apollo Steel had 30 percent of the stock?

Dr. Shapiro. I believe that is correct.

Chairman Udall. Did you personally invest any money in the operation?

Dr. Shapiro. No.

Chairman Udall. Can you give me the names of the two or...

Dr. Shapiro. I believe that is correct.

Chairman Udall. Can you give me the names of the two or three other principals, or four or five major principals that had invested besides Apollo Steel and anybody else?

Dr. Shapiro. There was an automobile dealer, Kittanning, or something like that. I believe it is all on record.

Chairman Udall. Okay. This isn't a vital point, but I had a question on it, and I thought maybe you could remember two or three of these people. Did the automobile dealer and others have their own money invested, to your knowledge?

Dr. Shapiro. Yes.

Chairman Udall. Substantial amounts? Could you give us a ballpark figure?

Dr. Shapiro. I really don't recall. The total investment at that time was, I believe, $250,000.

Chairman Udall. Did you receive stock for your expertise and management ability that you were going to devote to the company?

Dr. Shapiro. I received some stock, yes.[lxxvii]

Chairman Udall. Do you remember roughly how much, without going back to the records?

Dr. Shapiro. I don't recall what the initial number of shares were because they were split, and I just don't recall the actual numbers.

Chairman Udall. In the ballpark figure, would these people have a few hundred dollars, or few thousand dollars? Was it in that range?

Dr. Shapiro. It was in the thousand-dollar investment category. There were, as I said, about 25 all together, as I recall it, and Apollo was the principal through which the investment was made.

Chairman Udall. And did you go out and recruit the investors yourself? How did the group come together?

Dr. Shapiro. I helped to recruit some of them, but principally investors in Apollo Steel, as I recall it.

Chairman Udall. Who was the main person in charge of Apollo Steel?

[lxxv] Facilities funds were raised by Apollo Industries, into which Apollo Steel was merged, along with San Toy and American Nut and Bolt.

[lxxvi] Although Shapiro was repeatedly asked to identify investors by name, he declined to do so since it would reveal that many were highly active Zionists, or had dubious backgrounds, which would logically lead to questions about ideological motivations for investing in Apollo Industries.

[lxxvii] Shapiro did not mention he held almost 20,000 shares worth nearly half a million dollars under the terms of the Atlantic Richfield buyout.

Dr. Shapiro. The president at the time was Mr. Lowenthal, I believe.[lxxviii]

Chairman Udall. And who is he?

Dr. Shapiro. He was—

Chairman Udall. A businessman?

Dr. Shapiro. A businessman.

Chairman Udall. Had you had prior connections or business dealings with him?

Dr. Shapiro. I had met—no, I never had any business dealings with him, but I had met him through organizational work.

Chairman Udall. We have the name in the materials we have looked at of Raychord Corporation. Was Raychord Corporation connected with Apollo Steel?

Dr. Shapiro. Raychord was, I believe, a subsidiary of Apollo Steel, but it was formed subsequent to NUMEC's formation

Chairman Udall. Did you already have lined up in prospect government contracts when you formed NUMEC or were you forming the corporation and then looking about for business?

Dr. Shapiro. Right.

Chairman Udall. The latter?

Dr. Shapiro. Yes.[lxxix]

Chairman Udall. Okay. Can you tell us whether or not there were any foreign investors, or whether any foreign governments, including specifically the government of Israel, directly or indirectly was an investor in NUMEC?

Dr. Shapiro. Absolutely not.

Chairman Udall. You are certain of this?

Dr. Shapiro. I am positive.

Chairman Udall. This was local money, money of 25 or so investors, plus Apollo Steel?

Dr. Shapiro. Right

Chairman Udall. They were the only funds that went into this to your knowledge. When NUMEC got underway and founded the company, did you have a plant building assigned, equipment, or were you starting from scratch with vacant property?

Dr. Shapiro. No, Apollo Steel had a vacant building, a shipping and receiving building, and we occupied that, and we also occupied the top floor of the office across the street.

Chairman Udall. How soon was it before you began to get active contracts for the company?

Dr. Shapiro. I really don't recall.

[lxxviii] Shapiro's exhibited difficulty remembering Lowenthal, a legendary figure fighting for Israel's independence and a daring refugee smuggler, who was his close friend and neighbor, and the president of the company that allowed NUMEC to come into existence, reveals Shapiro's extreme evasiveness.

[lxxix] NUMEC's value as a corporation was largely derived from U.S. government contracts. NUMEC's sale to Atlantic Richfield and B&W also suggests that their motives for acquiring NUMEC were also largely tied to acquiring government contracts.

Chairman Udall. Fairly quickly?

Dr. Shapiro. Fairly quickly.

Chairman Udall. Months?

Dr. Shapiro. Months.

Chairman Udall. A few months after you were organized?

Dr. Shapiro. Yes.

Chairman Udall. Was the company formed with the idea of getting processing contracts with the AEC or the Federal Government to handle uranium and nuclear materials? Was this one of your objectives?

Dr. Shapiro. The principal objective was to get into the business of preparing fuel for power reactors, and we took other business to further our objectives in this because that business was at a low rate at the time.

Chairman Udall. You were recognized in those days as an outstanding expert in this field. Were there others in the organizational group that had expertise, or were they simply investors looking for a good business investment?

Dr. Shapiro. No, they had — the people in the community, you mean?

Chairman Udall. The original investors, incorporators and businessmen, when you put NUMEC together, were you the one person with expertise in this field?

Dr. Shapiro. As far as the business people were concerned, they had no expertise at all. As far as those of us who started the organization, we all had expertise in one aspect or another,[lxxx] except for the attorney that was also one of the few original members.

Chairman Udall. Do you remember his name?

Dr. Shapiro: Gray.

Chairman Udall. "Ray?"

Dr. Shapiro Gray.

Chairman Udall. Oh, Gray.[lxxxi] All right. In the Pittsburgh area?

Dr. Shapiro. No, he came from Washington.

Chairman Udall. Did you have an adequate capital base [to] begin original operations with these original contracts just from your stock sales, or did you go out and borrow money, make loans, in order to get a capital base?

Dr. Shapiro. Yes.

Chairman Udall. From whom?

Dr. Shapiro. Primarily from the Potter Bank.

Chairman Udall. That is a Pennsylvania institution?

Dr. Shapiro. Pittsburgh institution, yes, which was later merged into other banks.

Chairman Udall. Do you have a general idea of the magnitude of those initial loans we are talking about a few thousand, or a million?

Dr. Shapiro. Approximately a quarter-million.

Chairman Udall. And was this money secured in any way?

[lxxx] Shapiro did not mention the involvement of Leonard P. Pepkowitz, one of the initial three incorporators in 1956, who had extensive nuclear weapons experience in the Manhattan project. This would have provided valuable insight.
[lxxxi] Oscar S. Gray helped incorporate NUMEC and was Treasurer, among other roles.

Dr. Shapiro. It was secured against equipment that we purchased, for the most part.

Chairman Udall. Can you tell us whether any foreigners, foreign governments, anyone outside your group, was involved in making the arrangements for that loan, or those original loans that helped finance your business?

Dr. Shapiro. There were none.

Chairman Udall. There were none. I think I have already asked this asked this, but I want to make clear that to your knowledge the government of Israel was involved in no way in the original founding of the company.

Dr. Shapiro. Absolutely not.

Chairman Udall. Or in getting the loans that got the company underway?

Dr. Shapiro. Absolutely not.

Chairman Udall. All right, sir. NUMEC, your connection with NUMEC lasted what, ten years or so? You left NUMEC when?

Dr. Shapiro. In 1970.

Chairman Udall. So we are talking about a 13-year, roughly 13-year period?

Dr. Shapiro. Yes.

Chairman Udall. Give or take. During that time, what proportion of NUMEC's revenues and overall workload was involved in government contracts, and particularly the naval reactor program? Can you specify in general terms?

Dr. Shapiro. I really can't give you a—it would just be conjecture on my part at this point.

Chairman Udall. Less than half, or more than half?

Dr. Shapiro. I really don't remember.[lxxxii]

Chairman Udall. But you did have a number of contracts, and not just the processing of naval reactor fuel?

Dr. Shapiro. Yes, we did.

Chairman Udall. Can you give me a general idea of what the other kinds of functions NUMEC had, what other kinds of contracts you had?

Dr. Shapiro. Well, we processed a lot of reactor-grade material. We were one of the principal fabricators of uranium oxide for power reactors. We also made hafnium crystal bar for control materials, and for other purposes, and we made neutron sources and radiation sources and burnable poison materials, and did scrap recovery both for ourselves as well as for customers.

Chairman Udall. Some of your customers, I take it, then, were government agencies and some were private?

Dr. Shapiro. Some of our customers were government agencies and others were private firms, yes.

Chairman Udall. To what extent with regard to the Navy fuel processing program were you their sole supplier? Were you competing with other firms?

Dr. Shapiro. We were competing with other firms.

Chairman Udall. Many of them, or where there just a few of them in this business in the '50s and '60s?

---

lxxxii Shapiro should have known, and divulged, that the majority of NUMEC's revenue came from the government since the civil nuclear industry was still in a nascent stage.

Dr. Shapiro. There were quite a few compared to the business that was available.

Chairman Udall. Do you know Admiral Rickover?

Dr. Shapiro. Yes.

Chairman Udall. Did you ever have any controversy with him at NUMEC? There are some suggestions in all of this that Rickover was very unhappy with NUMEC's performance, which may not be surprising given Rickover's personality, but did you have—

Mr. Krash. Have you ever known him not to be unhappy with somebody's performance?

Chairman Udall. Have you had any conflicts with him, or did you have during that period?

Dr. Shapiro. Yes.

Chairman Udall. What did they regard, quality of work, or...

Dr. Shapiro. No, they did not generally—they were not generally with regard to the quality of our work. It was with regard to the price and with regard to other contractual issues.

Chairman Udall. Did he ever come to NUMEC; did you meet with him in Washington from time to time?

Dr. Shapiro. He never came to NUMEC.

Chairman Udall. And your—

Dr. Shapiro. My meetings with him were prior to—except in one instance—were prior to my corning to NUMEC.[lxxxiii]

Chairman Udall. Let me leave all this for a moment and hit something that I think is very direct, very fundamental, here. As you know, one of the charges that is kicked around, suspicions in the conversation over the years, has been that you were involved in diversion, may have been involved in diversion, of enriched uranium to Israel or some other country. Were you?

Dr. Shapiro. No, absolutely not.

Chairman Udall. And so I don't leave any loopholes, I am talking about not necessarily direct to Israel, but to agents of that country, to third countries who you understood were going to divert to Israel, anything of this kind?

Dr. Shapiro. I never diverted any material to anybody.

Chairman Udall. Do you know of anyone else who did?

Dr. Shapiro. No, I do not.

Chairman Udall. You are unaware of any other arrangement. Did you ever hear of any other scheme or plan of anyone else to use your plant or facilities, to divert to Israel, or any other source, enriched uranium?

Dr. Shapiro. No.

Chairman Udall. Would you like to speculate—this may be the kind of question you don't want to answer, why have these—this has bothered me—why have these rumors been so persistent these last 15 years?

Dr. Shapiro. I can only speculate.

[lxxxiii] Given a chance to discuss Rickover's letters excoriating NUMEC for its lax security and employing an Israeli national, Shapiro chose not to elaborate.

Chairman Udall. I am trying to get an understanding. Clearly, this doesn't call for factual information on your part—if you want.

Dr. Shapiro. You mean what have been the motives of these people?

Chairman Udall. Yes. Why is this thing so persistent, in newspaper reports, the memos back and forth between government agencies? Do you have any information?

Dr. Shapiro. First of all, congressman Udall, I don't believe they were persistent over the 15 years. They have been persistent over the past couple of years, I think, and I can only speculate as to why this whole matter arose again quite recently.

I believe there were probably two reasons. One, that it was a means of those who wished to demonstrate that nuclear power should be abandoned because of the possibility of diversion; and the other I believe, was because it suited some people, also, to use this as a means of embarrassing Israel.

Chairman Udall. Do you feel that you have been unfairly and unjustly treated in all of—

Dr. Shapiro. Absolutely.

Chairman Udall. —in all of these allegations?

Dr. Shapiro. Absolutely.

Chairman Udall. All right. Let me turn to something else. There is a memo, Feb. 27, 1962, and most of these memos I am referring to are documents or things that we furnished your attorneys. If you are unfamiliar with them, I will try to do something about it, but let me refer to one dated Feb. 27, '62, an Atomic Energy Commission memo from J. A. Waters, Director of Security, to Austin Betts, Director of Military Applications, indicating that NUMEC served as technical consultant and training and procurement agency for Israel in the United States. The memo also states that under this agreement one Israeli metallurgist worked at NUMEC's plutonium plant. Was there such a metallurgist, and do you recall his name?

Mr. Becker. Let me say that we don't have a copy of that document. If you have a copy, we could look at it.

Chairman Udall. Let me say by way of preface that I have been through a whole mound of materials myself.

Mr. Krash. What was your question, Congressman Udall?

Chairman Udall. Was there an Israeli metallurgist and what was his name?

Dr. Shapiro. His name was Dr. Bernard Cinai.

Chairman Udall. Can you spell that?

Dr. Shapiro. C-i-n-a-i. He worked at the plutonium facility.

Chairman Udall. At Apollo?

Dr. Shapiro. In Parks Township, not at Apollo.

Chairman Udall. What was that facility?

Dr. Shapiro. That was the facility located about six miles down the road from Apollo in which we were working on the development of fuel for, plutonium-bearing fuel for power reactors.

Chairman Udall. What was the connection of that concern, company to NUMEC?

Dr. Shapiro. It was part of NUMEC.

Chairman Udall. This was a subsidiary or…

Dr. Shapiro. It wasn't a subsidiary. It was just a separate operation.

Chairman Udall. A separate operation. I am going to come back to him in just a moment, but in your years at NUMEC were there other Israeli citizens that worked as technicians in your facilities?

Dr. Shapiro. Not that I know of.

Chairman Udall. He is the only one that comes to mind?

Dr. Shapiro. The only one that I know of that worked in our facilities.

Chairman Udall. Given the sensitive nature of the materials you were handling and so on, why was an Israeli citizen hired for this job as against a U.S. citizen?

Dr. Shapiro. Well, this was done under—with the knowledge of and permission of the Atomic Energy Commission. It was done under the Atoms for Peace program, and part of that program was for the training of personnel and this man was, as I say, a part of that program. He was, without question, an outstanding technical person.[lxxxiv] [389]

Chairman Udall. So you advised AEC or asked their consent to employ him at the plant?

Dr. Shapiro. Yes.

Chairman Udall. What was the procedure? Was this a written kind of procedure?

Dr. Shapiro. I believe it was written, but I can't recall at this point, but it was certainly done with their knowledge and consent without question.[390]

Chairman Udall. And he came to the United States specifically for the purpose of—

Dr. Shapiro. Working at that facility.

Chairman Udall. Working at that facility and learning and getting expertise and getting information about the peaceful uses of plutonium and nuclear power?

Dr. Shapiro. Of plutonium and nuclear power, exactly.

Chairman Udall. And he is the only Israeli technician you have said that you recall in your years at NUMEC—

Dr. Shapiro. Right.

Chairman Udall. …which came there for that purpose?

Mr. Krash [Shapiro's legal counsel]. If I could interject. I don't want to interrupt at all, but I would like to invite your attention to the fact that in the prepared statement, which I know you have not had a chance to see, Congressman Udall, at pages 7 and 8 we have listed the names of various individuals who were foreign nationals who worked at NUMEC. There was a Dutchman from South Africa, there were some Argentineans and there were some Japanese. In other words, there were a number of foreign nationals who were working there, about half a dozen actually, and they are all listed there.

Chairman Udall. They were here for the same purposes and same reasons?

Dr. Shapiro. No.

[lxxxiv] FBI interviews of the AEC concluded that the AEC disapproved of Cinai's employment at NUMEC

Chairman Udall. What was the difference?

Dr. Shapiro. The Japanese were there for the same purpose. The Dutchman was employed as an instrument expert. The Argentineans were analytical chemists who were working at our plutonium facility, as I recall it.

Chairman Udall. And these also were Atoms for Peace, helping to train other countries?

Dr. Shapiro. These were people who had gotten their degrees at the University of Pittsburgh, if my memory serves me correctly.

Chairman Udall. And would they typically seek you out and seek employment or were you looking for these kinds of people?

Dr. Shapiro. We were looking for analytical chemists.

Chairman Udall. And national origin was not of great concern so long as AEC would approve?

Dr. Shapiro. Right.

Chairman Udall. Mr. Cinai particularly, what was the nature of work that he performed and how long was he there? Do you recall?

Dr. Shapiro. He may have been there about 12 or 14 months and his specific project was on the degree of homogeneity achieved in the mechanical mixing achieved in the mix of uranium oxide and plutonium oxide for the purpose of making homogeneous mixed oxide for—

Chairman Udall. Nuclear fuel rods?

Dr. Shapiro. For nuclear fuel rods.

Chairman Udall. Did he seek you out or did you seek him out? Do you recall the circumstances of his employment?

Dr. Shapiro. I don't recall specifically how that came...

Chairman Udall. Over the years I suppose the work force varied from time to time, but during that 13-year period typically how many employees would you have total at NUMEC in all your different operations?

Dr. Shapiro. In all the different operations I think we peaked at 1,000.

Chairman Udall. And it was lower than that from time to time?

Dr. Shapiro. Yes.

Chairman Udall. Yes. The kind of work that Mr. Cinai was doing and I guess I would apply that to these other foreign nationals, was this an opportunity to gain expertise that could have been applied in nuclear weapons programs?

Dr. Shapiro. Not that I can see at all.[lxxxv]

Chairman Udall. The tasks he was doing, the technology involved was—

Dr. Shapiro. The technology was—

Chairman Udall. —specific to—

Dr. Shapiro. —was with uranium and plutonium oxides, specifically with regard to the fabrication of fuel elements for power reactors.

Chairman Udall. Would it be a fair summary with regard to these people—and I am talking now about various foreign nationals—that you are not aware of any

---

lxxxv Cinai was gaining experience handling quantities of plutonium not yet available in Israel. His skills could also come in handy in the future disassembly of Israeli nuclear weapons for use in mixed oxide, or MOX, fuel for power generation.

information or anything that came to your attention over the years that these people were trying to acquire information that might be helpful to them in a nuclear weapons program, particularly Israel's nuclear weapons program?

Dr. Shapiro. I certainly didn't give them such information.[lxxxvi]

Chairman Udall. I am going to refer now to an October 15, 1964 letter from [Charles A.] Keller of the AEC, Oak Ridge, to you, Dr. Shapiro, in which he states that "A crossover between different jobs has occurred." My first question is, this terminology, does this mean that materials assigned to one contract, nuclear materials assigned to one contract were mixed with materials from other contracts? Is that what you understand that allegation to cover?

Dr. Shapiro. If I could take a minute...

Chairman Udall. Sure, take your time.

Dr. Shapiro. I presume that he was referring to a job in which we had permission from the Atomic Energy Commission to utilize material from one contract which had been deferred for some reason and for which material was on hand be used in another and which permission, this permission was gotten from a New York operations office and apparently there was some misunderstanding and as a result communication with, as I understand it, Washington Headquarters, that was later rescinded and at the time that it was rescinded we stopped doing it.[lxxxvii]

Chairman Udall. In general your relationship, contractual relationship, with AEC was such that when you had material on one project, it was a violation of the contract for you to use that material in connection with some other project or contract. Is that what he refers to here?

Dr. Shapiro. There were two kinds of material. One was the leased material which we simply leased for the purpose of doing whatever jobs. Then there were specific contract materials and the specific contracts may have had on occasion a provision which indicated that that material should not be used for other materials, depended entirely upon the materials of the contract.

Chairman Udall. Some did permit crossover, some did not permit crossover?

Dr. Shapiro. I just don't recall.

Chairman Udall. All right. I am going to refer now to what is referred to in many of these materials as the WANL contract, which I take it is Westinghouse Astronuclear Laboratory?

Dr. Shapiro. Yes.

Chairman Udall. Can you describe briefly what that contract involved? What were they sending you and what were you doing with it?

[lxxxvi] Shapiro refused to speculate about Israel's nuclear program, admit any weapons-related discussions with Israeli officials, or any inside knowledge of Israel's program. Investigators were incredulous, given his expertise and ongoing contacts. His answer was also carefully hedged. If classified information stored at NUMEC was discovered to be compromised, he could still maintain his position that he hadn't participated in its disclosure.

[lxxxvii] In 1964 Keller was concerned about NUMEC taking material from the SNPO-C contract to cover material missing from the WANL contract.

Dr. Shapiro. They sent us uranium hexafluoride, which is a gaseous material which we got in cylinders from the diffusion plants. That was converted into a compound, uranium carbide, which was in a particular form of little spherical particles as large as essentially flyspecks, which had very stringent specifications associated with them from the standpoint of chemical composition, from the standpoint of size and distribution, and from the standpoint of density, and sporicity, and—

Chairman Udall. What was the end-use of these particles?

Dr. Shapiro. These were to be used in the fuel elements which were eventually fabricated by WANL.

Chairman Udall. For the Navy? For commercial fuel?

Dr. Shapiro. No, no, for the NERVA program, which was a nuclear rocket.

Chairman Udall. Now, this WANL contract, was that one in which crossovers were not permitted?

Dr. Shapiro. Yes.

Chairman Udall. Under that contract you had no right to use those materials in connection with any other contract that you were operating under?

Dr. Shapiro. As I recall it.

Chairman Udall. All right. Let me ask you then the obvious question: Was any of this WANL material to your knowledge ever used to satisfy obligations associated with other contracts?

Dr. Shapiro. If it did occur it occurred only in an accidental way and in scrap recovery. Now you have got to understand what scrap recovery entails.

Chairman Udall. I would like to.

Dr. Shapiro. To recover scrap you must dissolve the material and then it must be put through a chemical extraction column, liquid extraction column or columns, in fact. It is dissolved in a highly corrosive acid. It is then put into these columns where it mixes with an organic material and there is a partition that occurs such that the pure uranium goes into one phase and the impurities go into another phase. This material then comes out as uranyl nitrate, which is a uranium material which is insoluble. It must then be precipitated to form a solid, which was generally done with ammonia, and filtered, dried, and subsequently decomposed and then reduced to the uranium oxide form. There were a number of steps. Now, in the process of extracting the material it is necessary to fill the pipeline, so to speak, and to get an equilibrium established within those columns. And it is under those circumstances that you get the proper partition of the impurities and the uranium. It was essentially not only impractical, but impossible to take small quantities of material and put them through and segregate them and put them individually through those columns and achieve and maintain equilibrium. Therefore, the procedure that was involved was to dissolve the material, to assay that material, put it through—in other words, to establish the quantity, put it through and then to segregate a portion of the material coming through, depending upon what the assay is, make an estimate of what the losses in processing were, and—

Chairman Udall. Allocate?

Dr. Shapiro.—and allocate that material to that particular contract and so it went.

Chairman Udall. Uh-huh. So in that sense there may have been some mixing of WANL contract material, but only in the scrap recovery process that I hope you don't ask me to...

Dr. Shapiro. That's right. [lxxxviii]

Chairman Udall. —explain again. I think I understand enough about it to ask questions, but that is about all. Do you understand that your WANL contract permitted that kind of mixing? That this was a violation of AEC regulations or the WANL contract itself? Did they know you were doing this kind of thing? Did you do this —

Dr. Shapiro. I am sure they knew that we were doing it because eventually — I don't remember when the supply agreement was put into place, but the AEC recognized the necessity of doing it that way and adjusted their contractual arrangements to allow for that to happen in recognition of that fact. In general we were doing work on scrap recovery as it was being done, as we understood it, at the Oak Ridge National Laboratory. And I believe it was my understanding that they followed a similar procedure. So it was certainly not unknown to the AEC. [lxxxix]

Chairman Udall. And you feel you didn't violate your contract in scrap recovery with procedures of this kind?

Dr. Shapiro. There may be some who felt that we were violating it, but I believe that those who understood the practical aspects of the situation recognized the necessity for doing it that way. Otherwise, the losses would have, in fact, been considerably greater because it would have entailed disturbing of the equilibrium in the columns. You would be putting it through, establishing equilibrium and the material coming through first would be no good and the material that came through last would be no good. That would then have to be put back.

We would then have to clean up the columns between each one and under those circumstances obviously the losses that would have been incurred would have been greater in processing than they would have under these circumstances.

Chairman Udall. Okay. One of the things that has been kicked around by those who are suspicious is the suggestion that you were paying off old — that some kind of diversion or loss had been going on and that you would pay off old contracts by material from new contracts like running checks around to different banks, that you were doing this kind of thing and a lot of attention has been paid to this WANL contract. Can you tell us that you never engaged in that kind of a procedure with regard to this contract and that WANL material was not used to pay off obligations on old contracts?

[lxxxviii] Shapiro was now opening up a possibility of mixing material between contracts that NUMEC refused to admit in the early 1960s, even during AEC inspections and employee interviews.

[lxxxix] Shapiro was claiming credit for doing what NUMEC was contractually prohibited from doing. By the same logic, he could claim credit for heightening regulator concerns about diversion over NUMEC's record of the highest amount of MUF in the industry.

Dr. Shapiro. It became evident during the course of this contract that our losses were greater than we had estimated and, therefore, the amount that we had allocated to each contract and shipped back to the customer was greater than it should have been based on the actual losses that did occur versus those that were estimated. Is that clear?

Chairman Udall. Yes, I think I follow.

Dr. Shapiro. And, therefore, the only way that could have happened is that material which should have been charged to a subsequent batch was charged to the former batch and in that sense it could have happened.

Chairman Udall. But you don't see anything devious or sinister about—

Dr. Shapiro. Absolutely not.

Chairman Udall. —about this operation? I am about to leave this, Henry, and it is very clear in my mind. I don't want to come back to it if—all right.

Mr. Krash. I hope you will feel free to ask any questions here that—pursue any questions that you want to.

Chairman Udall. All right. I want to go through and do it just as quickly as possible and thoroughly as possible. [speaking to staff] If you have questions on items I have not clarified, whisper in my ear or speak up. Do you have anything to this point that is major?

Mr. Myers. No.

Chairman Udall. Let me move to another topic. I am going to be talking about section 6.13 of the AEC survey report conducted in November of 1965. I think you refer to that in the brief that your attorneys submitted to us some time ago.

The AEC survey report says something like this—it is just a paragraph: "In an attempt to establish yields and loss mechanisms directly applicable to this purchase order, the survey team requested NUMEC production control and process engineering data on this and other contracts. The data made available was of little or no value in this regard. Process lots or batches could not be correlated to points in time nor could a sequence of processing events be established. All efforts in this direction were negated when it was learned that many of the requested records had been inadvertently destroyed by supervisory personnel during a cleanup campaign at the time of an employee strike January 1 to February 25, 1964.

I wanted to ask several questions about that set of allegations. Tell me about this strike. Was it an actual work stoppage?

Dr. Shapiro. It was a very severe, serious strike that we had. It lasted, as I recall, for over two months.

Chairman Udall. Did it involve all your employees except for supervisory people?

Dr. Shapiro. It involved all the employees in the Apollo plant, except the supervisory personnel, and it occurred in the dead of winter, around the Christmas-New Year's period as I recall it, and into January and February.

Chairman Udall. When you are handling nuclear materials, does a stoppage of this kind cause any danger or particular problems for you? Can you shut down the operation without hazard?

Dr. Shapiro. Yes, you could shut down the operation without hazard.

Chairman Udall. Was there any violence connected with this strike?

Dr. Shapiro. There was violence.

Chairman Udall. Of what nature?

Dr. Shapiro. Some of our equipment was tampered with. For example, one of the trucks had sugar poured into its gas tank. The brakes of one of our supervisory employees were cut; the hydraulic lines were cut. This fellow could have, in fact, been killed because, as you know, we live in a hilly area and this man's car was located on a hill. I was assaulted, et cetera.

Chairman Udall. Okay. What can you tell us with regard to this statement here that the records they were trying to get to trace these various batches were inadvertently destroyed by supervisory personnel during a cleanup campaign? Did the cleanup campaign occur because of the strike or have connection with it? What can you tell me about destruction of these records?

Dr. Shapiro. First of all, the records were not destroyed per se, okay? I can explain that. These were discarded by the people who were doing the cleanup, not because I had ordered it, but because they had decided they were cluttering up area of the files and that they were no longer necessary and there was no necessity for retaining them. They were production type records. They had nothing to do with SNM control per se—

Chairman Udall. Give me an example. When you run a batch through or process a batch—

Dr. Shapiro. We have to give our employees instructions for—

Chairman Udall. I see.

Dr. Shapiro. —what they are to do and when they are to do it, et cetera. And these were processing and manufacturing instructions.

Chairman Udall. You had been in business for seven years at that time. You had an accumulation of these old things, is that what you are telling me?

Dr. Shapiro. Yes.

Chairman Udall. But none of these were of the kind of records that would help us to identify a MUF or a—

Dr. Shapiro. I don't believe that they were at all relevant to the situation, and Mr. Lovett, whom I talked to subsequently,[xc] felt that they had no relevance either. They were just grabbing at straws.

Chairman Udall. Did you indicate that your personnel destroyed cleaned them or, got rid of them, without your instructions on their own initiative?

Dr. Shapiro. Yes. The point was that during the strike supervisory personnel operated the plant to the best of their ability. I felt it was appropriate and desirable to clean up and maintain the plant during that shutdown, which they in fact did. And during the course of this the supervisory personnel in charge of production decided that these were no longer necessary to be kept and there was no need to keep them for any—we were not obligated to keep them for any reason and that they were not relevant to anything we were doing at that time. Consequently, they discarded them.

Chairman Udall. Wouldn't it have been customary or was there any requirement that you know of that you report the discarding of these records to AEC or anyone?

[xc] NUMEC hired away this key AEC inspector at the height of the MUF crisis.

Dr. Shapiro. No. There was no necessity for that because they were internal production and processing type records, and those records which we were obligated to keep for government, or other purposes were always kept.

Chairman Udall. How would you respond then to this statement, and I will just read a part of it, process lots or batches could not be correlated to points in time nor could a sequence of processing events be established? You are saying that five years after you processed something it is unimportant to anybody any more, the exact order or sequence in which you carried on an operation? Is that essentially what it is?

Dr. Shapiro. I would assume that this is what they had in mind, but I—really not having written this, I really don't—

Chairman Udall. I am not blaming you for their report, but I am trying to find out what it was they were complaining about, because they apparently took the loss of these records as being of some consequence. Let me pursue that just a moment. The brief you submitted to us at page 36 says that the lost records would not have been substantially helpful in determining the causes of material losses and you have just repeated essentially that to me today.

Dr. Shapiro. Yes.

Chairman Udall. The statement in the AEC survey that I just read seems to suggest otherwise, namely, that the survey team believed the records would have, in fact, been useful. I guess you have already commented on that conflict.

Dr. Shapiro. Yes. As a matter of fact, as I said, Mr. Lovett, who was responsible at that time for one of those who were engaged in the survey, felt that they would be of no use.

Chairman Udall. Do you have any knowledge of in what form the discarding took? Was it burning or shredding?

Dr. Shapiro. I really don't know.

Chairman Udall. What did you typically do with your waste paper?

Dr. Shapiro. Dumped it in the garbage, I guess.

Chairman Udall. But there was no particular procedure for these kinds of records to shred them or burn them or take precautions of that kind?

Dr. Shapiro. [Shaking head.]

Mr. Krash. If you shake your head, it doesn't show up on the reporter's transcript. You have to say something. The answer was "no," I take it?

Dr. Shapiro. No.

Chairman Udall. There was some reference—I can't remember what it was—of a fire that destroyed some records. Was this the only incident where records were discarded? Do you remember anything about a fire at the NUMEC plant?

Dr. Shapiro. We had a fire in our storage vault and there may have been some I don't recall, there may have been some involvement of paper at that time, but because we kept records of our storage in the vaults, since the man who had responsibility for incoming and outgoing material had to keep the records there—

Chairman Udall. This was the same vault in which you kept nuclear material?

Dr. Shapiro. Yes.

Chairman Udall. And the records related to them?

Dr. Shapiro. And it was a nuclear storage vault.

Chairman Udall. And the records relating to them were also, sometimes also kept in the vault?

Dr. Shapiro. Yes, and the important thing to recognize is that uranium carbide is a pyrophoric material. This is a material then that can spontaneously burst into flame.

Chairman Udall. Can you tell us about when that fire was?

Dr. Shapiro. I will guess around 1963 or thereabouts.

Chairman Udall. It was not connected with this strike that I referred to earlier?

Dr. Shapiro. No.

Chairman Udall. Now you think some records were destroyed in that fire that would have been of more consequence than the kind of batch processing records I was talking about a moment ago, would you?

Dr. Shapiro. Frankly, I don't remember whether there were any records involved at that time. I do remember that there was material involved and that what we had to do was to come in with special fire extinguishers that we had to use, and I just don't know whether there were, in fact, any records involved in that that might have been destroyed as a result of that fire.

Chairman Udall. Well, I had here that NUMEC notified AEC of the loss of records but not until 1966, which apparently was a couple of years or so after you had the fire in the vault. Is there anything sinister or suspicious about that?

Dr. Shapiro. Nothing in my mind.

Chairman Udall. Would you ordinarily—the kinds of records that the AEC required you to keep, had they been destroyed, would this have been an event that you would have been required to communicate with AEC on ordinarily?

Dr. Shapiro. If this were something which we were required to communicate with the Atomic Energy Commission, we would have, in fact, done so because we always tried to do what was required by regulation.

Mr. Terrell. Mr. Chairman, could I ask a question?

Chairman Udall. Sure.

Mr. Terrell. With regard to these records, were there any records that you kept proprietary in nature since you were dealing with Government and outside the Government, in the private sector, that would have been something that you would want to safeguard for the benefit of the contract?

Dr. Shapiro. Not for storage purposes. Not for material storage purposes.

Mr. Terrell. I mean production records that we are addressing. Would there be anything in there that would be of a proprietary nature? It would seem to me that their treatment, something of that nature, would be something greater than casual. In other words, you just wouldn't throw them out because they were excess in the everyday average trashcan if they were proprietary at that time?

Dr. Shapiro. You are talking about discarding of the production records?

Mr. Terrell. Yes.

Dr. Shapiro. I really don't know of anything that would cause us to treat them in any special way.

Mr. Terrell. Thank you, Mr. Chairman.

Chairman Udall. To return —maybe I can conclude this to the fire in the vault which you said was roughly in the '63 time frame, you can't recall now whether there were or were not any records destroyed?

Dr. Shapiro. I don't know.

Chairman Udall. You are saying there could have been the fact that we discussed, some were kept there at some time?

Dr. Shapiro. There were records kept there.

Chairman Udall. And if anybody is charging that you—that there were actually records that were destroyed in that vault fire that you didn't report for two or three years, you know nothing of that phase?

Dr. Shapiro. I don't recall anything about that.

Chairman Udall. I am now going to refer and quote from an August 2 '65 memo to the Commission by AEC Assistant General Manager Howard Brown. I believe your attorneys have been furnished with this previously. Let me quote from that: By the middle of July—

Mr. Becker. Could we have just one second?

Chairman Udall. Sure.

Mr. Becker. This is signed by Brown, to the Commissioners7

Mr. Myers. Yes; Howard Brown. I think you have it. Is it signed by Howard Brown at the end, the last page there?

Mr. Becker. We have another one with the same date.

Mr. Myers. I believe that's the one.

Mr. Krash. We have it.

Chairman Udall. I am quoting from that memo, Dr. Shapiro. He said: "By the middle of July the issue had not been resolved, and on July 21, members of the AEC staff" —and we better close that door—"members of the AEC staff met with Dr. Shapiro and members of his staff in his office at Apollo to discuss the situation and to ascertain what steps might be taken to satisfy the interests of the company and the Government. Shortly after the meeting began, Dr. Shapiro disclosed for the first time a new source of waste material at the plant which, he averred, would not only make up the dollar difference on the WANL contract, but would result in AEC owing NUMEC. Dr. Shapiro stated this new source of valuable waste was contained in about 800 drums of Kleenex, Kimwipes, et cetera, buried under four feet of earth on the company property. "We asked Dr. Shapiro why, in view of the numerous discussions and the almost daily contact on this matter for the past several months, he had not previously disclosed his estimate of the content of the material in the 800 drums. Dr. Shapiro simply said that the situation was embarrassing." That is the end of the quote.

In meeting with the Commission on August 10, you, Dr. Shapiro, proposed that the buried material be recovered over a 12-month period. You indicated at that time that NUMEC would experience difficulties in excavating the material quickly. It appears that NUMEC was urged to undertake the excavation on a more expedited schedule than you proposed. Dr. Shapiro eventually agreed, and exhumation was begun on October 11 and completed on October 21, 1965. The exhumed material contained approximately six kilograms out of the total of 52 kilograms missing."

So I guess what I want to ask you first is: What led you to be so firmly convinced in the first instance that these large quantities were actually buried in the disposal pile?

Dr. Shapiro. The fabrication of uranium carbide entailed a number of steps, and the material by nature—because of the particle sizes that we were involved in—was extremely dusty. It was necessary to, because of the pyrophoric nature of the material, to work in special atmospheric enclosures which had glass portholes.

In addition to that, there were lights inside to illuminate the work, and there were a number of pieces of equipment, and so forth, which were contained in these. It was necessary very frequently for the workers to clean off the surfaces of the portholes and equipment, and so forth, in order to continue operating. Those Kimwipes, rags, at cetera, plastic materials which were used to enclose certain things, accumulated this black uranium carbide material.

Chairman Udall. Uh-huh.

Dr. Shapiro. And these were then bagged and put in a drum and surveyed by our health physics man with an instrument, and it was his opinion, based on the survey, that the amount of material contained in these drums was really quite small and not worth recovering, so these were then buried.

When it became evident that our losses were greater than we had anticipated during the course—towards the end of the contract, I then became very concerned about where this material might have gone, and we looked for the various possibilities. Among the things that I asked our people to do was to take these Kimwipes and go through representative procedure, and then I had these things asked, so that we could concentrate and accumulate the material, and for the work that they did the average amount that seemed to accumulate per Kimwipe seemed to be approximately one gram.

Chairman Udall. So you extrapolated from that to the number of drums in the burial site?

Dr. Shapiro. Well, taking the number of Kimwipes that had been used over this period of time, I did just what you say; I extrapolated, and it appeared to me that we could have, in fact, accumulated a large quantity of material that had, in fact, been discarded by burial.

Chairman Udall. Why were you wrong on that? Did you get a bad sample, a bad cross-section?

Dr. Shapiro. I am not sure that I was entirely wrong. We did in fact—according to the GAO report—recover from the pits that we dug up, 7.4 kilograms of U-235, and, in addition to that, it was estimated as a result of sampling the earth from one of those pits that there was an additional 2.2 kilograms of U-235 for a total of almost 10 kilograms.

Chairman Udall. Were these drums metal?

Dr. Shapiro. The drums were metal. But some of the material was not put in drums, some of it was put in cardboard boxes.[xci]

Chairman Udall. Which had deteriorated?

Dr. Shapiro. Exactly. And in the process of digging this up, particularly during the period that this was done, we were in a rainy period, and as a matter of fact

---

[xci] Shapiro admitted NUMEC buried radioactive nuclear waste material that was toxic and flammable enough to require atmospheric chambers in cardboard boxes.

we had difficulty operating the bulldozer because of the character of the mud, much of this stuff was ruptured, and it was difficult to recover. Now, it was my feeling on the basis—first of all, you must understand that not all of the pits were dug up. I think I mentioned that previously.

Chairman Udall. Uh-huh.

Dr. Shapiro. It was my feeling that, in fact, that indicated that there was, in fact, considerably more material that had, in fact, been discarded in that way than we had estimated.

Chairman Udall. Do you then controvert the estimate that apparently the AEC and the other people, talking about six kilograms was all that they recovered, you feel there was quite an additional and larger amount?

Dr. Shapiro. I don't controvert that, Congressman Udall. I am merely quoting from the GAO report which was subsequent to that report, by which time we had, in fact, recovered all of the material that had been exhumed from the pits, and their number was 7.4 kilograms, and the amount of material that was estimated to have leached into the soil from that particular pit was 2.2 kilograms.

Chairman Udall. So we could account then for roughly 10 of the 50-some?

Dr. Shapiro. Ten of the roughly 50-some kilograms.

Chairman Udall. That were missing. Is that your judgment that that is about all there was now in the pit, or do you think that you —

Dr. Shapiro. I think there was more.[xcii]

Chairman Udall. How much? Ball park—would you care to make an estimate?

Dr. Shapiro. I really can't tell.

Chairman Udall. Considerably more?

Dr. Shapiro. I would say that there was considerably more by virtue of the fact that the conditions under which we exhumed the material were very difficult, and I am not sure that this was entirely representative, therefore, of what we could have gotten had it been done under more favorable conditions.

Chairman Udall. That leads me to the related question. Some might see something sinister in the fact that you said this would take a year to do—you were telling them it would take a year to do it and you wanted to spread it out and, in fact, they did it in ten days, which would seem to suggest that you were not anxious to have this done and were delaying it.

Dr. Shapiro. My recollection, Congressman Udall, was that it took considerably more than ten days. I don't recall exactly how much it took.

Chairman Udall. Henry, where did we get this figure that the exhumation began on October 11 and was completed October 21? Do you recall?

Mr. Myers. It's this October 22, '65 letter.

Dr. Shapiro. That may have been one pit.

Mr. Myers. The AEC staff said the excavation in the '63 burial pit was begun on October 11th and Mr. Newman reported that on October 21 the excavation was 98 percent completed.

Chairman Udall. Was there a pit for each year?

Dr. Shapiro. There were a number of pits. I think he was referring to just one.

[xcii] Shapiro would later claim in a letter to Glenn Seaborg (see appendix) that all MUF was recovered after B&W razed NUMEC's facilities. This was not accurate.

Chairman Udall. This says excavation of the 1963 burial pit; it's identified as the '63 burial pit.

Dr. Shapiro. Yes.

Mr. Myers. The impression from the documents is that's where it was expected that most of the material would be. Is that correct?

Mr. Becker. Just a moment, sir, so we can show him that document. Does your copy say 1/23rd of the material exhumed?

Mr. Myers. Right.

Chairman Udall. Do they have this?

Mr. Myers. I think so, yes.

Dr. Shapiro. First of all, the exhumation actually started earlier than October, as I recall it. I believe it was in process for probably a couple of months prior to that time.

Mr. Myers. It said that in a meeting on September 24th, I think, that you said that—this is a memo from APOLLO/NUMEC, dated September 30, 1965.

Mr. Becker. We do have it.

Dr. Shapiro. What do you—

Mr. Myers. In the first two paragraphs on page 4 it talks about the '62 pit.

Dr. Shapiro. I think it's obvious from this that the exhumation started before October.

Chairman Udall. Well, we have all been wandering around the documents. Do you recall suggesting it would take twelve months or so to completely dig up and burn and assay all of this material?

Dr. Shapiro. I don't recall that I said specifically twelve months, but I certainly felt that, in order to do this and do it properly under conditions that would allow us to recover the material properly, and to recover it through the scrap recovery so that we could, in fact, determine how much was in it, would certainly take a considerable length of time.

Chairman Udall. Not a ten-day job?

Dr. Shapiro. By no means.

Chairman Udall. Would you agree—the suggestion here in these documents is that you were looking for a place where you were going to find most of the material that might have been missing from the WANL contract which was highly enriched uranium, that the '63 pit was the obvious place?

Dr. Shapiro. I think I would agree to that, although material in the contract proceeded over a period of two-plus years, as I recall it.

Chairman Udall. The documents here reflect that one of the conversations with you was that while the '62 pit was most convenient, he admitted—quoting you here on this same document—that it would have been better from the standpoint of settling the WANL contract to have opened the '63 pit first. He stated that the '62 pit, however, did contain some material accumulated from the beginning of WANL and that he was encouraged that they had located some material in the 76-percent range.

Dr. Shapiro. Right.

Chairman Udall. He further stated that the '63 pit contains all of the remainder of material generated under the WANL contract. Does that sound right?

Dr. Shapiro. Not entirely because, as I recall it, the WANL contract was—we had gone from '62 through '64, and possibly even some of '65, but I don't recall exactly.

Chairman Udall. Of the materials going through your plant if I had wanted to get my hands on some stuff that was most useful in making nuclear weapons, was the WANL contract one of the most likely places I would have gone after, or were you handling, in other contracts, materials equally valuable?

Dr. Shapiro. We were handling other contracts that had highly enriched material, as I recall it, at that time.

Chairman Udall. But was this bigger than the others, the major source of that kind of material?

Dr. Shapiro. It was a large source.

Chairman Udall. Okay. Henry, I am kind of "let's see" on this. Do you have anything else on this subject, on the garbage pits?

Mr. Myers. No.

Chairman Udall. All right. Let me turn to something else. I am going to talk about contracts that NUMEC had, particularly foreign contracts. In briefing the Commission on February 14, '66, the AEC I am talking about, former Assistant General Manager Howard Brown stated that the data the AEC staff had on NUMEC's foreign shipments "was based only on NUMEC records that the present safeguards system did not provide for or require independent AEC physical checks of shipments."

Your brief, at page 40, submitted to us by your printers, suggested there were independent checks made on materials shipped from the NUMEC plant.

Let me ask you first, to get perspective here, how many in a typical year or month or some time frame, how many shipments would be going out of NUMEC to foreign contracts that you were fulfilling?

Dr. Shapiro. I really can't tell you specifically.

Chairman Udall. In the hundreds or a few dozen?

Dr. Shapiro. Oh, we had, as I recall it, a total of 30 foreign contracts, of which I think, there were only something in the 20's related to the shipment of special nuclear materials and only comparatively few of those pertaining to the shipment of highly enriched material.

Chairman Udall. Now, I think what we are talking about here is, when you got ready to make one of those shipments, was there any requirement that some outside person from AEC come by and physically check that shipment, or did you have authority to simply put it in a container and send it off?

Dr. Shapiro. I don't recall specifically. All I can recall is the following: That the foreign contracts were all with the knowledge, consent, concurrence of the Atomic Energy Commission, because they involved the bilateral agreement between those countries involved whether it was France or Belgium or whatever. And these arrangements were, therefore, made with the Atomic Energy Commission. The material was checked by a number of people in our own plant prior to going out, and I know that at least in some instances there were AEC people who were involved in the transfer. As a matter of fact, I have a photograph of an AEC man with me next to a bird cage—which was the container—showing that shipment next to the French person who was accepting it.

Chairman Udall. But there was no ongoing requirement or procedure at NUMEC whereby some official from AEC had to come in and check a shipment before it went to one of these foreign contracts that you had?

Dr. Shapiro. I don't really recall. I don't think so.[xciii]

Chairman Udall. I am trying to grasp what it was that was kind of the program you had ongoing there. How many foreign countries over the period of your association with NUMEC did you service? A few? A couple dozen?

Dr. Shapiro. I think there were less than a dozen.[xciv]

Mr. Udall. All of this is, in fact, a matter of record, and as a matter of fact I think Mr. Myers has a full, complete listing of all of the shipments that were made and the times and the quantities and the enrichments.

Mr. Myers. Yes.

Chairman Udall. Okay. I don't want to get into a lot of great detail, but I was just trying to get some idea of the procedures. What percentage of your business over the years was with Israeli firms or with the Israeli Government, the shipments which you made abroad?

Dr. Shapiro. Comparatively small. Are you referring to special nuclear material? Or are you referring to business in general?

Chairman Udall. I am referring to business in general at NUMEC.

Dr. Shapiro. Comparatively small.

Chairman Udall. They were not—the Israelis were not your largest customer, you would say?

Dr. Shapiro. By no means.

Chairman Udall. You can't really recall or help me very much with regard to the procedure, whether AEC would come to your plant and verify different shipments? Sometimes they did, sometimes they didn't; but you can't—

Dr. Shapiro. At this time I can't recall. I am really sorry, but I would only be speculating as to that.[xcv]

Chairman Udall. Is there any way that the AEC could tell today or that this committee or anybody else could determine that highly enriched uranium was not shipped from NUMEC in containers that purportedly had other contents or

---

[xciii] The most troubling Israeli shipments detailed by former NUMEC employees involved directly trucking material from NUMEC for loading onto ships waiting on New York piers, with no AEC inspection, warehousing or overt customs procedures. This stood out in the shipping manager's memory since it was so unusual.

[xciv] NUMEC claimed only ten foreign destinations of U-235 shipments to the AEC between 1957 and 1965, Australia, Canada, France, Germany, Italy, Japan, Netherlands, Sweden, Switzerland and France. NUMEC claimed irradiator shipments to Israel contained only non-weapons-grade source material such as Cobalt.

[xcv] Shipments observed by NUMEC employees on the loading dock during the 1960s revealed no overt AEC presence, so Shapiro should have candidly admitted that not all foreign shipments were overseen by the AEC.

low enriched uranium? How would we go about establishing this today? Do you have shipping records that would tell us?

Dr. Shapiro. There were certainly shipping records. These records were not only kept at NUMEC, but they were also, we were obligated to send copies of these records to the Government. They had their own set of records.[xcvi]

Chairman Udall. But the signature of your company and the shipping officials — there is all the proof that we would have or the AEC would have that the carton contained what it says it contained?

Dr. Shapiro. I really don't know whether this is all of the proof. I suspect that there was more because I do know that in at least a couple of cases they intercepted shipments and sampled material and had them sent to the New Brunswick laboratory for analysis and consequently there must have been other means of determining —

Chairman Udall. This was AEC you are talking about?

Dr. Shapiro. Yes.

Chairman Udall. They would make spot checks from time to time, apparently?

Dr. Shapiro. Apparently. I really didn't know whether they made spot checks from time to time or all the time.

Chairman Udall. But you know some of your shipments were intercepted and checked?

Dr. Shapiro. I do know in a couple of cases shipments were intercepted and checked.

Chairman Udall. Typically, maybe there isn't a typical case, but your shipping to France or Belgium nuclear materials, how would they go? Would they go by Apollo, by truck? Do you call United Parcel? Do you put them on an airplane? Give me, as a layman, an idea of how you would get a shipment of material you had processed to your customer in Belgium or France, or wherever.[xcvii]

Dr. Shapiro. First of all, it's necessary to understand that when a shipment went out, any shipment went out, and particularly a foreign shipment went out, these were very carefully checked. They have to be packed in special containers, bird cages. They were checked by a number of people in the organization for, one, whether we had in fact fulfilled the order, terms and conditions of the order.

Secondly, whether we had fulfilled the requirements with regard to specific quality, whether they — whether the shipments were properly packaged, whether there was any residual radiation on the package, and so on.

So there were a number of people who were involved in checking each specific shipment. These papers were dispatched not only to the receiver, but to the Government, and to the shipper, that is, to the carrier, and it was trucked

xcvi The correct answer is that there was no easy way to determine whether diverted HEU was shipped in irradiators or other equipment such as remote weather stations from either NUMEC or AEC records.

xcvii Shapiro never volunteered information on NUMEC's New York shipping agents, such as Wolf & Gerber, Inc., how they were chosen, or their role in expediting the highly unusual shipments which never passed through warehouses.

generally to the airport and transfers were generally made by air directly to the country involved.

Chairman Udall. Suppose I had had evil purposes, is there some place along this route of shipment that I could— a private citizen, or someone could intercept and maybe substitute high-enriched uranium for low-enriched uranium or make some other changes in the contents of the package?

Dr. Shapiro. I think this would be very difficult to do. First of all, these were sealed; and, secondly, the receiver would recognize that there had been some mistake made both in terms of the enrichment or the weight or whatever the case may be.

Chairman Udall. I noticed just this one, down your list of shipments here, U02, I guess uranium oxide pellets?

Dr. Shapiro. Yes.

Chairman Udall. U308 fuel pellets, U02 powder pellets. How big would a typical one of these shipments be, the kind I have been reading? Is this a box I could carry? Is it a huge container? Give me some idea.

Dr. Shapiro. I don't know if there is anything here that is really typical because, as you can see, the quantities varied very significantly.

Chairman Udall. Give me the biggest and the smallest that you just note there. The largest one would be how big?

Dr. Shapiro. The largest would be a number of bird cages, which were probably 6 feet tall and several feet in diameter, and these were the containers that would be within this bird cage and they were bolted in place and the containers themselves were sealed.

Chairman Udall. What do they weigh when you get all through with it?

Dr. Shapiro. Well—

Mr. Becker. With the bird cage and everything?

Chairman Udall. Yes, the shipping container and the contents.

Dr. Shapiro. Oh, several hundred pounds or more.

Chairman Udall. So you would need a forklift or crane?

Dr. Shapiro. Oh, yes.

Chairman Udall. I see several to Germany here, U02 pellet. Is that some hospital shipment of some kind?

Mr. Becker. We may not have the same list.

Mr. Myers. We have a slightly different list.

Chairman Udall. I am trying to get an example now of a small shipment. You have given me the big one.

Mr. Krash. You mean the size or weight of the shipment.

Dr. Shapiro. This would also go in the bird cage containing this. There were some that did not require bird cages depending upon the enrichment. If they were very low enrichment no, but I believe that unless we had natural or depleted material, it was necessary to use a bird cage to assure that there was no possibility of accumulating critical configurations because of the possibility of other shipments that might be on the same carrier.

Chairman Udall. I don't want to take too much time, but give me dimensions and weight for a small shipment. Find one there that is quite small and tell me how big that would be, 6 inches square or what-have-you. Give me an example.

Dr. Shapiro. Well, here is oils and powder amounting to only 4 grams of material apparently. That would be a container probably packed in a container which might be like a pail, a 5-gallon pail or something of that sort.

Chairman Udall. All right. One foot to 2 foot in size, something of that order?

Dr. Shapiro. I guess.

Mr. Krash. That would be for what quantity, 4 grams?

Dr. Shapiro. That would be just 4 grams.

Chairman Udall. Does the size of the contain[er] have any connection with the column "Percent Isotope," for example, here is one to France, U0 2 powder, 89.2; is that the level of enrichment indicated there?

Mr. Myers. 11/23/62 to France.

Chairman Udall. 4,000.

Mr. Becker. I don't think we have the same document.

Mr. Myers. Yes, this is from your documents.

Dr. Shapiro. 11-23-62, 89.2.

Chairman Udall. That is the percent isotope?

Dr. Shapiro. That is the enrichment.

Chairman Udall. Would that take a smaller or larger carton or does that matter? Does it have any bearing on it?

Dr. Shapiro. It is primarily the total quantity of the material and there were apparently somewhere in the[amount] of 4,000 grams, 4 kilograms, and in a bird cage. So it would be a large container.

Chairman Udall. All right. Let me proceed to a couple of other things here. I asked a moment ago if I were evil-minded, how difficult it would be for me to intercept at some point along the shipment track one of these containers and substitute highly enriched materials for low-enriched materials or low-enriched material for high-enriched material and you gave me an answer on that. How difficult would it be after the container was put together in the plant for me or someone else to get a hold of the container and do the same thing, make the same kind of substitution?

Dr. Shapiro. I would say it would be quite difficult because my recollection was that those were stored in the vault until shipment.

Chairman Udall. Over the years—and again you had a great variety, but I just want to get a general approximation—what portion of NUMEC's revenues came from foreign governments?

Dr. Shapiro. Comparatively little.

Chairman Udall. The majority of your work was for the U.S. Government and domestic?

Dr. Shapiro. For domestic, yes.

Chairman Udall. And I may have asked this previously. You said your shipments to Israel and to Israeli companies were a relatively small part of your foreign operations.[xcviii]

[xcviii] Shapiro claimed to the AEC that the Israeli government owed NUMEC a significant enough sum that it required an airport meeting with an Israeli government official.

Dr. Shapiro. I think it is important to note, Congressman Udall, that we never shipped special nuclear material to Israel other than possibly—and I don't even know whether that is true—a neutron source, which would contain only a very small amount of SNM.

Chairman Udall. Nothing in the range of bomb-grade quantities?

Dr. Shapiro. No material other than that. What we shipped to Israel was equipment, an instrument or something of that sort.[xcix] [391]

Chairman Udall. All right, Henry, I am going to leave this subject. Anything else?

Mr. Myers. What is your feeling with regard to the confidence we can have that high-enriched uranium was not shipped to foreign recipients in containers with labels indicating other contents? I think you may have answered that.

Dr. Shapiro. I think there is a high degree of confidence

Chairman Udall. It would require for something of that nature to occur, it would require collusion at both ends, both at the shipping end and the receiving end?

Dr. Shapiro. Yes.

Chairman Udall. Because in an ordinary contract the receiver would know he was not getting what he had contracted for?

Dr. Shapiro. Absolutely.

Chairman Udall. All right. Let me turn to something else related. It is of great importance to us in this inquiry whether a diversion could have occurred as it is whether diversion actually did occur. I want to pursue that a moment. I want to talk about some of these matters. At the NUMEC plant did you have a 24-hour operation, three shifts, most of the time?

Dr. Shapiro. Yes.

Chairman Udall. And give me a brief outline of the layout. How big an operation did you have there?

Dr. Shapiro. The plant itself, I think, was 450 feet long or 500 feet long.

Chairman Udall. Okay, I am not trying to pin your down, just a rough idea.

Dr. Shapiro. 70 feet wide.

Chairman Udall. Big parking lot for your employees?

Dr. Shapiro. Yes.

Chairman Udall. Chain link fences around the whole operation?

Dr. Shapiro. Yes.

Chairman Udall. Security guards and gates and that whole business?

Dr. Shapiro. Yes.

Chairman Udall. What kind of screening did you do to people coming in, visitors and employees?

[xcix] Under the United States – Israel Agreement for Cooperation, NUMEC shipped 320 grams of plutonium to the Department of Nuclear Science at the Israel Institute of Technology via carrier's agent Wolf & Gerber of New York City. They were allegedly sources for four Plutonium-Berrylium irradiators fabricated by NUMEC and part of a 600 pound El Al shipment sent June 30, 1963.

Dr. Shapiro. Well, the employees for the most part were cleared either Q or L[c], and all visitors had to be announced through the guard or the receptionist, and there was not free access, if that is your question.

Chairman Udall. Yes, that is what I was getting at.

Dr. Shapiro. No, there was no free access.

Chairman Udall. And you had round-the-clock guards?

Dr. Shapiro. Round-the-clock guards.

Chairman Udall. And sentries at all places of entry?

Dr. Shapiro. Yes.

Chairman Udall. An outsider could not wander in?

Dr. Shapiro. When you say "round-the-clock guards at all places of entry," the answer to that is "no." We had round-the-clock guards at the place of entry.

Chairman Udall. I understand. I didn't state it correctly.

Dr. Shapiro. And they did also regular rounds.

Chairman Udall. And in addition, did you have more than one vault for the safekeeping of nuclear materials?

Dr. Shapiro. Yes.

Chairman Udall. Several?

Dr. Shapiro. At Apollo I believe we had two, but I don't recall.

Chairman Udall. Did your plant consist of several buildings or just one building?

Dr. Shapiro. We had — do you mean the principal plant at Apollo?

Chairman Udall. Yes.

Dr. Shapiro. That consisted of one major complex of buildings with some outlying buildings in which we had some storage.

Chairman Udall. I want to talk a moment about an outside group gaining entry to your operation. Were the measures, security measures, you had adequate in your judgment to protect against the clandestine entrance of two or three people, a small group, who might have known the layout?

Dr. Shapiro. Yes.

Chairman Udall. This would be extremely difficult and undoubtedly would have been detected by your people in the way you operated?

Dr. Shapiro. Yes.

Chairman Udall. If a group of this kind maybe aided by some insiders had gained entry to the plant, what difficulties would they have had in trying to get to highly enriched uranium in your vaults or your safe storage places?

Dr. Shapiro. First of all, the vaults were locked and those who had access were the vault custodians, so they would have had to have overpowered a custodian or forced him in some way to open the vault or he would have had to have been in collusion with them.

Chairman Udall. Sure.

---

[c] Q and L are non-military security clearances. Q gives access to information classified as "Top Secret" such as nuclear weapons design information. L is a lower level access to confidential and secret classified information, and unescorted access to protected areas.

Dr. Shapiro. So certainly that would be difficult. Furthermore, he would have to go through the change room. Otherwise it would become quite obvious that somebody who didn't belong was there. All of our people had badges.

Chairman Udall. So this would be highly unlikely that any group could have penetrated from the outside—

Dr. Shapiro. I believe so.

Chairman Udall.—and gotten through whatever security you had, to the uranium materials themselves? All right.

Mr. Myers. Was there also a custodian at the vault 24 hours a day?

Dr. Shapiro. If material were in process—and it was required that material be shipped in and out of the vault—then we had a vault custodian. Otherwise, the vault was locked and the only access then would be via the guard himself, who at one time punched a clock in the vault and later on it was decided by, I believe, the security people and the AEC to eliminate the clock punch from the vault.

Chairman Udall. I have talked about outside penetration. Let me talk about inside. How difficult would it have been for some person with an evil design to seek work at your plant where he would have access and then to spirit away quantities of SNM?

Dr. Shapiro. First of all, the person would have had to go through some clearance procedure. And, therefore, if the person had evil designs and had a history of that sort, presumably that would have been picked up by the security check. So that is certainly a major inhibitor to that, to start with. Secondly, people were not isolated in their work. They worked with other people. So what that person would be doing would certainly over a period of time be noticed by neighbors.

Chairman Udall. There is no way that a person by himself could go in and seize or get his hands upon substantial quantities of special nuclear material?

Dr. Shapiro. I should think that would be highly unlikely

Chairman Udall. You would almost need, from the standpoint of what you have told us about the weight of shipments and the places that your materials were kept, it would be almost impossible for one person by himself, even if he were in the plant and had fooled or had the confidence of his co-workers, to do anything much by himself?

Dr. Shapiro. I think that is true. In terms of material that was prepared for shipment, I think it would be highly unlikely that a person could ever do a thing like that by himself.

Chairman Udall. You have said that both of these instances, the outside penetration and the person who gets inside, are very unlikely. But assuming that in either case that some diversion had taken place. Were your records such that a shortage would have shown up in the record?

Dr. Shapiro. I believe they were. We kept records in accordance with the requirements of the Atomic Energy Commission, and I feel they were adequate to have done it, to have determined whether there was any major amount of material that might have suddenly disappeared.

Chairman Udall. My notes here talk about skimming. Henry do you mean getting material in process and—

Mr. Myers. Taking a small amount out from time to time.

Chairman Udall. And gradually would increase to a large amount. Is this possible or difficult given the nature of your processing, your operations?

Dr. Shapiro. I would say that that would be highly unlikely because what you are implying here, I think, is that this would have to be an operation that would have to go on over a period of time?

Chairman Udall. Right.

Dr. Shapiro. And repeated many times.

Chairman Udall. Yes.

Dr. Shapiro. And given the set of circumstances I would say that that would be highly unlikely.

Chairman Udall. Was the high-enriched uranium scrap that was associated with the WANL contract any more or less accessible to a skimming process than other materials that you had?

Dr. Shapiro. I would say it would be—as far as accessibility to skimming, I think in general the answer is the material is really not accessible to skimming. I would say that in the case of the WANL contract by virtue of the nature of the material it is even worse, because of the pyrophoric character of the product and the necessity to keep it under an inert atmosphere and to package it under inert atmospheric condition.

Mr. Krash. You mean it would be even harder to skim?

Dr. Shapiro. Yes.

Chairman Udall. The word you used is a word to describe a material that can burst into flames spontaneously?

Dr. Shapiro. Yes.

Chairman Udall. In the papers we have and in the submission you made there is talk of some of the missing uranium at NUMEC being discharged with effluents. Could you tell me how this would happen?

Dr. Shapiro. Yes.

Chairman Udall. How was your plant locked up and to where would the effluents, liquids coming out of your plant, go? Was it connected to somebody's sewer system, or your own someplace? Discuss that with me so that I understand.

Dr. Shapiro. First of all, you realize there are a number of operations which involve fluids, water, acids, et cetera?

Chairman Udall. Yes.

Dr. Shapiro. These were collected and put into large tanks, holding tanks, where they were assayed. They were neutralized or whatever in order to assure that the chemical nature of the material was suitable for discard. The assay was done and then these—as I recall it now the material was then pumped into another holding tank so that the check could be made and then dumped out to the river.

Chairman Udall. Just dumped in a pipe running down to the river?

Dr. Shapiro. Yes.

Chairman Udall. Was there radioactive material in that effluent?

Dr. Shapiro. There were small quantities of radioactive material in the effluent, but these were in the parts per million quantities which were within the limits, allowable limits, for discard.

Chairman Udall. What kind of quantities?

Dr. Shapiro. Thousands of gallons.

Chairman Udall. Every day?

Dr. Shapiro. Yes.

Chairman Udall. And this was in compliance with the Clean Water Act?

Dr. Shapiro. Yes.

Chairman Udall. And whatever Federal regulations?

Dr. Shapiro. Whatever compliance was required with whatever acts were in existence at that time, yes.

Chairman Udall. You have agreed, I think, in all of this that there were probably several tens of kilograms of nuclear material missing over the period you operated at NUMEC?

Dr. Shapiro. There were materials—

Chairman Udall. —unaccounted for.

Dr. Shapiro. There was material that—yes, which was unaccounted for, not what I would say was "missing" per se.

Chairman Udall. All right. Well, we have talked about some of it and I have asked you to give me a general estimate of what went into the garbage dump and how much was leached out. How much do you estimate of the missing material went out through the effluent into the river? Could you quantify that in any range?

Dr. Shapiro. I am sorry, I can't. Obviously there were kilograms over a long period of time.

Chairman Udall. Is this the major, in your judgment as an operator of the plant and with all your experience, is this the major point at which we have lost uranium?

Dr. Shapiro. I really don't know whether that was the major point. There were a lot of places where uranium is normally lost, some of which are susceptible to measurement, and others are not susceptible to measurement. The materials that we handle were not only in liquid form, but they were in powder form and consequently they became airborne rather easily as well. So you had material that would collect on the ceilings, on the walls, and as a matter of fact then we had special—in order to assure that people were not breathing the material, we used to have hoods which were constantly pumped and therefore material went into filters and again some went through the stacks. The plant contained three huge ventilating fans because we had lots of furnaces in the plant, and it became rather hot, especially in the summertime, and it was necessary to have a large volume of air flow through that plant in order to keep it cool and certainly there was material that must have been sucked up through that and left the plant through that mechanism. In addition to that, material spilled on the floors in spite of attempts to prevent it and it was necessary to clean it up and we had janitors constantly mopping the place to assure that we—the mop waters were monitored to assure that there wasn't an excessive amount, but over a period of time it is quite obvious that even if you had a small amount, you would accumulate quite a bit of material over the years this way and later on a study was done by Mr. Lovett, and he indicated that we were, in fact, losing more material through mop water alone than we had ever anticipated.

I can't give you numbers. As a result of the nature of the mop water as being heterogeneous and, consequently, was not susceptible to good measurement, and the determinations by our analytical laboratory were therefore not as good as they should have been. Later on, of course, instrumentation [and] better methods were devised. But in any event, there was a great deal of material that was lost by mechanisms of that sort. So—

Chairman Udall. There are several ways this move can—

Dr. Shapiro. There are several ways in which material could be lost and not accounted for.

Chairman Udall. Some in the effluent, some in the leaching in the garbage pile, some in the mop water, some blown out by the fans?

Dr. Shapiro. Some even in the concrete, because the concrete is porous and we used to have—this is highly acidic material and when they spill it on the floor, it would leach into the concrete.

Chairman Udall. This talk—there is a lot of loose talk in some of these documents about losing this stuff in the pipes. By that, do you mean or do these people mean it is physically in the pipes or are you talking about these kinds of drainages we just mentioned, the effluent, the mop water—

Dr. Shapiro. You can lose it two ways in terms of the pipes: You lose it from the effluent and you also have material that coats the walls of the pipes just as you would in the case of household plumbing.

Chairman Udall. Sure.

Dr. Shapiro. If you have ever opened up a pipe, you would notice there is a good deal of sediment that collects, material that collects in the thing.

Chairman Udall. Sure.

Dr. Shapiro. So, in addition to that, we had a great deal of maintenance on the plant as a result of the corrosive nature of the material, and they would remove pipes and/or valves, pieces of equipment, and we found that on occasion the workers would not follow instructions properly and they would discard that and bury it without having properly washed it and leached off the material as they were instructed to do.

Chairman Udall. So it is your best judgment that the accumulation of all these little things we have talked about, the slippage of each one of these points, could add up to the kind of MUF that we had at Apollo?

Dr. Shapiro. Absolutely.

Chairman Udall. What would then be your answer to some of those who point the finger of suspicion here that the MUF at Apollo, quantities of that were much larger than comparable other facilities used by the Government. Do you challenge that assertion or do you have some explanation of why your losses would be larger than other plants?

Dr. Shapiro. Our losses were not larger than those of other plants doing similar work. The report that was put out by the Atomic Energy Commission, NRC, DOE, whatever it was, last August—

Mr. Becker. Of 1977.

Dr. Shapiro. August of 1977, indicated that there was no statistical difference between the losses which were incurred in our plant versus those that were incurred in other plants doing similar work.[ci]

So it is true we did not take credit for some of the losses that were through the stacks which we probably should have, and did later take credit for.

Chairman Udall. All right.

Dr. Shapiro. We had over 100 stacks in the plant.

Chairman Udall. I have run by a lot of theories; let me try one more on you. Could a workman intent on diverting uranium have used low enriched uranium to dilute high enriched uranium and then skim off some of the high enriched material?

Dr. Shapiro. All of the product had to be assayed from the quality point of view. One of those assays was the enrichment, and we couldn't ship the product unless it met all those specifications. Furthermore, if a product were shipped to a customer he does his own assays on the other side. He had to determine whether our weights were correct, whether the assay of U-235 itself were correct, and if they were not in accordance with the specifications I can assure you we would have heard.

Chairman Udall. I am going to refer to a memorandum of November 27, 1973, of Paul Gaughran, Director of the Division of Security of the AEC wrote that in examining the history of NUMEC's security procedures, that such "procedures were not designed or geared to prevent an employee from removing quantities of strategic nuclear material from the facility." Would you agree with that assessment?

Mr. Krash. Excuse me just a moment, Mr. Congressman.

Mr. Myers. This is page 8. That's [indicating] it.

Mr. Becker. Background and history, page 8.

Chairman Udall. Yes, 4th line down. I just read one sentence. It's the 5th line down.

Dr. Shapiro. First of all, this refers to a period of inspection after I had left the facility and had nothing to do with it. I mentioned I had left in 1970. This refers to 1971 when this was under either Babcock & Wilcox, or Atlantic Richfield, I don't know. I certainly can't speak for them. What I can say with regard to our security is if you will refer to another document that was produced by the inspection people of the Atomic Energy Commission, they had indicated that during this period that we did in fact have adequate security measures and not only that, but that our personnel were highly conscious of the security aspects. I believe you will see that among the documents that you have.

Chairman Udall. I think what they are getting at here—and I want to leave this subject in just a minute—they are saying in effect that while you can have adequate protections against the people who with violence and force come in an seize something, and while you can have adequate protections against the clandestine invader who tries to take something, that the procedures were not

---

[ci] The numbers were revised by a 2001 Energy Department audit that revealed NUMEC had almost twice the normal industry losses.

really designed to prevent an inside job, some official or somebody in the management of the facility from diversion. Isn't that what they are trying to say?

Dr. Shapiro. If that is what they are trying to say, I would say that that certainly was not the case at NUMEC as far as I was concerned.

Chairman Udall. Could you have made a diversion if you had wanted to? Could you have arranged a diversion of substantial material yourself if you had wanted to with your knowledge of the process? Would you have required a number of people helping you, and this sort of thing?

Dr. Shapiro. I would have required many people to help me.

Chairman Udall. People inside the plant who knew something was otherwise wrong?

Dr. Shapiro. Absolutely.

Chairman Udall. All right. Let's try another facet of this. We are making a little progress here. I think we will finish in due time. You had, NUMEC had a subsidiary in Israel, Israel Isotopes and Radiation Enterprises Ltd., or sometimes called "ISORAD", and it was allegedly half-owned by the Israeli Atomic Energy Commission. Is that correct?

Dr. Shapiro. Not "allegedly," it was half-owned by the Israeli—

Chairman Udall. Alright, strike "allegedly". Why did you form this subsidiary? What was it supposed to do? What was your relationship with it?

Dr. Shapiro. Do you mean NUMEC's relationship with it?

Chairman Udall. Yes.

Dr. Shapiro. We owned 50 percent of the company. The purpose of the company was primarily to irradiate food for the purpose of pasteurization to prevent spoilage. Also, to kill fruit fly infestation in citrus, and for export purposes.

Chairman Udall. Is this a fairly common procedure in many countries?

Dr. Shapiro. No.

Chairman Udall. This is a new kind of technology, is it?

Dr. Shapiro. No. It is not a new technology, it was not a new technology at the time. What we were doing was adapting the technology which was under development in the United States at the time primarily at NADIC[cii] which is an Army supply center and the Davis Campus of the University of California where they were doing development of procedures for the irradiation of citrus for preservation purposes.

Chairman Udall. How highly enriched does material have to be that you use this process?

Dr. Shapiro. I think you misunderstand, Congressman Udall. These processes do not employ special nuclear material. They employ a radiation source, generally cobalt.

Chairman Udall. Okay, how big an operation was this in terms of people, offices, including ISORAD, and others?

Dr. Shapiro. We had our facilities initially at Soreq which is the Israel Atomic Energy Commission's research center and there were several irradiation facilities,

---

[cii] He was likely referring to the Natick Laboratory Army Research, Development and Engineering Center in Massachusetts.

one of which was in the pool of the pool reactor; and we constructed the facilities for this. We constructed the equipment and the tools and worked with the technical people there in the development later on. There were other radiation facilities that were put in operations.

Chairman Udall. At peak, how many employees would ISORAD have had?

Dr. Shapiro. Oh,

Chairman Udall. A few dozen?

Dr. Shapiro. Not even that many.

Chairman Udall. Were any of those—were most of the ISORAD employees Israelis or was it a mix of—

Dr. Shapiro. They operated in Israel so they were all Israelis except for an occasional person from NUMEC who went there to either install equipment or to assist them in a particular problem.

Chairman Udall. Was there any need or requirement that ISORAD employees who were Israelis come to NUMEC in Pennsylvania from time to time?

Dr. Shapiro. Yes, on occasion.

Chairman Udall. Generally for what purposes?

Dr. Shapiro. First of all there were discussions with regard to the business itself; secondly, there were technical—when we shipped equipment there were technical discussions with regard to the design of the equipment and et cetera for the utilization of it, and the installation. And it was some sort of—

Chairman Udall. So there was an occasional visit. It was not unusual to have ISORAD Israeli employees visiting you at NUMEC from time to time?

Dr. Shapiro. There were occasional visits of this type.

Chairman Udall. Did they have free access to the NUMEC premises, and to your ongoing operations there?

Dr. Shapiro. Their access was restricted like the access of any other alien.

Chairman Udall. They could not go through your equipment or your files or your processes?

Dr. Shapiro. No.

Chairman Udall. All right. The *Washington Star*, in a story published last November 6th—

Mr. Myers. That means '77, of course.

Chairman Udall. —in '77, yes, reports that Ephraim Lahav,[ciii] [392]then counselor on scientific matters to the Israeli Embassy in Washington, repeatedly visited with you at the NUMEC Apollo plant in the early to mid-1960s. Did he, and why? Why did he visit you, if indeed he did?

Dr. Shapiro. Congressman, I don't even recall who Ephraim Lahav is. There were people who visited us but I don't recall Lahav.

Chairman Udall. Was it unusual for scientific people from the Israeli embassy or offices in New York to come to Apollo on visits?

[ciii] Dr. Ephraim Lahav had been involved in early negotiations with the AEC over the acquisition of ten tons of heavy water. At that time, Shimon Peres had already begun negotiating with France to produce the Dimona reactor, but France could not supply heavy water Israel needed for a plutonium-based facility.

Dr. Shapiro. It was not unusual for the counselor, scientific counselor to come to NUMEC.

Chairman Udall. How about Mr. Hermoni?

Dr. Shapiro. I don't recall his ever visiting Apollo, although he came to Pittsburgh; he may have. I don't recall.

Chairman Udall. All right.

Dr. Shapiro. But one of the people that did visit that I do recall on two, possibly even on three occasions, was David Pellig who is the administrative director of the Israeli Atomic Energy Commission who had specific responsibility for our joint operations of ISORAD, and in addition to that, the business relations were generally carried out by the scientific attaché in Washington.

Chairman Udall. Since the government itself was a half partner.

Dr. Shapiro. Yes.

Chairman Udall. These people were acting and visiting you on behalf of the government.

Dr. Shapiro. Yes, they acted on behalf of the government.

Chairman Udall. Fine. In the January 9, 1977, I guess that means, *Newsweek* — you say '78.

Mr. Myers. No, that one is '78.

Chairman Udall. [Continuing] *Newsweek* refers to meetings you held with Avraham Hermoni, a scientific counselor at the Israeli Embassy in Washington. Who is he and what kind of relationship did you have with him while you were at NUMEC?

Dr. Shapiro. Well, you describe him as the scientific counselor, which he was indeed, and during the time that he was scientific counselor he conducted the business, the ISORAD business on behalf of the government of Israel. And in addition to that he was interested in technical assistance from time to time.

Chairman Udall. Did you ever have any indication that he was associated with the Israeli nuclear weapons program, if they had one?

Dr. Shapiro. I don't.

Chairman Udall. This was just never called to your attention, or you had no reason to believe that that might be the case?

Dr. Shapiro. No, as a matter of fact, I first met him at the University in Israel as I recall it.

Chairman Udall. How frequently did you see him during the time he was in Washington, and you were operating NUMEC?

Dr. Shapiro. I don't recall exactly. Maybe, probably less than half a dozen times.

Chairman Udall. The British Broadcasting System, in a documentary on the question of Israeli nuclear capability, called you a frequent privileged visitor to Israel. Were you frequent, and did you feel privileged?

Dr. Shapiro. I am flattered.

Chairman Udall. I think one of the things they mentioned was that you were given a special flying tour of the Egyptian frontier or something. On the occasion of your visits did you get attention that you thought was special or privileged?

Dr. Shapiro. Oh yes, but I got attention that other visitors have gotten, I know that. I have been to Israel I believe some 17 times. If that is "frequent," yes, I was a frequent visitor. Insofar as "attention" is concerned, I am a member of the local board of Israel bonds; I am a member of the Jewish National Fund; I am a national officer of the Zionist Organization of America, and, therefore, I am met and greeted by my counterparts and to the extent that I had business with the Israel Atomic Energy Commission I am met and greeted by them.

Chairman Udall. Do you recall an event where you got a flying tour that was unusual?

Dr. Shapiro. A flying tour that was unusual? It was very unusual for me because what happened was that I was there in '67, and I took a tour that others took, too, and then we flew over—we flew over the Sinai, landed in Al Arish and Tiran.

Chairman Udall. I don't particularly want to beat this one to death, but there was nothing particularly special about this; had you requested it, or did the government want you to do something, or see something, or was this a special opportunity to see the front?

Dr. Shapiro. This was right after the '67 war; and I wanted very much to see the area.

Chairman Udall. Okay. One of the documents says that—a memorandum from Tharp, Deputy Director of the Division of Security in the Albuquerque Operations—Tharp quoted Dr. Shapiro as having present intentions of moving to live in Israel. Did you ever intend to emigrate or consider it or take any steps toward it?

Dr. Shapiro. I never took steps towards it, but I certainly considered it because I think any Zionist considers what is called "aliyah," and I certainly considered it.

Chairman Udall. Have you ever had at any time that you can recall any discussion with Israeli citizens who you knew or believed were involved in the Israeli nuclear weapons program?

Dr. Shapiro. Mr. Congressman, I don't know that they have other than what I have read in the newspapers. Consequently, I really would not know, therefore, who would be involved in such a program if they have such a program.[civ]

Chairman Udall. Were you ever involved or do you recall any relationships or contacts with people who might have been known to you to be an Israeli intelligence operator or in Israeli intelligence operations in this country?

Dr. Shapiro. In this country?

Chairman Udall. Yes.

Dr. Shapiro. No.

Chairman Udall. Or in Israel?

Dr. Shapiro. In Israel I met the head of the military intelligence there.

Chairman Udall. Have you ever been asked to provide classified information to people who were not authorized to receive it?

Dr. Shapiro. No.

[civ] Shapiro's adoption of the language of "strategic ambiguity" contrasts with his earlier interviews with the AEC, in which he refers to Israel's nuclear weapons program without qualifiers.

Chairman Udall. On December 11, 1971 the AEC Security Director Riley wrote that Dr. Shapiro had told him that the Israelis had told Dr. Shapiro—we are getting third- or fourth-hand removed here—the following information: that an AEC Commissioner told officials of a foreign government other than Israel that the Commission suspected Dr. Shapiro of having helped divert bomb-grade uranium to Israel. Do you remember saying this to Riley?

Dr. Shapiro. I don't recall it.

Chairman Udall. On June 20, 1969, there apparently was a meeting between you and someone from the Israeli Embassy at the Pittsburgh airport—let me back off here. On October 30, 1970, Robert Tharp—was this the material give to us?

Mr. Myers. Yes.

Chairman Udall.—then Deputy Director of Security at the AEC, wrote a memorandum to H. C. Donnelly, manager of the AEC Albuquerque Operations Office, in which Tharp described a meeting between Dr. Shapiro and AEC staff. He said that during the course of this meeting that Riley had advised Shapiro that he, Riley, had indicated to the Commission that Dr. Shapiro had been less than candid in clarifying the derogatory implications, particularly the airport incident. Did you have a meeting with an Israeli official in the Pittsburgh airport on or about June 20, 1969, at the airport?

Dr. Shapiro. I had a meeting but the date I don't recall, with an Israeli official—specifically the scientific counselor.

Chairman Udall. What was his name?

Dr. Shapiro. I think his name was Epharaim Lahav, but I don't really recall.[cv]

Chairman Udall. How did this meeting come about? Who asked for it?

Dr. Shapiro. I asked to meet him because as I had indicated earlier, we had in fact shipped some equipment to Israel and I had—they were delinquent in payment; and I had repeatedly asked for payment and this was getting very embarrassing to me because in view of the fact that we had already merged with Atlantic Richfield, and we were about to have an audit, they wanted to have payment made as quickly as possible.

Chairman Udall. For what? Was this a contract?

Dr. Shapiro. This was for equipment that we had shipped.

Chairman Udall. ISORAD, or another contractor?

Dr. Shapiro. To Israel. I don't know whether it was in connection with ISORAD or whether it was in connection with the research establishment or whatever.

Chairman Udall. What was the amount of the claim you had on them in indebtedness?

Dr. Shapiro. It was quite a large amount. Many thousands of dollars.

Chairman Udall. Tens of thousands?

Dr. Shapiro. Yes.

Chairman Udall. You don't remember anything more than that?

Dr. Shapiro. I don't remember precisely. And I had asked to meet with him.[cvi] And it was a matter of convenience for both of us that we met at the airport. I

[cv] It was Jeruham Kafkafi

had offered to come out to Washington as I recall it, and he said he was actually going to be passing through Pittsburgh on his way to I think it was Dayton, and asked in view of the fact that he was stopping over, could we meet; and I agreed to do so. It was a convenience to him because otherwise he would have had to travel an hour and a half or two hours from the airport to Apollo so it would be, you know, a three or four hour trip just in travelling. And it was a convenience to me because it means that I didn't have to go into Washington.

Chairman Udall. So you drove to Pittsburgh, and met his flight?

Dr. Shapiro. Yes.

Chairman Udall. By yourself?

Dr. Shapiro. Yes.

Chairman Udall. Did the two of you meet?

Dr. Shapiro. Yes.

Chairman Udall. For how long?

Dr. Shapiro. Between planes. I don't recall exactly how long.

Chairman Udall. A few minutes; an hour?

Dr. Shapiro. Oh, it may have been an hour, thereabouts.

Chairman Udall. He was by himself?

Dr. Shapiro. He was by himself.

Chairman Udall. Where did you meet him?

Dr. Shapiro. I think I met him at the gate and then I think we went to the Ambassadors Club, the TWA Ambassadors Club.

Chairman Udall. Can you tell us what occurred there?

Dr. Shapiro. I told him I was very embarrassed and that I had requested payment on a number of occasions, gently, and that for some reason or another I had been reassured that it was on its way, but it was — we had not yet received payment and I felt that it was desirable for him to urgently request Israel to remit.

Chairman Udall. What did he say to that?

Dr. Shapiro. He arranged for it and we got payment very shortly thereafter.

Chairman Udall. Within a few days?

Dr. Shapiro. I don't know whether it was a few days but it was shortly after.

Chairman Udall. That payment was by check, a draft through a bank or something?

Dr. Shapiro. I would assume that that would be the way it was paid, but I don't remember exactly how it was paid. I am sure it was not in cash.[cvii]

Chairman Udall. Was anything exchanged at that meeting between you, cash, objects or anything else?

Dr. Shapiro. No.

Chairman Udall. You are certain of that?

[cvi] According to Gilinsky and Mattson, it is likely that the Israelis called the meeting to inquire about which U.S. officials would be traveling to Israel to inspect Dimona.

[cvii] There is no declassified record of available FBI or AEC files that any attempt was made to see if invoices or payments matching Shapiro's description were processed or receiving during this time frame at Atlantic Richfield.

Dr. Shapiro. Positive.

Chairman Udall. No package or parcel or object or paper or anything that you can recall?[cviii]

Dr. Shapiro. I certainly don't recall any exchange of anything.

Chairman Udall. It's probably unfair to ask you to comment on other peoples' motivations or conclusions, but do you have any idea why this fellow Riley would have said that you had been less than candid in clarifying the implications of this particular airport meeting?[cix]

Dr. Shapiro. I really don't know.

Chairman Udall. You do recall specifically that he was going on to Dayton somewhere beyond Pittsburgh?

Dr. Shapiro. Yes.

Chairman Udall. Not as the information was that that I thought I had, that he turned right around and went back to Washington?

Dr. Shapiro. My recollection was that he was going to meet his daughter who was visiting a friend.

Chairman Udall. Did you take him back to his departure gate or did you leave him?

Dr. Shapiro. I probably took him to the plane, but I don't really remember.

Chairman Udall. You don't recall whether it was a Dayton plane or Chicago, or Washington plane?

Dr. Shapiro. I don't remember.

Chairman Udall. Let's talk about secure communications. Various John Fialka articles in the *Washington Star* have talked about a secure communications system. He has talked about scramblers and encoding device. Without getting bogged down in terminology, was there any kind of device or system that you would call "secure communications" system at Apollo or at NUMEC?

Dr. Shapiro. Not that I know of.

Chairman Udall. You would have known it as the chief executive officer.

Dr. Shapiro. I was told in a story I heard, Mr. Congressman, on the air about a congressman who was talking to Indians—

Chairman Udall. Okay.

Dr. Shapiro. That's it precisely, it's pure unadulterated goombah.

Chairman Udall. All right. Let me not limit my question to the NUMEC plant at Apollo. Let me ask you about Raychord Steel, or Apollo Steel. They were adjacent to your premises?

Dr. Shapiro. Yes.

Chairman Udall. Was there a secure communications system at either one of those operations that you know of?

Dr. Shapiro. Not that I know of.

---

cviii FBI surveillance photos of the meeting have not been declassified and released.

cix The initial reason was that although the June 20 meeting had taken place quite recently, Shapiro "did not volunteer to disclose the fact…until pressed" and then seemed surprised that the AEC even knew about it. The AEC was advised of the meeting by the FBI, which had Shapiro under surveillance.

Chairman Udall. And if there was, if you had used it, or had access to it, you certainly would remember.

Dr. Shapiro. I certainly would have remembered it. I never used such a device.

Chairman Udall. This is one of the troubling things, go ahead.

Mr. Terrell: No, that's alright.

Chairman Udall. This is one of the troubling things about this whole endeavor we have been engaged in is these rumors and reports get started and gain credence, and they are repeated.

Mr. Krash. We would like to have Mr. Fialka here to cross examine him.

Chairman Udall. This is one of the most persistent ones that we have heard about.

Dr. Shapiro. It makes for a good story, I am sure.

Chairman Udall. As far as you know, you never heard of it, never saw it, or never used it either at your company, Raychord, or Apollo Steel?

Dr. Shapiro. That is right.

Chairman Udall. All right. Let me refer again to the October 30, 1970 memo of Robert Tharp. The memo indicates that you were informed that the Commission might decide whether the Commission's Administrative Review procedures should be used to resolve the question of whether you should be granted a Q security clearance.

Do you know why the AEC had questions and I guess for the record we ought to say we are talking about the period when you sold NUMEC, after you sold NUMEC to Atlantic Richfield and you went back to work for Westinghouse and you were seeking a security clearance to do some work there; is that correct?

Dr. Shapiro. This was at Kawecki Berylco.

Chairman Udall. Okay. Alright. What seemed to be the problem? Do you have any idea why they were giving you difficulty with regard to the security clearance? Can you enlighten us on that?

Dr. Shapiro. My clearance pertained to reactor applications. When I went to Kawecki Berylco they were working on other security matters, and it required an enhanced security clearance. Furthermore, it is customary that there isn't a simple transfer of security clearance, it is necessary to reinstate the clearance after one leaves one installation for another. So I went through the usual application for the clearance, and I then expected that it would take the usual time which was maybe two or three months and when I did not receive notification of clearance, I called to determine what the problem was and I was told that it was necessary to reinvestigate the situation because of the enhanced clearance requirements, but that it was a matter of time, and to be patient; and that happened on a number of occasions.

I don't know whether I have answered your question.[cx 393]

[cx] According to Glenn Seaborg, an AEC official met with Shapiro in April of 1971, telling him that the Justice Department was blocking his security clearance upgrade. This is likely something Shapiro would not forget, but he likely did not want to discuss it at the hearing and later claimed he had no idea of the magnitude of resistance.

Chairman Udall. Did you get the feeling that the delay, inability to get a fairly prompt security clearance again related to the various investigations of NUMEC and the MUF and the circumstances surrounding this, or did you feel that you believed it had regard to something else?

Dr. Shapiro. As I recall it, as a result of my conversations with Riley, I believe that one of the factors that he talked about was the Israeli—my Israeli connections.

Chairman Udall. All right. One of the questions raised here and I will raise it with you directly, you apparently had a right to go before the Commission's Administrative Review Procedures Board and get this matter cleared up, and you decided apparently not to press the clearance question with the Atomic Energy Commission and instead moved into a position with Westinghouse. What were your motives, your reasons for doing that?

Dr. Shapiro. It wasn't quite that way.

Chairman Udall. Alright, explain it to me.

Dr. Shapiro. I was in fact pressing for clearance; however, without seeking a position at Westinghouse I was called by the President of the Power Systems Company and asked to come to see him about the possibility of a position. I was offered a very attractive position at a much more attractive salary and I decided to accept that position.

Chairman Udall. It did not involve a security clearance, or this level of clearance?

Dr. Shapiro. It did not involve this level of clearance and when I accepted the position, the whole matter, of course, became irrelevant to the Westinghouse position.

Chairman Udall. The implication in all of this is that you were afraid you couldn't get clearance and were fearful of further investigations and, therefore said "to heck with it" and took the Westinghouse position.

Dr. Shapiro. That may be the implication, but that is not—

Chairman Udall. That is not the case.

Dr. Shapiro. —That is not the case.

Chairman Udall. I wanted to clear that up. Related to that, and not very important, there is a letter from you to then-Commissioner Ramey in June '71 which suggests that Ramey had provided you assistance and advice that led to your taking the Westinghouse job. Did he in reality assist and advise you?

Chairman Shapiro. I would hardly say that he assisted me but we discussed it. And as a result of not only discussions with him but in discussions with others and my family and all other considerations, I finally decided to accept the job.

Chairman Udall. One final one on this security clearance problem; there is a memo by Riley of December 11, 1970 which says that you consulted Mr. Hermoni, the science advisor I mentioned earlier, concerning your inability to get your clearance renewed with the AEC, which would raise an inference that you were aware that somehow your problems on the clearance were bound up with your relations with the Israeli Government. Did you think that Hermoni could help you with regard to that?

Dr. Shapiro. No, sir, I just stated, you had asked me did I have any feeling for why there might have been some holdup in the clearance, and I indicated that I, as a result of conversations with Riley, it became clear that this was one of, if not the principle, issue. This was one of the issues.[cxi]

Chairman Udall. Yes.

Dr. Shapiro. And I must have—although I don't recall specifically—must have discussed the, he may have discussed the Hermoni issue with me, and that is how this whole thing arose.

Chairman Udall. I think the thing that puzzled me was if indeed your connections and affection for Israel was part of your problem with the clearance, why one would assume that Hermoni could help you get the clearance. It seemed a little bit odd. I guess you have explained it.

Dr. Shapiro. I don't recall the specific circumstances of that but it undoubtedly led from the discussions I had with Riley.[cxii]

Chairman Udall. I don't know whether this is important enough to take your time, but let me run it by you quickly. In some of the documents that we provided you, there are suggestions that people on the AEC staff felt that you went over their heads from time to time to the Commission itself when you felt the staff had been overly scrupulous in enforcing contract conditions.

Did you appeal to the Commission to make known such concerns from time to time and why, and which ones? Do you care to comment on that? I recognize it's broad and kind of shotgun.

Dr. Shapiro. I really have no recollection of...

Chairman Udall. Did you ever complain to the Department of Justice that the FBI had harassed you or had you under surveillance or had given you trouble?

Dr. Shapiro. No.

Chairman Udall. I have about concluded the range of things I wanted to cover. Let me suggest, Abe, that we talk about a five minute break here and let me check with my colleagues and check my notes.

Mr. Krash. Sure, fine.

Chairman Udall. Bob, for Congressman Bauman, do you have anything you wanted to cover?

Mr. Terrell. No.

Chairman Udall. Let's take a break here.

Mr. Terrell. Would you go through the statement?

Chairman Udall. Yes, go ahead. Ready?

Mr. Terrell. Is the machine running?

On page 4, you note, "Apart from what I have read in the press, I have no knowledge whether Israel does or does not have a 'nuclear weapons program'." Is that correct?

Dr. Shapiro. That is what I stated.

Mr. Terrell. You really have no knowledge one way or the other, other than the extent of this?

[cxi] At the time, Shapiro's Israeli intelligence connections were a major issue holding up a higher security clearance.

[cxii] The fact that Hermoni was closely tracking Shapiro's move into a position involving advance nuclear weapons designs was the way LAKAM operated. It is far more likely that Shapiro was helping Hermoni rather than the reverse.

Dr. Shapiro. That is true.

Mr. Terrell. With your connections in Israel as far as military intelligence, there has never been any discussion with you at any time relative to their capability?

Dr. Shapiro. My discussions with the military intelligence people pertained to a long-lived battery to be used in intrusion detection.

Mr. Terrell. But it never got outside of that area of discussion; you never — your concern for Israel's defense, it would seem to me that that would be a natural thing you would want to know, what military capability they would have, particularly in the nuclear weapons end of the thing. I know you were concerned in 1967 to take the tour that you did after the war with your concern, it would just seem that you have a little bit more knowledge than I.

Dr. Shapiro. I would say that many others have taken the tour.

Mr. Terrell. I am not suggesting that you were alone on that tour. I am just suggesting that from your concern —

Dr. Shapiro. Others were interested in taking the tour, and did take the tour in '67.

Mr. Terrell. And you don't know any more than I know from reading the press.

Dr. Shapiro. I would know exactly what you know if I had read the same press articles.

Mr. McNulty. I have a question.

Chairman Udall. Yes, go ahead.

Mr. McNulty. In the BBC documentary they carried this possibility of the diversion of NUMEC materials by something called "plumb bat." Does that ring a bell? Are you familiar with that at all?

Dr. Shapiro. I have heard the name.

Mr. McNulty. It's supposedly some scheme to divert some unknown tons of material—

Mr. Krash. There is a book on the subject.[cxiii]

Mr. McNulty. Yes.

Mr. Krash. Published by the *London Tines*, *Sunday Times*, which I assume you have seen.

Mr. McNulty. Right. And I just wonder if you had known anything about it. This book fairly convincingly portrays the ability of Israeli intelligence to — well, I guess "divert" is a word, to divert a whole shipload of yellowcake from the middle of the Mediterranean. You talk about how extremely difficult it would be to divert uranium at NUMEC. If they are so skilled as to be able to carry out this "plumbbat" affair, it just seems that that same skill could be used. I just wanted to throw that out. I guess you really couldn't comment on that.

Chairman Udall. That was pretty far afield.

Mr. McNulty. Yes, I guess.

Chairman Udall. I have only one or two more questions and I am nearly through. Why did you sell out the NUMEC facility to Atlantic Richfield, just as a matter of curiosity?

---

cxiii *The Plumbat Affair*, by Elaine Davenport, Paul Eddy and Peter Gillman, published in 1978

Dr. Shapiro. Well, I had been looking for the possibility of selling the company to a large company for—ever since the new law, I guess it was '64, was passed, which allowed for private ownership of special nuclear material. The reason for that was that the cost of special nuclear material was so high that private ownership[cxiv] required large amounts of capital and, also, the ability to borrow money at prime rates, and we were not in the prime rate category, and consequently we were at a serious disadvantage from a competitive viewpoint with companies which did have the ability to borrow at prime rate. And I

felt it was necessary to rectify that situation by becoming connected with an organization which, in fact, could have that kind of capital; and, as a matter of fact, I fought hard against the private ownership bill for, as I recall it, two years and gave testimony in connection with that and managed to hold it up, I think, for a year, with a little conflict, but nevertheless I did that, and so I wanted somehow or other to rectify that problem that we were in, and this was not simply a progression of events that occurred.[cxv]

Chairman Udall. Had you kept the same group of investors over the life of NUMEC, the same original group that went into it? Had there been some changes?

Dr. Shapiro. Oh, a number of changes, because we had a number of financing arrangements subsequent to that original group. The original group, as I indicated earlier, was somewhere in the neighborhood of $250,000, and that was a pittance compared to what was eventually required for the nature of the business that we in fact built up.

We needed a multimillion-dollar investment, and therefore we had followed that original investment not only by the borrowing, but then floated a $295,000 fund that is, sale of stock within the state, and that was in May subsequent to, as I recall it, subsequent to our going into business. Then there were...

Chairman Udall. You didn't go public in the sense that you went out and solicited investors from a large area?

Dr. Shapiro. We did go public.

Chairman Udall. You did go public?

Dr. Shapiro. Yes, we did go public, and that was handled by one of the investment bankers for us.[cxvi]

---

cxiv Even normal contracts with no private ownership component required the AEC to estimate whether a contractor could cover any MUF, which NUMEC simply couldn't.

cxv NUMEC's more immediate need to sell itself to Atlantic Richfield was that its $2.6 million loss in 1966, a direct result of the MUF and AEC fines, put it into technical default on its loan agreements. Atlantic Richfield spun off NUMEC close to the end of its first lucrative AEC facilities contract. An AEC official interviewed in 1979 characterized the NUMEC sale as "forced" by the AEC in order to precipitate a change in its management.

cxvi NUMEC was a "closely held" corporation. Most of its voting stock was held by a small number of stockholders and was not traded on any major national stock exchange. When NUMEC's letter to stockholders announced the sale to ARC in 1967, there were only 323,390 outstanding shares.

Chairman Udall. At any time as you expanded and got large investors, were you aware of any investment by the Israeli Government or people acting in behalf of them?

Dr. Shapiro. No. No.

Chairman Udall. When this large MUF finally was discovered or identified at NUMEC, did this cost your company a good deal of money; were you liable, was the company liable, for missing —

Dr. Shapiro. We were always liable for missing material.

Chairman Udall. Was this a very heavy burden?

Dr. Shapiro. It was certainly a heavy burden, but as you may recall from reading the reports, we had already declared losses, as I recall it, of 149 kilograms of material, and we had to pay for those, so this was the difference between 149 and 178 so we had already been paying heavily for the material. But in every contract what we tried to do was anticipate what those losses might be and to put in the contract price those anticipated losses so that they would cover it. Naturally, if we underestimated those losses, then, of course, it would hurt us.

Chairman Udall. Were you ever in serious financial trouble as a company because of the losses that you had to pay the Government for?

Dr. Shapiro. Well, we certainly had — we certainly had a severe cash flow problem as a result of these losses. But in terms of financial — serious financial difficulty we had arranged prior to this time a fairly substantial line of credit with the Mellon Bank.

Chairman Udall. When you finally sold out to Atlantic Richfield, did you and the other investors make a profit on your original investment of money? Was this a profitable sale for you?

Dr. Shapiro. Well, I had not invested any money, and I got some stock, and therefore anything I would have gotten out of it —

Chairman Udall. You were ahead.

Dr. Shapiro. I would have been ahead except for the fact that I put my heart and soul into this whole thing, and I don't know how much one can pay for that in money. And insofar as other investors are concerned, as I mentioned, it depends upon when.

Our original sale of stock was equivalent, as I recall it, to ten cents per share, and what we call our automobile dealer friend sold out at $135 a share. So I think he made a substantial profit.[cxvii]

Chairman Udall. Yes. I am through. Questions, anybody?

Mr. McNulty. Not here.

Mr. Myers. I have a brief one. While you were announcing along the way that these losses were occurring, it seems somehow not to have been gotten to the Commission itself. Because it seems that it wasn't until the summer of '65 that they seemed to demonstrate a lot of concern, they had meetings and deciding

---

[cxvii] Shapiro was able to remember the share price obtained by a lone hold-out, but not his own profit which would have been close to half a million dollars under Atlantic Richfield stock swap arrangement if he sold immediately, and much more if he held onto the shares until the Atlantic Richfield began managing the Hanford facilities.

what they were going to do about all this. Somehow it seems to have escaped their consciousness until then.

Dr. Shapiro. I don't believe that that was so.

Mr. Myers. I mean like this August 2 memorandum starts out by saying, from Brown to the Commissioners, starts out by saying

Mr. Becker. Just a moment. August 2?

Mr. Myers. August 2. It's the one we had out before.

Chairman Udall. For the record, what year is it?

Mr. Myers. This is August 2, '65. It starts off by saying, "We are faced with a potentially serious problem involving the possible shortage ... "and so forth. I don't think this has—may not have—anything to do directly with you, but the fact is there seems to be starting about a this time a major concern demonstrated by the Commission about all this and how they are going to handle it, and they have a meeting where you attend, on August 10, and then they have their own meeting—other meetings—along the way, and it just seems hard to understand why this concern suddenly gets to them and why it had not been there before. What was it that happened at this time that made them demonstrate this worry? Maybe we should ask them, but—

Dr. Shapiro. I think that that would be most appropriate.

Mr. Myers. But do you agree that that's the case? There seemed to be—Howard Brown seems to be notifying the Commission for the first time that we are faced with a potentially serious problem?

Mr. Becker. He is talking about a particular contract and the documents related to it.

Mr. Myers. There were 60 kilograms associated with that contract, but, if you had been announcing losses along the way—which I gather you had been—why was it that the Commission, Howard Brown is suddenly saying "We are faced with a potentially serious problem"? Because most of that 60 kilograms—

Dr. Shapiro. Would you care for me to speculate? Is that it?

Mr. Myers. Sure. Well anyway, that's one of the big questions about all this, of course, which...

Mr. Becker. I think you would have to ask the people who wrote the document what they had in mind at the time. That is what your question is.

Mr. Myers. I think what they would say is that, "Well, yes, maybe it was being reported, but somehow it didn't, or, we weren't, you know, it didn't strike us. We were not really aware of what it really was." I think that's what they would say.

Mr. Becker. I am lost. Is there a question?

Mr. Myers. The question is that here, if losses were being announced along the way, and they were being reported along the way, why is Howard Brown suddenly saying "We are faced with a potentially serious problem?" Since the problem had been existing for a long time, why is he suddenly on August 2 informing the Commission of this?

Mr. Krash. Do you have anything to shed any light on that, Dr. Shapiro?

Dr. Shapiro. I think it is clear that what they are referring to here is a particular contract which entailed the loss of approximately 60 kilograms of material which would—on that basis approximately six percent of the total material—and I would guess that was the reason for their concern.

Mr. Myers. Because that would be a higher percentage than on most contracts?

Dr. Shapiro. As I pointed out earlier, the reason for the higher percentage of loss, processing loss in this contract than in other contracts is because of the nature of the material that needed to be fabricated and the low yields—I didn't mention this previously—and the low yields that we got which required the constant recycling of materials. So what you had was the accumulation of, repeated accumulation of, losses as a result of the recycling.

Mr. Myers. All right.

Dr. Shapiro. Nobody previously had ever attempted to make in quantity this flyspeck material having these stringent specifications. In retrospect, I think we were fools to have undertaken it.

Chairman Udall. Bob?

Mr. Terrell. I have only one question and I think I was making a phone call when you were talking about it. You mentioned, and I think it was verified later, that your assay on the Kimwipes and whatever other materials were used, was inaccurate. This was only discovered as a result of digging up and doing some sort of backward counting to find out that you, in fact, not only varied two kilograms, in fact it was considerably more than that.

What I am wondering is: Is the same technique applied to the effluent that leaves the plant that goes into the river, and can we assume that there could be some inaccuracies there as to how much actually went out through that effluent?

Dr. Shapiro. Let me explain. The problems with regard to assaying material, the amount of material depends upon the character of the materials to a large degree. Whether it's homogeneous, whether it's inhomogeneous, and whether it is in fact distributed on material.

When it is inhomogeneous and distributed on a lot of material, it is extremely difficult to get an accurate assay. And in this particular instance the assay was done by radioactive counting of the package, and that indicated by virtue of the counts that we got off of it, that there was very little material associated with it. Consequently it was buried.

But, when we tried to do a more accurate estimate, by actually burning the material, then it became apparent that there seemed to be much more associated with that than we had dreamed of.

Now, in connection with the liquid effluent that we are talking about, there are also problems in that again there is a homogeneity problem, whether or not the material is suspended, or whether it is in solution. Obviously, if it is in solution, it is much easier to get an accurate estimate. If it is suspended, then it depends upon how you take the sample and many other aspects as to whether or not you get an accurate—as to how accurate the assay is.

Did I answer your question?

Mr. Terrell. Yes; to the extent that I don't think that anybody knows, including yourself; depending upon how much confidence you put in the analysis you did on effluent, I don't think that you have any confidence, or anyone else, as to how much actually left the plant in the effluent that went into the river. You are saying you were within certain Federal guidelines, certain parts per million, but how do you know that you were actually within those 'stats'?

Dr. Shapiro. By taking the assay that we did.

Mr. Terrell. Using the technique at that time?

Dr. Shapiro. That was available to us at that time.

Mr. Terrell. That is what I am saying. What you are saying today — obviously to the extent of your knowledge — would you say today the techniques being utilized at the receiving and shipping end of this sort of thing is more refined to the extent that we know more clearly how much material is being received, how much is being brought in solution, how much is being concentrated out of the process?

Dr. Shapiro. You could continuously find variance during the period of our operations.

Mr. Terrell. And did you adjust the procedures when you went back through and had it burned and assayed? Did you do anything to rectify your assay procedures?

Dr. Shapiro. Well, for—

Mr. Terrell. You know, for future work.

Dr. Shapiro. I don't recall specifically.

Mr. Terrell. There were things that you probably — I would think — that you would do something along those lines.

Dr. Shapiro. It was our practice that if, obviously, we learned a lesson, that we apply that lesson.

Mr. Terrell. Yes.

Chairman Udall. Let me lay one more question on you, and then we are really going to quit. This will just take a minute. Henry, correct me — I am trying to find my reference, and I can't find it. With regard to this famous June, 20 meeting at the Pittsburgh airport, there was something in the file that the AEC had interviewed you about that in August. The meeting was in June, and that someone at the AEC called upon you to interview you about in August, and that the information I am searching for was to the effect that, when they first contacted you about this, you had no — expressed the feeling you had no — memory of the meeting, and then later recalled a lot of details and called them back with supplemental information about the airport meeting. Do you recall this?

Dr. Shapiro. No, I cannot.

Chairman Udall. There is no way you can give me an explanation of that, why you would forget that meeting at the airport, if indeed you did?

Dr. Shapiro. My recollection is that I have always been forthcoming about that.

Chairman Udall. About that meeting?

Dr. Shapiro. Not only that meeting, but everything pertaining to this whole issue.

Chairman Udall. Were there other airport meetings at that time? Any other time?

Dr. Shapiro. Not that I recall.

Chairman Udall. That you met an Israeli official or citizen at the airport that you can recall?

Dr. Shapiro. I may have—

Chairman Udall. In connection—I am talking about a pre-arranged meeting where, during the time NUMEC was in operation, where you had gone to the airport to meet someone?

Dr. Shapiro. I have certainly gone to the airport to meet people. I have often done that. And I would generally go out of my way for an overseas visitor, whoever that was, whether he from England, France, Belgium, or wherever. And it was likely that I would have done the same for an Israeli visitor.

Chairman Udall. I don't question that. The puzzling thing about this to some people has been the idea that he made a special trip to the airport and you made a special trip to the airport and apparently a long conversation was held. That was why I was—

Dr. Shapiro. Certainly that matter has been discussed and explained many, many times.

Chairman Udall. Okay. Well, I want to thank you for coming. You have been very good to be with us and answer questions not only relating to your own knowledge, but some of the speculation that we have asked you to indulge in. I must say I don't know where we go from here. I want to ponder what has occurred today. You are a very effective witness in your own behalf and, if what you told us today is true, you have got every right to feel aggrieved about the way you have been treated. That is all I will say in conclusion.

Mr. Krash. I want to thank you, Congressman, for let the record reflect—the most courteous treatment the witness has received here, and we appreciate that very much. We do stand ready and willing to cooperate with you in any way we can in responding to any questions that you may have.

Chairman Udall. While the record is being transcribed, as I said in the beginning, feel free to supplement additional material or statements.

Mr. Krash. We understand that the reporter will give us a copy of the transcript and we will have a chance to review that.

Chairman Udall. You will, indeed.

Mr. Krash. Thank you very much.

Chairman Udall. All right. We stand adjourned.[cxviii]

Although Udall's final comments conveyed empathy with Shapiro's situation, the former prosecutor clearly did not believe Shapiro's version of events and seemed dumbfounded that Shapiro would not even admit the existence of an Israeli nuclear weapons program. Just six months after the hearing, a BBC interviewer asked Udall whether Israel had removed HEU from NUMEC. Udall responded in drastic terms. "If someone had me write in an envelope whether a diversion occurred or did not occur and that I would be put to death if I answered wrong, I suspect I would have to put in the envelope that I believe there was a diversion."[394]

According to Michael McNulty, Udall's chief counsel, Udall's "staff was pretty convinced that fissionable material had been spirited out of the country. But to get to the bottom of it required a law enforcement agency rather than a congressional committee."[395] But unknown to Udall, the FBI was only a few years

---

[cxviii]At 5:01 p.m. the meeting was adjourned.

away from truncating the investigation once again, after receiving stunning new employee accounts of diversion.

There were many questions that Udall didn't have the background information or perspective to ask that Shapiro probably would not have been able to credibly answer. Why had Shapiro invited Israeli covert operatives Ephraim Biegun and Rafael Eitan into the plant in 1968? Exactly when had he first met David Lowenthal, at what "organizing events" and what was the true nature of their relationship? Why was he running a facility that was cutting so many corners, seemingly oblivious to the health concerns of Apollo and Parks residents and the environment? Given his lack of concern over the 1969 toxic spill, why should he be entrusted with a security clearance? Had he ever visited the Dimona plant? Reviewed in context, Shapiro's highly selective memory and evasive answers to FBI agents, the AEC and congressional investigators all seem to point to a very cool and collected man carrying very big secrets.

13. NUMEC investigation terminated

Publicly available documents from the FBI's NUMEC investigation files generated in the late 1970's mostly consist of NUMEC employee and management interviews and interrogations of those former AEC and other government officials willing to make themselves available. Not all did. Leaked content from the classified GAO report *Nuclear Diversion in the U.S.* and titillating press accounts of NUMEC diversions to Israel to some observers pointed either to LBJ-sanctioned CIA involvement in the diversion, or a clumsy government cover-up fueled by establishment political party desires for future Israel lobby campaign funding largesse.

Adding fuel to the "with LBJ approval and CIA assistance" theory was a January 9, 1979 article in *Newsweek* titled "Mysteries of Israel's Bomb" by David C. Martin. *Newsweek's* piece was an anomaly among other NUMEC press coverage in assuring readers that the latest FBI investigation already concluded that Shapiro had committed "no provable illegal act." This was not true since declassified records from the investigation revealed that although the FBI clearly wanted to terminate it, the investigation was still underway. The article, published the same year as the Mossad-inspired Andrew Young stories, strongly suggests Shapiro had intimate ties to the CIA. Presumably referring to some period after LBJ ordered CIA Director Richard Helms to drop Carl Duckett's warning about a growing Israeli arsenal fueled by diversion, government officials were "told by the Justice Department there was no need to pursue because there would be no prosecution." Shapiro, *Newsweek* asserted, was "close to the CIA" and had even helpfully arranged for the agency to debrief Westinghouse employees just back from visiting the Soviet Union.

David C. Martin, the son of a senior CIA official, served as *Newsweek's* defense and intelligence analyst from 1977-1983 and wrote the book *Wilderness of Mirrors*. The book, which covered the career of CIA counterintelligence desk and Israel liaison James Jesus Angleton, was considered an unusually accurate inside account by another CIA scholar.[396] But Martin had a number of material omissions in his coverage of NUMEC. He failed to reveal Rafael Eitan's visit to NUMEC, Hermoni's role in the Israeli nuclear weapons program or that not only were traces of HEU picked up in Israel, but that they matched the Portsmouth signature of the kind of 90-plus percent enriched U-235 mainly reprocessed at NUMEC. Most improbable of all, Martin wrote that "others wonder if the whole story of Israel's early nuclear capability was actually 'disinformation' — fabricated by a sympathetic CIA to worry Arab leaders."[397] Dimona technician Mordecai Vanunu's soon released detailed photos of Israel's weapons and plutonium plant which were printed in *The Sunday Times* October 5, 1986. This unprecedented exposure made such talking points about nuclear capabilities rather than an actual arsenal, often advanced by Israel supporters in the establishment press, seem silly if not deceptive. But in CIA-NUMEC story was highly useful, since the agency had no motive to reveal its sources and methods in Israel and could therefore neither confirm nor deny any ties to Shapiro.

Another executive, likely NUMEC's CFO Fred Forscher, also seemed to cryptically advance the CIA collusion angle when interviewed by the FBI in his Pittsburgh home on November 11, 1977. There was "no illegal activity" and "no cover-up" at NUMEC, claimed Forscher. Forscher further tried to clear himself stating that while there had been an inventory issue with the AEC over SNM at NUMEC, he personally had no involvement in such side businesses as ISORAD. He was aware of all the news stories about diversion, and felt they were being promoted by "anti-Zionists or people opposed to the nuclear program."[cxix] Or just perhaps, according to Forscher, they were the work of something entirely different. "Some intelligence agency may be trying to divert attention from its own shortcomings."[398] The FBI did not document whether it sought any clarification about Forscher's strange intelligence agency speculation.

Neither did the FBI pursue this line of inquiry with a former AEC division of security inspector interviewed by the FBI on June 29, 1978. He too seemed to support the theory of a lavishly funded covert operation. He couldn't understand why former government employees were granted permission to work for NUMEC, or why Shapiro was often allowed to visit U.S. weapons labs for no obvious purposes. During the time preceding the official conclusion of extraordinary materials loss, there had been a "lot of Israeli contact with NUMEC." When this all became public in 1966, according to the inspector, suddenly the "Israelis avoided the plant." Overseas shipment controls were lax, and NUMEC "steered" his own inspection visits. William Riley, the AEC's director of security, had finally just been "told to shut up about NUMEC." This led the inspector to conclude that NUMEC was "bigger than [the] AEC" and somehow "out of [its] control or jurisdiction." Although he recommended the FBI thoroughly review NUMEC's foreign visitor records which "had to be maintained indefinitely" there is no indication that the FBI ever followed up on this lead.

No others interviewed by the FBI in 1977 advanced the investigation with new leads. Leonard Pepkowtiz underscored that while Apollo was a "defunct" facility Zalman Shapiro was an honest, trustworthy, and loyal American. Other former managers advanced similar views. A woman affiliated with NUMEC interviewed on March 21, 1977 attributed "problems in infancy" as "inevitable." Sylvester Weber, a manager of NUMEC engineering and accountability testified there had been no cover-up, even when the FBI asked if employees had been "pressured" into hiding losses. [399]

However Charles Keller, who had recorded a blow-by-blow account of how the AEC handled the NUMEC crisis in the 1960s, again interviewed with the FBI from his new position as assistant manager for manufacturing and support at the Department of Energy's Oak Ridge facility. As he turned over his notes, Keller asserted that if he were planning to steal nuclear material, he would have exactly the sort of operation that existed at NUMEC. Sloppy handling of SNM, lax

---

[cxix] This was the same response Zalman Shapiro made during his Congressional interview.

accounting procedures, all done under the pretext of a "hand-to-mouth" operation. Keller repudiated the AEC's official processing loss position. "No one [was] willing to state an actual diversion had taken place, but we are in no position to show that it wasn't."[400]

Undaunted, on August 8, 1979 the Justice Department furnished its latest list of 50 individuals for the FBI to interview. Categorized into two groups, individuals in the first were to provide "any knowledge they may have concerning the alleged diversion of special nuclear material from the NUMEC plant." Members of the second group were to be asked whether they knew of "any attempt by anyone in the executive branch to prevent or impede an investigation into this alleged diversion, or to withhold any information regarding this alleged diversion from any investigative body."[401] FBI agents had earlier discovered that two of the most important officials, Hyman Rickover and Glenn Seaborg, were uncooperative. There is no record the FBI ever successfully interviewed former CIA Director Richard Helms, who was prosecuted and convicted of lying to Congress in 1977. But the FBI did try a new legal strategy to find out everything the CIA knew — and hadn't told them — about NUMEC.

Citing new authorities in an executive order signed by Carter in early 1979, FBI Director William Webster formally solicited an "exchange of information" about the NUMEC nuclear diversion. The FBI would give the CIA directorate of operations information about individuals of high counterintelligence interest, in exchange for CIA information developed on allegations of U.S. persons involved in diversions. [402] The authority for this exchange, Executive Order 12036, was an intelligence reform effort which expanded the U.S. ban on assassinations while restricting intelligence operations targeting U.S. citizens "unless the President has authorized the type of activity involved and the Attorney General has both approved the particular activity and determined that there is probable cause to believe that the U.S. person is an agent of a foreign power." It also banned the CIA from conducting electronic surveillance in the U.S., leaving the FBI as the sole member of the intelligence community allowed to perform physical searches in the United States. However, there is no record that the CIA cooperated with the FBI request.

The FBI reported on March 10, 1978 that Admiral Hyman Rickover — upholding his reputation for being difficult — at first refused to cooperate with the FBI and even hung up the phone after telling agents "I have nothing to say." Rickover later called back the agents demanding their list of questions in advance, claiming that all information about NUMEC had already been sent to the Department of Energy. Rickover finally relented and allowed the FBI to visit, which star-struck agents noted had produced "good dialogue."[403] But nobody thought to ask the celebrated pioneer of the nuclear Navy whether its contracting clout had been used to pressure B&W's acquisition of NUMEC in 1971, or other obvious questions that only Rickover likely could have answered.

William Riley, the AEC's Director of Security between 1967 and 1972, provided highly derogatory insights about Shapiro to the FBI on October 19, 1979. Shapiro had frequently used Rickover's "name and fame" when dealing with the AEC. Their mutual admiration and friendship blossomed at Westinghouse and bootstrapped NUMEC by providing a steady stream of revenue. Riley assured the agents he had personally recommended that all government contracts to NUMEC be terminated due to "poor security." He was also alarmed by the implications of Shapiro's ideology, asserting that "helping Israel is helping the United States," declarations that U.S. and Israeli interests were one and the same, and stated intention to one day emigrate to Israel. According to the FBI interview report, Shapiro had once implied to Riley the angle trumpeted by *Newsweek* that full knowledge of his activities was simply beyond Riley's pay grade. "Riley advised that Shapiro indicated that his actions were with the approval of some unnamed Federal agency not further expanded upon by Shapiro during the interview."404

On March 3, 1979 the special agent in charge of the San Francisco FBI field office reported to headquarters that Glenn T. Seaborg had outright refused to be "re-interviewed" about NUMEC and Shapiro.405 As the major proponent of the AEC's official "processing loss" theory, Seaborg would have found it increasingly difficult to respond to ever more informed interrogators. The FBI's Washington Field Office had already confirmed that "U-235 can be traced back to the facility which processed it, based on isotopic differences." Seaborg's compelled testimony to Energy Department investigators who told him flat out that traces of Portsmouth material of the kind primarily sent to NUMEC for WANL and other contracts were discovered in Israel is still secret, but not the fact that Seaborg was forced to answer. How would Seaborg have answered informed, more pointed questions about why he urged the Justice Department not to release Shapiro's FBI file to the JCAE? Or why he defended Shapiro even after AEC interrogators constantly found him less than forthcoming over questions about his activities within a network of Israeli covert operatives? Or whether the AEC had encouraged the Atlantic Richfield buyout in exchange for its lucrative Hanford contracts? He could not, and so he did not.

Earle Hightower, a former AEC assistant director of policy plans in the office of safeguards and security noted "a lot of suspicions" about NUMEC among the management ranks. "I think it might have been intentionally diverted" Hightower told FBI agents on October 23, 1979. From his perch, he could see a broader pattern among U.S. scientists quietly working on critical Israeli projects. "We also had a number of scientists leave the U.S. on sabbatical....and go to Europe during the same time span..." Hightower thought that NUMEC developed contacts in the nuclear weapons program, while noting Shapiro's extensive Zionist credentials. 406

Despite documenting such suspicions, an interviews report submitted in November of 1979 by the FBI's Washington Field Office stated confidently that "there has been no information presented which supports the alleged diversion of special nuclear material from the NUMEC plant, nor has there been any indication of anyone in the Executive Branch attempting to prevent or impede an investigation..." But it also noted that "Several individuals remain to be interviewed."407 Then the FBI suddenly hit the jackpot.

A group of NRC officials interviewed by the FBI said they had recently been documenting complaints emanating from Apollo that the NUMEC plant "was causing health problems, particularly cancer..." The NRC was not yet pursuing toxic NUMEC pollution complaints in earnest, claiming it would require epidemiological studies as well as action by Pennsylvania's state government. [408]

One NRC fuels inspector interviewed by the FBI notified them of a troubling February 1980 phone call about contamination caused by NUMEC facilities. A former NUMEC worker living close to the Apollo facility complained of dizziness, nausea and lightheadedness. A growing number of worker health problems were flooding into the Pennsylvania Department of Environmental Resources, the State Health Department, congressional representatives, the University of Pittsburgh and recorded on the Governor's "action" hotline. [409] In desperation an anonymous NUMEC worker, referred to as the "Ammonator operator" called the NRC.

In the early 1980s, the FBI finally obtained a solid lead from NUMEC employees with still-vivid eyewitness accounts of a nuclear diversion to Israel. A NRC fuel facility inspector with the Office of Inspection and Enforcement in King of Prussia met with an Apollo resident who worked at NUMEC to discuss his health concerns on March 5, 1980. The NUMEC employee went to the NRC as part of a last ditch effort since "both the company and the union consider the Nuclear Regulatory Commission as the 'enemy'". On March 11, 1980 the NRC official sat down with the FBI to recount second-hand the aggrieved NUMEC employee's story.

After listing a litany of health complaints and a contentious dismissal which led to a union grievance filing, the NUMEC worker had told the NRC that when he read a newspaper account linking "losses of nuclear material when Mr. Zalman Shapiro was here, he had to laugh." Perhaps fearing retribution, the employee asked that his NRC interview be entirely "off the record."

Early in the employee's employment, 1965 or 1966, he and a co-worker walked near the NUMEC loading dock, encountering people he was unable to identify, "loading cans into some equipment." The cans were of "the approximate size and dimension that would contain high enriched uranium." He noticed that "the shipping papers for the equipment indicated that the material was destined for Israel." An armed guard ordered them to vacate the area, which they did. Identifiable NUMEC personnel at the scene included a foreman and truck driver. The employee had not gone into too many details with the NRC about diversion "since the original purpose of his interview was to discuss any health problems." [410]

Activists and victims seeking compensation for toxic emissions during Shapiro's tenure had (and have) little incentive to talk about illegal SNM diversions. Criminal activity is not covered by many liability insurance policies. By providing credible evidence of such activities under Shapiro's reign at NUMEC, former NUMEC employees could diminish the possibility of future compensation. Activists leading the compensation drive, such as Patty Ameno[cxx],

---

[cxx] Patty Ameno's "Linked In" website states: "Of most significance is that Patty has been able to establish to a much higher degree and with documented evidence, that the alleged diversion of nuclear material from NUMEC to Israel never happened. Rather, the Material Unaccounted For (MUF) were all process losses into the environment. Something that has been commonplace with many

exonerated Shapiro of any diversion (mainly citing misinformation or obsolete data). Such exonerations strengthened their compensation claims against Atlantic Richfield, B&W and the U.S. government.[411]

In 1980 the FBI's special agents were intrigued and interviewed the NRC's James Devlin, Chief of Security and Investigations for Region One and two other employees, Deputy Director James Allen and Chief of Safeguards Walter Martin on March 11. All agreed that while NUMEC shipped a great deal of irradiation equipment, it was simply not legitimate for NUMEC to have inserted any SNM containers into equipment that was being shipped. Equipment and source material were always shipped separately.[412] Finally, with solid leads, the FBI interviewed the NUMEC employee who had tipped off the NRC, as well as a former NUMEC truck driver and shipping and receiving foreman.

The former employee sat down with FBI agents on March 21, 1980. He confirmed joining NUMEC in February of 1965 and that he had made health - related grievances with the NRC. Like many employees, he had received limited training at NUMEC, only three days of schooling on the equipment he operated, along with briefings by personnel and enrichment facility managers on health and safety. His job as Senior Ammonator Operator in the Low Enriched Operations area was immediately adjacent to the NUMEC Apollo plant's loading dock. He also sometimes he worked overtime in the High Enriched area.

He was working a 3PM to midnight swing shift in March or early April of 1965, when his ammonator[cxxi] shut down. The plant was unusually hot, so he walked over to the loading dock, just twenty feet from his equipment, for "a breath of air."[413]

He noticed on this particular evening a flatbed truck "backed up to the loading dock with some strange equipment on it." The equipment resembled steel cabinets with gauges on the front section. Other equipment resembled lathes. He saw Zalman Shapiro "pacing around the loading dock" while the shipping and receiving Foreman and a NUMEC truck driver "were loading 'stove pipes' into the steel cabinet type equipment" already on the flatbed. This was unusual, since these particular employees, "never loaded the trucks themselves" and normally put "other workers" to such tasks. [414]

The "stove pipes" loaded into the equipment were cylindrical storage containers used to "store canisters of high enriched materials in the vaults located at the Apollo nuclear facility." Each stove pipe contained three to four HEU canisters "highly polished aluminum with standard printed square yellow labels, approximately three inches in diameter by six inches tall, that normally were

---

of the legacy sites in the U.S. It is the only answer to: What is more likely than not. The loss of nuclear material continued at NUMEC under the new ownerships of ARCO and Babcock & Wilcox, long after the former founder and president of NUMEC, Dr. Zalman Shapiro left the company. An NRC Task Force Report dated April 25, 1977, on Material Unaccounted For confirms this.
Patty believes that Dr. Shapiro has been wrongfully investigated and accused of the alleged diversion."
[cxxi] Ammonate is the same as ammoniating, or forming compounds with ammonia.

used to store high enriched uranium products...defined as 95 percent uranium."
415

Two workmen were bringing the stove pipes over to the loading dock from the High Enriched vault area, located 150 feet away from the Ammonator operator's loading dock. They opened the stove pipes and withdrew the canisters while a man with a clip board checked each HEU canister's label and checked it off against a shipping order. Each canister was approximately three by six inches of brightly polished aluminum with yellow labels bearing typewritten information and the universally recognized nuclear fan warning symbol on the upper corner.

The canisters were then reinserted into their stove pipes, which were "loaded into the cabinet type equipment after being wrapped with a brown paper type insulation." The Ammonator operator observed "one cabinet being loaded and that the 'stove pipes' were placed one in each back corner of the cabinet, and one in the front center of the cabinet directly behind the door." 416

The Ammonator operator was sure these were HEU containers being loaded into the equipment "due to the size and shape of the container and the labeling." Low enriched containers he commonly processed were much larger and used different labeling. The Ammonator operator had never seen stove pipes used as shipping containers, much less within equipment. Previously, HEU canisters would be removed and loaded into large cement-lined steel drums. The Ammonator operator also questioned why the workmen carrying the HEU were using a different, less-observable route through the plant than they normally took for carrying such material.

Curiosity piqued, the Ammonator operator walked over and read a clipboard, which had been laid down on an empty drum on the loading dock. The equipment was to be transported by truck to a ship, and then to Israel. Having served in the as a radio man, 3rd class, at the Naval Radio Facility in Londonderry, Northern Ireland from 1956 to 1960, the Ammonator operator believed the ship's long name was written in Greek.

The clipboard quickly snatched from his hands, he was informed that "the shipping order was confidential and not for his eyes." Although he had been on the dock for approximately fifteen minutes, and Shapiro had never asked him to leave, it was at this moment that an armed guard suddenly appeared and ordered him off the loading dock.

How could the Ammonator operator remember this incident in such vivid detail a quarter century after it occurred? In his words, it had been "highly unusual to see Dr. Shapiro in the manufacturing section of the Apollo nuclear facility; it was unusual to see Dr. Shapiro there at night; and very unusual to see Dr. Shapiro so nervous as to pace around." As so many others noticed, Shapiro was "a very calm, cool and collected man who never got upset." 417

The Ammonator operator never again saw such equipment as appeared on the dock that night, nor did he ever see such equipment again within NUMEC's fabrication facilities. When he became aware of nuclear diversion stories in newspapers, it caused him to think back to this incident. "Everyone at the plant knew there were losses of materials from the High Enriched area, but nobody seemed to care during the time it was owned by NUMEC." Just prior to the strange incident, it was an open plant rumor that "Shapiro had just returned from an extensive vacation in Israel." [418] But there were plenty of reasons to keep quiet about it.

The employee had a large family to support and the following day the plant personnel manager came forward threatening to fire him if he "did not keep his mouth shut concerning what he had seen on the loading dock the night before." He was again told to shut up by some Kittanning "union goons" after he complained about the armed intimidation to his union steward. "The prevailing attitude at the plant in 1965 by management, union and the employees was that the Atomic Energy Commission was the enemy looking for a reason to shut the facility down with the resultant job losses." Moreover, the Ammonator operator "did not know how or who to contact in authority who would take action." [419]

On April 3, 1980 the FBI interviewed by telephone the truck driver identified by Ammonator operator. He was a 6 foot tall, 220-pound former Army veteran who had joined NUMEC as a laborer in 1961 and was transferred to shipping and receiving within two months. The truck driver still worked at NUMEC. His memory about the suspicious shipment was hazy, but he added a few important details.

The truck driver remembered cast metal, radiator-like equipment being crated into a four-by-four foot shipping container. After storing it at the NUMEC Park Township facility for one or two days, he received instructions to drive it to New York City, for immediate loading onto a docked ship waiting on Pier 34. Although warm and rainy in Pittsburgh, he encountered icy road conditions on the way to New York City. At a New Jersey Turnpike toll booth he received an urgent message to call NUMEC and "explain why he was late with the delivery." There was no delay in "off-loading it from his truck directly onto the ship" when he finally arrived. At the prompting of FBI agents, the truck driver could not provide the name of the ship or verify if "anything was contained inside the cast metal equipment."

The truck driver did recall that Shapiro roamed freely throughout the plant twice a week, at "any time of day or night" but rarely spoke with any of the employees.[420] The truck driver suggested the FBI talk to the shipping clerk.

On April 15, 1980 the FBI interviewed one Mr. Desmond at his NUMEC office at the Apollo facility. Desmond had worked at NUMEC since February 15, 1960 when he signed on as a quality control supervisor. He rose to transportation supervisor in the early 1960's until June of 1979 when Atlantic Richfield spun off the firm to Babcock and Wilcox, Inc and he became transportation specialist.

Desmond oversaw all aspects of air, sea and land transportation for NUMEC shipments. He could not specifically recall shipments in early 1965, but recalled that "high-enriched nuclear products were loaded by the vault handlers into special packaging in the 'hot cell,'" a clean room where only manipulator arms handled the materials. Products would be loaded into 'casks with cooling fins' and closed in the hot cell before they were ever delivered to the loading dock. "The only function performed on the loading dock was to place a final seal on these casks and then load them aboard the trucks."

According to Desmond, storage of nuclear materials on the loading dock was strictly prohibited. "They had to be immediately loaded and dispatched to their destination...under no circumstances were the casks opened on the loading dock to verify either the type of substance or the amount." Only the shipper's list was consulted to verify what material was contained in the casks.

Desmond recalled that "some of the high-enriched materials, which he described as 20% uranium, arrived at the loading dock in temporary storage containers known as stovepipes, which are made out of corrugated sheet metal." Any 3" x 8" HEU containers removed from a stovepipe should have been shipped in a single steel container, called a 2R inner container, which was then placed into a 55 gallon insulated steel drum. According to Desmond "this is the only time he or his personnel come in contact with any high-enriched uranium products." [421]

Desmond did not remember any shipment to Israel fitting the Ammonator's early 1965 date. But by virtue of a birthday or other personally meaningful date, he recalled a shipment on October 26 (of either 1962 or 1963). He loaded one cask with cooling fins containing what he believed to be Strontium 90 onto a three-ton NUMEC company truck. After a false start, he rushed the shipment to New York using phone communications to toll booths to monitor the trucks progress with Dr. Shapiro. It was loaded at Pier 32 onto an Israeli Zim steamship bound directly for Haifa.

Between October 1964 and September 1965 he shipped a "remote weather station" (another large cask with cooling fins and hot material) to Israel via Zim. As was procedure, he had to trust the vault clerk's account about the exact material content "insamuch as opening this type of container would expose himself and his crew to extreme danger." [422] In late summer to early fall of the same year, he was personally dispatched by Dr. Shapiro to take a three and a half foot square crate to the Israeli Defense Ministry in New York City via TWA. He believed it may have contained radioactive material, and Shapiro advised him to contact the Israeli Consul if "he needed any help." [423]

Shapiro dispatched Desmond yet again in September of 1966 to personally accompany two large crates bearing radioactive symbols to New Jersey. Escorted to the Pennsylvania-New Jersey border by State Police, the shipment was delivered to the Maheir Steamship Lines at Newark onto the ship Yafo destined for Haifa. Desmond did not recall any other shipments to Israel, and stated how "it was highly unusual that materials would be taken directly from the backs of trucks on to ships...it was his experience that materials are usually removed from incoming trucks and placed in a warehouse prior to being loaded on a ship." Desmond could "not align Dr. Shapiro's trips with any shipments made to Israel, nor could he recall any phone calls received from Dr. Shapiro while he was in Israel." Desmond did recall requests from time to time for items from NUMEC's machine shop from a NUMEC engineer in Israel. He also gave the FBI a contact at Zim-Israel Shipping lines while pointing out shipping records for 1962-1966 were still likely maintained by the Accountability Department at the Apollo facility.[424] There is no record the FBI ever sought or obtained such records, even though the Justice Department was truly intrigued by the FBI's solid new leads. The DOJ urged the FBI to "please follow all other possible leads that may develop."[425] However, in the Carter administration's final months before a devastating November election defeat by Ronald Reagan, the NUMEC diversion trail began to grow cold.

In May a NUMEC employee knowledgeable about shipping said it was common to ship 5 by 3 inch cans encased in boxes, but these were lead casks of Cobalt 60, not highly enriched uranium.[426] On August 12, 1980, the Washington Field Office pushed back, complaining to the FBI director of the "never ending flow of leads" being developed in the DIVERT investigation. After wondering aloud about the statute of limitations, the WFO questioned "if the final objective is finding the location of the missing uranium, office wonders if this is in the prevue of the FBI."[427]

The FBI interviewed in December of 1980 an executive, undoubtedly Oscar Gray, who had left NUMEC to work in a Washington DC legal practice. Gray dismissed the mysterious loading dock accounts given by lower level employees. All shipments were rushed "emergencies" in that they were "usually late in departing" because having trucks standing by was expensive. NUMEC did not ship at night, so any equipment, claimed the executive, would surely have been shipped in the afternoon.[428]

Still, none of the NUMEC executive interviews could explain away the explicit threats made to employees, or warnings by armed guards that they not repeat what they had seen on NUMEC's loading docks. Why was Shapiro opening and double-checking stove pipes on the loading dock before clandestinely loading them into equipment?

Three days after Carter left office, the Justice Department ordered the FBI to "place the matter in closed status" until further instructions. The FBI specified that there should be "no destruction of files" without FBI headquarters authority.[429] As the Reagan administration swept into office, the NUMEC investigation quietly ended. What happened?

Data provided by the Ammonator's loading dock observations and Desmond's description of "stove pipes" in equipment may provide a chilling answer. If all three "stove pipes" loaded into equipment in early 1965 were full of highly enriched U-235, they would have easily accommodated some of NUMEC's known MUF from the WANL contract.

Volume of a 3 x 8 inch NUMEC HEU "can"	**56.57**	**Cubic Inches**
Stovepipes shipped within equipment to Israel spring 1965	3	Stovepipes
Cans shipped to Israel spring 1965	9	Cans
Total can volume	509.1	Cubic inches
Weight of 1 cubic inch of U-235	.6828	Lbs
Potential Spring 1965 diversion weight	347.6	Lbs
Weight of 1 cubic inch of U-235	309.7	Grams
Potential Spring 1965 diversion weight	157.7	Kilos

Figure 17 Estimated 1965 HEU diversion to Israel.[430]

NUMEC did not manufacture Cobalt-60. The metal is used for medical radiotherapy and as a gamma radiation source for food irradiation and other industrial uses. The synthetic isotope of Cobalt is created by placing slugs of 99.9% pure cobalt in capsules into a nuclear power reactor. After 18–24 months of neutron absorption a small percentage of the atoms in the cobalt slug are converted into Cobalt-60 atoms. The half-life[cxxii] of Cobalt-60 is only 5.27 years, so legitimate industrial users need to replace their sources every 20 years. [431]

Since NUMEC did not have a reactor to create Cobalt-60, its only value added as a reseller of AEC licensed material was the fabrication of irradiators, or consulting on the fabrication of overseas commercial irradiation facilities. Today Cobalt-60 is widely used in industrial-park type facilities for commercial purposes. None of the legitimate uses of the material would have warranted threatening an employee at gunpoint, when a simple explanation would have eliminated all doubts. The added cost of rushing Cobalt-60 shipments to freighters on standby in New York piers or special El Al flights also doesn't make any sense. NUMEC's brief entry into the irradiator business created a cover for many secure shipments of tamper-proof containers to Israel. Unlike Haganah smugglers embarrassed when boxes marked as innocuous shipments but really containing high explosives broke and spilled onto New York piers, NUMEC's dispatches couldn't spill.

Many declassified FBI files about high-profile investigations released to the public have better documented "closure" than NUMEC's. At times the Justice Department orders the FBI director to shut down by providing reasons why investigation is no longer warranted, citing expired statutes of limitations, or forecasting the impossibility of a successful prosecution. The FBI's NUMEC case files provide no documented Justice Department closure memos. In fact, they seem to indicate that just when investigators finally broke the case with credible eyewitness accounts of diversion at the time of greatest HEU loss, the investigation was quietly shut down in handover between administrations. Freedom of Information Act requests for documents better justifying the closure of NUMEC filed with 20 separate units of the Justice Department have yielded no release of solid justifications for closing down DIVERT.

However some conclusions are now obvious despite the truncated investigation record. Was there an "attempt by anyone in the executive branch to prevent or impede investigation into this alleged diversion?" There clearly was. LBJ certainly impeded the investigation when he ordered his CIA director not to publish or circulate Carl Duckett's analysis about the Israeli nuclear program and illicit source of HEU. Richard Nixon impeded the investigation when he promised Golda Meier the U.S. would be "standing down" opposition to Israel's clandestine nuclear activities — many of which included Israeli covert operations in the U.S. Since acquiescence to diversion extended to the very top of the executive branch, what was the point in further FBI investigations? Was Jimmy Carter capable of prosecuting Zalman Shapiro and his collaborators under the Atomic Energy Act where LBJ, Nixon and Ford could not? Was DIVERT simply a final attempt to maintain appearances that rule of law would be upheld and

---

[cxxii] The amount of time it takes for half of the material to decay.

justice pursued even when the criminal activities were perpetrated by elite operatives at the margins of established lobbies?

The NUMEC investigation, though not its toxic fallout, ended with a whimper in the Carter Administration. No documents pertaining to Shapiro or NUMEC are to be found in the Ronald Reagan archives, indicating that Carter, unlike Gerald Ford, failed to successfully hand-off the affair. The Reagan Administration, which heavily courted the Israel lobby with huge military and overseas covert operations spending that benefitted Israel in Central America and later Israeli-brokered U.S. weapons sales to Iran during the Iran-Contra scandal, soon faced its own espionage and covert operation challenges. Al Schwimmer, a convicted felon for his role as a Haganah smuggler, was intimately involved in Iran-Contra, a crisis that could that could have taken down the administration. In 1984 Israeli Minister of Economics Dan Halpern passed stolen classified business secrets and official U.S. trade preference negotiating documents to the American Israel Public Affairs Committee (AIPAC) to lobby and conduct public relations.[cxxiii] Rafael Eitan, who visited NUMEC in 1968, began tasking Navy analyst Jonathan Pollard to loot America's most precious intelligence until he was arrested for spying in 1985. By failing to adequately conclude the NUMEC investigation, the Reagan administration signaled it would be an easy target for Israeli covert operatives and their parastatal organizations in the U.S.

Despite the lack of warranted law enforcement, the NUMEC affair is not over. Shapiro demanded the federal government formally restore his reputation. In calling such attention to himself, Shapiro inadvertently raised public awareness about the huge toxic legacy he and NUMEC quietly had transferred to U.S. taxpayers, renewing public interest in NUMEC's true origin and purpose.

[cxxiii] See the author's book "Spy Trade: How Israel's Lobby Undermines America's Economy"

14. Shapiro rehabilitation derailed by year 2001 loss estimates

Since 1948 every major Haganah smuggler convicted of a felony in the U.S. has received a presidential pardon. Most were never even charged because of the prosecutorial immunity network and its visible leaders, including Nahum Bernstein and Abraham Feinberg. Others received indirect support from the Israeli government—and the Israel lobbying network—to quash the possibility of prosecution even before indictments were ever handed down.

Zalman Shapiro has never been charged of a crime even though many people inside and outside of the U.S. government are convinced that he diverted SNM to Israel. "I was a smelly dead fish" Shapiro once lamented. "Contracts were pulled away and given to others."[432] Late in life Carl Duckett seemed to disavow any direct knowledge of Shapiro's involvement in nuclear diversion. He was likely unaware of FBI interviews conducted nearly a half decade after his retirement, Duckett's statement also may be yet another sign of the absence (and perhaps impossibility) of coordination between the CIA and FBI.

"With all the grief I've caused," Duckett told investigative reporter Seymour Hersh "I know of nothing at all to indicate that Shapiro was guilty. There's circumstantial information, but I have never attempted to make a judgment on this. At no point did I have any vested interest in this whole process. It was a matter of trying to be sure when you had information that you passed it along. Ultimately," Duckett said, "you have no control over the information. I never met Shapiro and at no point was I interested in peddling the story." Although Shapiro and his supporters would claim this statement absolved Shapiro, Duckett never recanted CIA assessments that NUMEC as a source for diversion.

Shapiro and his supporters have also periodically made assertions that all of the MUF at the heart of the diversion investigation has now been accounted for. However, these "the lost material has been found" assertions are easily debunked by the latest Energy Department data. NUMEC continues to reveal itself in subtle ways. If it were really were a front company modeled on the Haganah experience, the Israeli government and its lobby would be quietly operating in the background to obtain the equivalent of a presidential pardon for Shapiro. This is precisely what is happening.

In 2011, Shapiro received his 16th patent for growing diamond crystals from "seed diamonds" at a lower temperature and atmospheric pressure than other artificial processes. His previous patents included a related method for growing diamonds (2009) a thermoelectric generator with nuclear fuel rods as the primary heat source (granted 1974), a removable thermoelectric module for such generators (granted 1976) and a process for chemically creating a "uniformly porous body" (1975). [433] In 2009 Shapiro claimed his diamond process invention would lead to more artificial diamonds being produced for jewelry and industrial uses in the United States, "slashing the annual $40 billion import bill for cut and uncut diamonds."[434]

However by filing such diamond process patents, while not using or licensing them, Shapiro is more likely trying to help sustain Israel's lead in natural diamond exports to the U.S. A defensive patent strategy involves filing patents in order to exclude competitors from exploiting a new market. Those who want to license the technology might find themselves prevented from doing so if Shapiro's intention is really to protect Israel's position as the dominant U.S. market supplier. Or Shapiro could be "trolling" by filing patents with no intention of manufacturing or using the patented invention, but positioning himself to be able to sue others for infringement.

Whatever Shapiro's strategy, he is late in life again positioned to affect the future of an industrial process of vast importance to Israel. According to Shapiro, the process would allow the "large-scale, low-cost synthesis of large gem-quality diamonds."[435] If his forecast market potential for artificial diamonds is true, the impact for Israel would be immense. Under 1985 trade preferences granted to Israel, tariff and quota free access to the U.S. market massively increased the Israel exportation of pearls, precious stones, and metals. The category grew 13 percent annually between 1989 and 2007. Over the same period, Israel's revenues for the category grew from $1.5 billion to $9.8 billion per year. Israel supplied half of total U.S. import demand ($19 billion) for such precious objects in 2008. The Israel Diamond Controller's office of the Ministry of Industry, Trade, and Labor reported that total 2008 diamond exports reached $6.2 billion. LLD Diamonds Ltd., owned by Israeli American Lev Leviev, topped the list of exporters at $417 million.[436] But from a professional standpoint, Shapiro's latest invention was most important for—with Israel's hidden help—the attempted restoration of his tarnished reputation.

On July 29, 2009 Deborah Shapiro nominated Zalman Shapiro for a National Medal of Technology and Innovation (NMTI). The medal is the highest honor given to America's leading innovators and personally awarded by the President of the United States. But Shapiro had to overcome a high hurdle in the process. All nominees selected as finalists are subjected to FBI security checks and all information collected through the security check is considered in the final selection of process.[437]

Shapiro's submission of his contributions to America in his NMTI application was all-encompassing. Winning the medal would serve as de facto exoneration after decades of suspicions, a mountain of FBI, CIA and intelligence agency documents, and tell-all accounts from former U.S. and Israeli spies.

The filing exalted "his continuous, diverse, and profound contributions to the defense and welfare of the United States, including his pioneering innovations in nuclear-related materials processing and equipment development. For his seminal contributions to our nuclear-powered navy and commercial nuclear power industry, long-lived cardiac pacemakers, and prospective ground-breaking applications of synthesized diamonds." Deborah argued that Zalman's key development work on the USS Nautilus alone justified recognition, gushing "For just these innovative contributions to the success of the Nautilus and paving the way to our nuclear Navy and much-enhanced defense capabilities, Dr. Zalman Shapiro is worthy of this honor."

At the end of a lengthy list of accomplishments, Deborah Shapiro wrote that awarding the NMTI to Shapiro would help realize President Obama's own vision for America. "Dr. Zalman Shapiro is the very embodiment of what President Obama has been advocating to get the U.S. back on its feet, proceed to regain economic leadership, and to remain on the forefront of global technology. Dr. Shapiro is truly a compelling example of what any American can achieve with education, innovation, hard work and perseverance, and at any age. Since the National Medal of Technology and Innovation is awarded by the President, what could be more fitting than President Obama recognizing and thus highlighting my father's lifetime of enduring and significant contributions to the welfare of the United States and in so doing advance his presidential imperatives."[438]

As required by the application process, Shapiro also assembled an impressive array of recommendation letters. Chairman Emeritus of the Defense Nuclear Facilities Safety Committee John T. Conway wrote not only on his own behalf, but on that of the deceased Hyman Rickover. "During my career I interfaced with most of the early pioneers in the field including Drs. Oppenheimer, Seaborg, McMillan and Admiral Rickover. If Admiral Rickover were alive today, I am certain he would also be among those recommending Zalman Shapiro for the award." [439]

Senator Kirsten E. Gillibrand tied Shapiro's accomplishments to her state's electrical power generation. "New York State, as of January 2009, ranked fourth in the country for next generation of megawatt hours provided by nuclear power plants. Dr, Shapiro's breakthrough achievement in mass-producing high quality nuclear fuel paved the way for our current multi-billion dollar nuclear power industry."[440] Consultant James Roddy eloquently waxed on about Shapiro's personal mannerisms, perhaps without realizing they were also a testament to Shapiro's powers of evasion during FBI interrogations and congressional inquiries and highlighted his cool ability to lead complex operations. "I became acquainted with Zalman Shapiro approximately 25 years ago. We served together on a committee to raise funds for environmental projects in Israel. Several things about Zalman impressed me. He makes everyone with whom he has contact feel that they are an important person for him to know. This is true whether he is speaking to a restaurant waiter or a corporate CEO. His inquisitive mind, attention to detail, gentle manner, and zest for whatever project happens to be before him create a unique and irresistible persona. He can control a conversation while making everyone in his presence feel that they are 'in charge.' He has the power to motivate by deed and through logic. Simply put-he is brilliant, modest, generous, kind, and one of the most pleasant and pleasing persons I have ever known."[441]

Former NUMEC Secretary and Treasurer Oscar S. Gray penned a lengthy recommendation from his perch as the Jacob A. France Professor Emeritus of Torts at the University of Maryland School of Law. Gray argued that Shapiro greatly advanced U.S. national security. "Dr. Shapiro's technological contributions have advanced the public interest in a number of important ways. First, as to our national security interests, the development of the nuclear-powered Navy, for which his development of production processes for ultrapure zirconium and, later, for hafnium crystal bar, was a key element, was one of the great success stories of the twentieth century."

But given the decades of documented national security concerns raised by his activities — necessitating costly investigations — Shapiro needed a much more influential legal advocate than Oscar Gray. It is unsurprising that K Street lobbying giant Arnold & Porter LLP again stepped in to represent Zalman Shapiro in his bid for presidential redemption. According to the U.S. Department of Justice, the firm is the longest-running active registered foreign agent of the Israeli government in the United States, claiming ongoing representation since June of 1964. The firm claimed to be working for free "on behalf of our longtime pro bono client and friend, Dr. Zalman M. Shapiro."[442] In reality, there were plenty of Israeli government slush funds available to Arnold & Porter to restore Shapiro's — and by Israel's — reputations. In 1978 the U.S. Department of Energy warned former AEC Commissioner Glenn Seaborg that Zalman Shapiro had contracted Arnold & Porter and that they might try to contact him. This occurred during the same visit when they told Seaborg "some enriched Uranium-235 which can be identified as coming from the Portsmouth, Ohio plant has been picked up in Israel."[443]

Arnold & Porter represented several Israeli government officials in U.S. courts by arguing that sovereign immunity mandates provide blanket protection from legal liability for their actions. In 2007 the firm won dismissal of war crimes and crimes against humanity claims brought by Palestinians against former General Security Service head Avraham Dichter. In 2005 the firm won dismissal of similar claims against Israeli Prime Minister Ariel Sharon and other senior officials, Israeli military forces and an intelligence agency. In 2006 Arnold & Porter also won dismissal of similar claims focusing on a single Israeli official's actions that resulted in civilian casualties in 1996. In 2008 Israel's Treasury paid Arnold & Porter $483,401 to defend such actions.[444] In the year 2010 the firm signed a renewable contract with Israel for a $10,000 per month retainer for legal and advisory services and "special projects" with $8,000 in allowed travel expenses. [445] Arnold & Porter reported $1.2 million in fees from the government of Israel for the year 2010.[446]

On August 27, 2009 Senator Arlen Specter formally requested that Rebecca Schmidt of the Nuclear Regulatory Commission's Office of Congressional Affairs intervene on Shapiro's behalf. Specter asked the NRC to "issue a formal public statement confirming that he [constituent Zalman Shapiro] was not involved in any activities related to the diversion of uranium to Israel." Specter forwarded a five page report compiled by Hadrian R. Katz of Arnold & Porter documenting why NRC clearance of Shapiro was so important to his legacy. But Arnold & Porter's submission was an embarrassing panoply of errors, misrepresentations and selective retrieval. "In 1965, the Atomic Energy Commission sent a team of nuclear material management personnel to NUMEC to conduct an audit and determine, if possible, the reason for the processing losses, a cumulative total of approximately 100 Kg (220 Ibs) of enriched uranium. The investigators concluded that there was no indication of any diversion, and that a diversion would have been as a practical matter impossible." A careful review of history reveals the AEC never claimed diversion was impossible. While the AEC did settle on emphasizing "processing losses" as its public statements and institutional position before the JCAE, the AEC documentary record and St. Valentine's Day meeting accurately reflects internal awareness that diversion could have occurred, since the entire SNM accountability regime was at best an honor system in desperate need of a makeover. When that makeover finally occurred, it was largely because of NUMEC.

Arnold & Porter tried to enhance Shapiro's credibility claiming "In the course of our representation of Dr. Shapiro in the 1970s, we spoke with every significant individual involved in these investigations personally, and all of them repeated their conclusion that there was no diversion." [447] The reason for selectively focusing on the 1970s is obvious. Whatever Arnold & Porter discovered in the 1970s, the most compelling testimony on record was not made until the early 1980s when the FBI finally had its "smoking gun" eyewitness of Shapiro supervising unknown individuals covertly loading of SNM containers into export equipment. The Arnold & Porter version of history is also discredited by the revealed identities and nuclear weapons specialization of Shapiro's guest visitors to NUMEC and his home during critical periods of Israel's nuclear program. But Katz continued undaunted "following the AEC investigation, the FBI and the Joint Committee on Atomic Energy sent in their own teams to investigate. Various possibilities were investigated, but no evidence indicating diversion was ever found. One CIA analyst concluded that Dr. Shapiro had diverted uranium to Israel, and that view was picked up by journalists. One journalist would rely on another's misinformation, and the diversion suspicions were treated as fact. The distortions snowballed; books and articles magnified and embellished damaging falsehoods. The more these maligning assertions were repeated in print and online media, the greater the perception of credibility." In reality Carl Duckett, the "one CIA analyst" was far from alone at the CIA. CIA Tel Aviv station chief John Hadden and CIA director Richard Helms were both convinced that a diversion from NUMEC had taken place. Shapiro's known contacts with Avraham Hermoni, Rafael Eitan and Efraim Biegun weren't fictionalized embellishments, and have never been adequately explained by Shapiro. Shapiro's evasiveness and seeming inability to explain his many curious movements and clandestine meetings all point to the truth. That Portsmouth signature U-235 used by NUMEC was picked up outside Dimona is also an established fact. Arnold & Porter's brash demand that the NRC clear its client was oddly reminiscent of Shapiro's earlier demand for a top security clearance to work on advanced weapons designs.

This time, however, Arnold & Porter did not seek the complete formal exoneration of NUMEC, but rather only Shapiro by arguing "Dr. Shapiro has never had an opportunity to obtain a formal statement from any government agency clearing him of the false accusations made long ago. We respectfully suggest that the time has come for the NRC once and for all to confirm that Dr. Shapiro committed no diversion. An attack on Dr. Shapiro is necessarily an attack on the Atomic Energy Commission as well, and the NRC would itself benefit from putting the stories of diversion to final rest. The NRC's unequivocal statement that Dr. Shapiro did not divert nuclear material to Israel, that the material has been accounted for, and that he is, and has always been, a loyal citizen of the United States who has contributed significantly to its defense should be conveyed to the FBI, with a recommendation that this statement be given a prominent position in the files on Dr. Shapiro."[448]

While Glenn Seaborg probably would have agreed that an attack on Shapiro was an attack on the AEC, the Nuclear Regulatory Commission to its credit did not. In a droll November 2, 2009 letter to Specter, the NRC's Executive Director for Operations R. W. Borchardt refused to exonerate Shapiro, or accept the inaccurate assertion that all NUMEC MUF had been properly accounted for. He wrote "The Atomic Energy Commission (AEC) and other Federal agencies investigated whether Dr. Shapiro played a role in the possible diversion of nuclear material. The AEC concluded that it had no evidence that diversion had occurred. However, the AEC also determined that it did not have sufficient evidence to conclude, unequivocally, that uranium had not been diverted. In the time period that these statements were made, the conclusions reflected the minimal nature of the material control and accounting features that were in use to detect the loss or diversion of special nuclear material. Your request is based on information that during the decommissioning of the facility, an amount of uranium equal to the amount alleged to have been diverted had been recovered at the facility. NRC staff has reviewed agency documents related to the 'material unaccounted for' (MUF) discovered at the site and investigated by the AEC in 1965, including those pertaining to additional inspections and MUF evaluations during subsequent operations, and the decommissioning activities for the facility. Accordingly, after a thorough review of its records, NRC found no documents that provided specific evidence that the diversion of nuclear materials occurred. However, consistent with previous Commission statements, NRC does not have information that would allow it to unequivocally conclude that nuclear material was not diverted from the site, nor that all previously unaccounted for material was accounted for during the decommissioning of the site."[449]

Shapiro did not win the 2009 NMTI announced in 2010 or the 2010 award given in 2011 for which he was automatically made eligible. Shoring up the NRC's private ruling on Shapiro with a public analysis, former NRC commissioner Victor Gilinsky and staffer Roger Mattson penned the best documented and most damning scholarly article ever written about NUMEC in the April/May 2010 issue of the *Bulletin of Atomic Scientists*. "Revisiting the NUMEC Affair" was brutally frank, debunking and labeling Seymour Hersh and Zalman Shapiro assertions that all MUF had been recovered as "canards." Their final prescription for resolving all outstanding questions on NUMEC unsurprisingly did not involve pressuring the NRC to issue retractions. "The bottom line of the article" coauthor Dr. Roger Mattson later told a Federation of American Scientists blogger "is that it is time to end FBI and CIA secrecy on the now 40 plus year old Apollo/NUMEC affair."[450]

According to the two experts "A frequently encountered canard espoused most notably by Hersh in his book *The Samson Option* is that the missing HEU was recovered when the Apollo pant was taken apart. Beginning in 1978 B&W did carefully dismantle Apollo, recovering HEU containing 95 KG of uranium 235 from equipment and structures, and estimating that another 31 KG was left unrecovered in the concrete floor and walls, for a sum of 126 kilograms. But the cumulative 'material unaccounted for'—the unexplained missing amount—for the entire 1957-79 period of HEU operations at Apollo was 163 kilograms. That leaves 337 kilograms as the cumulative amount of HEU still unaccounted for— more than three times the amount unaccounted for in the 1965 inventory. In other words, the fact that about 100 kilograms of uranium 235 in the form of HEU was found during post-1978 decommissioning does not eliminate the possibility that the 100 kilograms of HEU that the AEC could not account for in Apollo's 1965 inventory, or the larger amount that went missing during the 1966-68 period, was diverted."[451]

Timeframe	1957-1968	1969-1978
Total NUMEC HEU loss (Kilograms)	269	76
MUF as a percentage of total NUMEC throughput	2.0 percent or more	0.2 percent or less

Figure 18 Year 2001 Energy Department NUMEC MUF estimates.[452]

According to the 2001 Energy Department report the final tally of total cumulative NUMEC loss from 1957 to 1968 was 269 kilograms of U-235. While this accounts for the 100 kilograms of material discovered in the inventory of 1968, the data suggest that heavy losses occurred between 1966 and 1968. The year 1968 was when Rafael Eitan and his team were hosted by NUMEC while working undercover as scientists. Only after Shapiro and his management team departed NUMEC did MUF as a percentage of total throughput return to industry standard levels. [453]

Shapiro's bid for NRC exoneration and a presidential award have only made NUMEC look even more like a typical Israeli smuggling front. Did Shapiro insist that lawyers on the Israeli government payroll work to clear him in the 1970s and 2009? It certainly appears that way. But Shapiro's quest to force a sitting U.S. president to exonerate unprosecuted crimes committed in the name of Israel's security are not the final episode of the NUMEC saga. Newly declassified evidence of Shapiro's criminal negligence—affecting the environment rather than the world nuclear balance of power—could finally fully expose Israel's elite nuclear materials smuggler.

15. NUMEC's toxic legacy

U.S. acceptance of Israel's policy of total secrecy over its nuclear weapons program has undermined public debate and governance. In the United States, enterprising reporters have continually tried to pull U.S. presidents and high officials out of Israel's "ambiguity" policy swamp by publicly asking pointed questions about the Israeli nuclear arsenal. The last significant reporter to do so was veteran White House correspondent Helen Thomas. When she asked the new President if he knew of any countries in the Middle East that had nuclear weapons, Obama said that he did not want to "speculate" about such an issue, but that preventing a regional nuclear arms race, as well as reducing nuclear arsenals internationally, was an important goal.

Since the 1969 Nixon-Meir summit, U.S. presidents have remained under constant pressure from the Israel lobby and its campaign finance network. They have refused to engage in any bona fide public discourse about the developmental history and implications of Israel's nuclear arsenal. One convenient reason for the stale policy is that a truly informed discussion would reveal the breakdown of U.S. governance that occurred as a result of NUMEC and its support network. Open discussion would confirm to Americans the disturbing reality that powerful parastatal interest groups have so much unwarranted influence in America that they can steal even the most precious weapons materials and secrets with utter impunity. It can deliver them to a foreign state, and then demand cover-ups, secrecy, and, when the time is right, pardons and normalization. Since that influence is so undemocratic, unwarranted and harmful, it by nature is maintained only through secrecy and cover-ups. Today however, Americans can observe a spectacle even people living beneath the shadow of Israel's nuclear arsenal cannot—a cleanup of Israel's nuclear waste. Unfortunately, they are also paying for it in many different ways.

Perhaps the most disturbing file, declassified and released late in 2011, reveals how Zalman Shapiro's negligence polluted Apollo. A wire tap conducted on Shapiro's phone picked up news of a major radioactive spill on the NUMEC property on May 5, 1969. By then NUMEC had already been sold to Atlantic Richfield. Shapiro and another NUMEC employee, likely Frederick Forscher (or possibly David Lowenthal, the subject of the FOIA filing), were unperturbed about the spill, quickly changing the subject to new acquisition targets. "9:18 PM....reported on a spillage at the plant. They have the area roped off and it will take some pick and shovel work to dig up the contaminated areas...they are getting 100,000 counts. CENSORED said, 'oh god.' They are dampening it down to avoid dust and will cover it if it looks like it may rain. CENSORED asked if there is anything on AAI?...he heard that American Instrument Company up for sale. He said he heard this from the former sales manager."[454]

For perspective, Japan which used 6,000 counts per minute as its standard radiation level necessitating mandatory decontamination raised it to 100,000 only after the Fukishima disaster.[455] In a phone call 2 minutes later it was confirmed that "it's not only a bad spill but 'actually they are operating outside compliance.' They had the drums all together. They have about 200 drums and estimate that about 6 a day will corrode through. The trouble lay with a fluoride which was put in to help the decay and this was not checked...they are also about $230,000 over on their construction costs for the scrap plant. Z [Zalman Shapiro] said if they could get other people, there would be a lot of firing."

Later in the call it was confirmed that material "ate through the stainless steel drums. It got out into the field and it is a real problem...they have a very serious problem and are running up a bill of $100,000 on use charges..."[456]

Although Shapiro was still a year away from leaving NUMEC, the ecological disaster of a plant which only performed one function particularly well— materials diversion—is evident from the transcript. If Atlantic Richfield and subsequent corporate parents of NUMEC had any inkling of what the FBI then knew, they would have avoided NUMEC like the toxic waste it so freely spewed into its environs. Yet, in another sense, the spill is analogous of the jurisdictional turf wars that undermined the entire diversion investigation.

According to a former congressional investigator who reviewed 1968 Senate and House intelligence committee files on Shapiro, FBI director Hoover fought bitterly with CIA Director Richard Helms about NUMEC for a year. "The CIA was saying to Hoover 'You're responsible for counterintelligence in America. Investigate Shapiro, and if he's a spy, catch him.' Hoover's answer was, 'We don't really know if anything's been taken. Go to Israel and get inside Dimona, and if you find it [evidence of the Shapiro uranium], let us know.' It was kind of a game...The memos were hysterical—they went back and forth."[457] Carter's CIA Director Stansfield Turner seemed wholly uninterested in compromising the CIA's intelligence assets in Israel in order to assist prosecutions for crimes in the U.S. during the renewed investigations. This could have jeopardized sources and methods. The FBI, in the midst of an exciting counter-espionage investigation, failed to alert environmental regulators or affected citizens of Apollo or Parks Township that it had wiretap evidence of a criminally negligent spill. If it had relayed the wiretap to the AEC, the question of ever issuing another security clearance to Shapiro would have been moot. The AEC could have forever banished Shapiro from the industry over criminal negligence, and his preference for firing workers after they cleaned up avoidable toxic spills. However cover-ups, infighting, and a total lack of concern for the environment are integral to NUMEC's legacy ever since the first toxic fire of the early 1960s vented out of Apollo's smokestacks.

B&W, like Zalman Shapiro and David Lowenthal, would ultimately extricate itself from liability for NUMEC. At first it received matching taxpayer subsidies to decommission NUMEC. In late 1990, Congress passed a military appropriations bill that mysteriously included $30 million for NUMEC remediation to be matched by B&W funds. NUMEC's Apollo HEU processing area, located on the second floor and East bay of the main building, underwent complete "remediation" between 1978-1991 in which all processing equipment was dismantled and sent for decontamination to Chem-Nuclear. NUMEC's low enriched facility was remediated between 1983 and 1984. The official Apollo

Decommissioning Project began in earnest in June of 1992 when the rest of NUMEC's buildings, including offices, laboratories, elevated mezzanines and visitor areas were decommissioned and some radioactive materials were recovered.[458] The demolition concluded in 1995.

Figure 19 Waste removal during NUMEC decommissioning.

The razing was costly and handled with care. When NUMECs facility for packing shipments such as the irradiators—known as the "box shop"—was demolished in 1990, all the masonry was crushed, pelletized and then shrink-wrapped in polyethylene before being sent to a processing plant. Shapiro's administrative office just east of the plant wasn't spared. Though mainly used for management and engineering, it also contained a laboratory to analyze both non-radioactive and radioactive products. The building's floor boards and drainage lines contained elevated levels of uranium, so the building underwent a controlled demolition; the lot was graded, re-vegetated and sold to an adjacent business in 1996.[459]

In 2009 Apollo attempted to seize by eminent domain the riverfront parcels where NUMEC first opened for business. In the end Babcock and Wilcox agreed to simply hand the land over to the borough. Although it has been cleared of chemical and nuclear waste, no structures can be built and no digging can take placed on the land. Apollo plans to build an access road through the property using $340,000 in federal grant money in a bid to redevelop this severely restricted center of Apollo. According to Borough Manager Lori Weig-Tamasy "we want to prove a point: Apollo Borough may have fallen down and seen better times, but Apollo Borough is capable of coming back."[460]

Figure 20 Razed NUMEC site at Apollo, PA.

NUMEC's toxic waste dump, located in Parks Township, failed to attract federal cleanup attention until much later than the Apollo site. NUMEC chose a particularly unsuitable geography for its waste dump. In the early 1900s, companies exploiting the Upper Freeport coal seam mined a warren of subsurface voids and shafts 60 to 100 feet below NUMEC's buried waste. [461] The Department of Energy performed "historical research" into NUMEC and its "Shallow Land Distribution Area" where radioactive toxic waste was stored. On May 25, 2000 it determined that the former NUMEC site was eligible for a federally-funded cleanup under its "Formally Utilized Sites Remedial Action Program" or FUSRAP. FUSRAP is a project to manage and cleanup contamination at thirteen early USACE activity areas.

The Parks Township Shallow Land Disposal Area site, located in Armstrong County, encompasses 44 acres of private land now owned by B&W spinoff BWX Technologies, which was licensed by the Nuclear Regulatory Commission to maintain the site and ensure the protection of workers and the public. The SLDA site's ten known trenches contain contaminated soil and other waste materials. The estimated quantity of contaminated waste material from the trenches is approximately 24,300 cubic yards. This equates to the area of a football field twelve feet deep. Any additional waste material encountered outside of the trenches is also slated for removal. uranium, thorium, americium and plutonium contaminated waste has already been identified. Uranium and thorium contaminated waste material consisting of process residue, equipment; scrap and trash from the nearby Apollo nuclear fuel fabrication facility were disposed of within the trenches at the SLDA site only between 1961 and 1970. The uranium in the trenches is present at various levels of enrichment ranging from highly depleted to highly enriched. Americium and plutonium, the presence of which is attributed to storage of equipment used in the adjoining Parks Facility has been detected in surface soils adjacent to a single trench.[462]

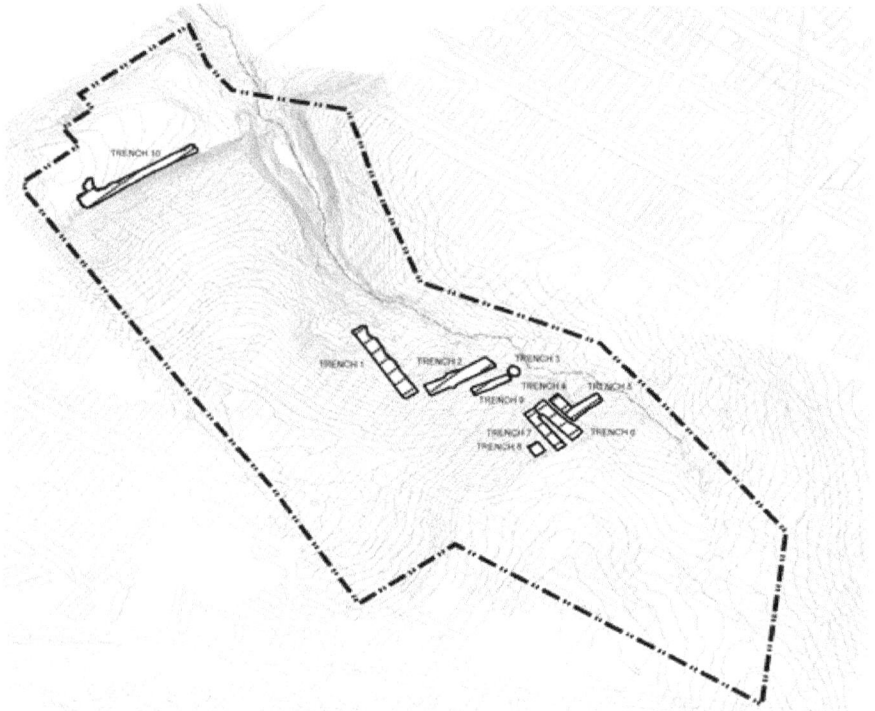

Figure 21 NUMEC Shallow Land Disposal Area topography.

On September 6, 2007 U.S. Army Brigadier General Bruce A. Berwick approved the U.S. Army Corps of Engineers decision to remedy toxic waste at the NUMEC waste dump through a formal and open "Record of Decision." The Corps of Engineers received its authority for such cleanups under 2002 Defense Appropriations. The USACE initially proposed five options. The first three ranged from doing nothing, containing the area with fencing and monitoring it for 1,000 years, to full containment. The ACE found that NUMEC's waste trenches "contain significant concentrations of radioactive contaminants, and these materials could pose a potential risk to human health in the future." Options 4 and 5 involved either decontaminating materials onsite, or trucking them offsite for disposal. In its 2007 decision, USACE chose the most costly option "Excavation, Treatment and Off-site Disposal." But even this costly option failed to cover all contamination, including polluted groundwater or isolated non-radiological chemical contamination. B&W identified 12 private wells within 2 kilometers of the SLDA during a 1995 survey. In 2007 the estimated cost of the remedy was $44.5 million.[463] This had spiraled to $170 million by 2010. In 2011 the USACE had to stop its cleanup after unexpected contractor difficulties in moving old toxic waste barrels.

If there is any constant in the story of NUMEC site management it is how effectively NUMEC functioned as a "free rider"—transferring costs to other government contractors and insurers which were gradually passed directly to the U.S. government and taxpayers. NUMEC's founders and corporate parent, Apollo Industries, managed to slough off all liability for future cleanup costs and criminal penalties for diversion through timely, liability-free divestment. The USACE study noted that Atlantic Richfield acquired NUMEC's stock in 1967, but had wisely discontinued disposal into the SLDA by 1970. ARCO and subsequent owners nevertheless assumed all legal liability for a decade of diversion-related "sloppy" administration by Zalman Shapiro. It remains to be seen whether the previously "government-classified" spill in 1969 detected by the FBI due to improper waste storage practices ever results in direct civil or criminal action against Shapiro.[cxxiv]

Although no state or federal authority has officially designated NUMEC's Armstrong County neighborhood a "cancer cluster" expert analysis presented in a series of court cases based on reviews of state health department data found rates that were above and "outside the normal range." An early lawsuit against B&W in 1998 won a $35 million judgment for cancer victims. Ultimately it was delayed when a judge found errors in the trial and declared there should be a new one.[464] However, by 2009, Babcock & Wilcox and Atlantic Richfield had paid out $80 million to 365 health claimants over the course of a 14-year lawsuit filed in 1994.[465] In 2012 additional massive lawsuits against the NUMEC's corporate parents (excluding Apollo Industries and Zalman Shapiro) appear to be in the works.

The federal government, beyond Medicaid, Medicare and Social Security benefits for NUMEC workers, is also now obligated to make payments under the Energy Employees Occupational Illness Compensation Program Act of 2000.[466] Members of a defined "Special Exposure Cohort" (SEC) can obtain compensation for one of 22 specified cancers.[467] After a petition process, all staff members of NUMEC for the period January 1, 1957 through December 31, 1983 automatically became eligible as SEC members to file claims since all worked at jobs which "should have been monitored [for radiation] since all NUMEC workers had the potential to receive exposures to onsite releases of radioactive material."[468]

[cxxiv] ARCO sold NUMEC to Babcock and Wilcox Company in 1971. BWX Technologies Inc. (BWXT) has assumed ownership of the site since 1997.

16. An Israeli operation from the beginning

When questioned by an over-eager reporter about NUMEC diversion in 1986, Zalman Shapiro testily responded. "Do you think that if there was any truth to any of this stuff that I'd be walking the streets?"[469] Only after carefully reviewing how NUMEC fits into the broader history of individuals and organizations committing U.S. crimes in the name of securing Israel can that question be properly answered. Statistically speaking, most individuals and organizations caught committing serious crimes in the name of Israel continued walking the streets — compounding rule of law and governance problems in America and its international standing. But undue government secrecy contributed as well.

That the CIA is still actively keeping its information about NUMEC out of the public domain is not difficult to prove. Reporter John Fialka's numerous FOIA inquiries to the CIA triggered no release. National Archives and Records Administration presidential libraries from LBJ to Jimmy Carter process multiple Mandatory Declassification Reviews every year from researchers seeking release of the key set of memos, reports and presidential decision about NUMEC containing CIA and other intelligence "equity." The CIA fully denies or renders released files useless through over-redaction.[cxxv]

Yet the cover-up has not kept top-level CIA analysis and covert observations from Israel completely secret. During an exhaustive BBC program comparing and contrasting the NUMEC diversion to Israel's yellow-cake theft "Plumbat Affair" CIA Tel Aviv Station Chief John Haddon expressed an almost blasé attitude toward Israel's SNM extraction from NUMEC.

John Hadden: "These gentlemen have been extraordinarily adept at removing things at long distance." (Hadden may have been using the word "gentlemen" as a jab at Shapiro's references to his frequent Israeli covert operative visitors.)[470]

BBC: "Which gentlemen are we talking about?"

Hadden: "The Israelis, and they are gentlemen. Just imagine to yourself how much easier it would be to remove a pound or two of this or that at any one time, as opposed to — which is inert material — as opposed to removing all at one blow. 150 pounds of shouting and kicking Eichmann. You see, they are pretty good at removing things. So I would have no argument with that kind of a judgment without knowing anything about it."

BBC: "You mean it would be quite consistent with Israeli practice to clandestinely go about getting any materials they needed?"

[cxxv] In April of 2010 the author tried to end-run the CIA by asking the Chief Archivist at the National Archives and Records Administration (NARA) for direct release of NUMEC files in possession of NARA under the authority granted by Section 11 of the National Archives Act of 1934. The gambit did not work, but did trigger an irate phone call from CIA FOIA officer Susan Viscuso, who refused to accept the author's offer that she become the public's champion for future NUMEC document releases.

Hadden: "Well, there were those ships out of Cherbourg, and there was that...garage full of Mirage plans. I don't think it's unusual for them to have removed things or acquired things. "[471]

Hadden's comparison of Israeli spy Rafael Eitan's snatching Nazi war criminal Adolph Eichmann from the streets of Buenos Aires for prosecution in Israel with a diversion of HEU from NUMEC was revealing in ways not broadly recognized at the time. It would be almost another decade before the American public became aware that Israeli spy Jonathan Pollard's spy master, Rafael Eitan, had also visited NUMEC during the peak period of uranium loss. Hadden was clearly implicating Eitan — from the CIA's perspective — in the NUMEC losses.

The CIA station chief may have considered instances of well-placed Jewish insiders helping out Israeli covert operators to be inevitable. Each of the other cases cited by Hadden, the Israeli theft of French Mirage jet fighter blueprints Israel wanted in order to build its own copycat Kfir fighter, or the Plumbat affair, all involved sympathetic foreign nationals. The BBC report noted with curiosity that suspects and detained individuals never seemed to suffer any consequences for their activities. Dan Ertisch (AKA, Ert, AKA, Erbel) who was a straw ship buyer for the Plumbat affair was convicted of murder when another Mossad plot went bad. Yet — as the BBC noted — he was back in Israel along with his co-conspirators within 19 months despite the murder charges.

Hadden wasn't alone in speaking very carefully about the Israelis, or in the midst of such criticism trying to positioning them as the "good guys." The other highly vocal CIA observer of the NUMEC affair was just as reticent about assigning harsh — but obvious — judgments. Carl Duckett, the biggest NUMEC leaker in the entire agency, later seemed to absolve Zalman Shapiro to Seymour Hersh for his 1991 book about the Israeli bomb "The Samson Option." "I know of nothing at all to indicate that Shapiro was guilty," claimed Duckett.[472] Though it was among his last public observations about NUMEC before he died in early 1992, Duckett never actually cleared NUMEC or Israel of diversion. What thread binds the CIA's ongoing cover-up and the CIA leakers' unwillingness to be critical of Israel? Political reality, powered by the ongoing volume of Israel lobby campaign contributions that can make or break American politicians from the presidents on down. Behind all the investigations into the ties between NUMEC and the Israeli nuclear weapons program lay political calculations about how much publicity and political harm could be unleashed by the Israel lobby.

Limiting the distribution of intelligence reports about Israeli nukes has never been of any real benefit to U.S. national security. Yet by blowing the whistle on NUMEC, the CIA could damage the relationship between the President and the Israel lobby with all of its attendant campaign contribution, media and societal power. From LBJ's vantage point, the CIA's discovery and any proposed actions were of vastly less importance than preserving access to Abraham Feinberg's network of cash contributors.

In the intervening decades, this largely opaque political reality has continued to function unabated, fortifying U.S. acquiescence toward maintaining "ambiguity" toward the Israeli arsenal. No sitting president will probably ever allow the CIA to release final secrets it holds about NUMEC. To do so would reveal how the influence of the Israel lobby has so corrupted U.S. government that even sitting presidents cannot protect the most powerful technologies unlocked by American science from immediate proliferation to Israel.

In 1995 the United States approved the sale of powerful Cray supercomputers to a network of Israeli academic institutions. Although supercomputers are used for many civilian functions, they were invented primarily to design U.S. atomic and hydrogen bombs. Supercomputers can run simulations of the implosive shock wave that detonates a nuclear warhead, as well as modeling forces in effect during a ballistic missile's launch, flight and impact.

The university network that sought to acquire the supercomputers included Technion and the Weizmann Institute.[473] A 1987 Pentagon-sponsored study[cxxvi] conducted by the Institute for Defense Analyses found Technion University participated in designing Israeli nuclear missile re-entry vehicles.[474] U.S. officials also claimed that Technion scientists worked at the Dimona complex which led to a 1989 denial of a supercomputer export license. The Pentagon study discovered that Weizmann Institute scientists developed a cutting-edge high energy physics and hydrodynamics program "needed for nuclear bomb design" as well as the most advanced methods for enriching uranium to weapons-grade through the use of lasers. A U.S. official worried that by networking supercomputers, Israel would be able to boost computing power enough to design reduced-size thermonuclear warheads that could fit on Israel's long range missiles. A survey of authorized U.S. supercomputer sales to Israel revealed that the Weizmann Institute had the second highest processing capability (3937 CTP[cxxvii]) trailing only Tel Aviv University (12,021 CTP). [475]

Supercomputer	Composite Theoretical Performance	Buyer
Cray Research	5,225.0	Tel Aviv University
Cray Research	1,325.0	Weizmann Institute
IBM	6,796.1	Tel Aviv University
IBM	1,421.0	Hebrew University
IBM	1,421.0	Bar Ilan University
IBM	1,278.1	Technion Institute
IBM	1,278.1	Weizmann Institute
Silicon Graphics	1,334.0	Weizmann Institute
Silicon Graphics	1,071.0	Bar Ilan University

Figure 22 U.S. supercomputers approved for sale to Israel thru Nov. 1994.

When questions arose in Congress about the wisdom of supplying such advanced computing power to a non-NPT signatory, the American Israel Public Affairs Committee (AIPAC) quickly stepped up charges that Saudi Arabia's national oil company had a more powerful supercomputer than any proposed for sale to Israel at that time. Although Saudi Arabia openly used its computing

---

cxxvi Although not classified, the IDA denies the report's content and even its full title to the public. The DOD has not authorized its public release.

cxxvii Composite Theoretical Performance: a performance measure of millions of theoretical operations per second.

power for energy exploration, AIPAC charged "There are no real safeguards to verify the formal Saudi commitment that the computer be used...solely for oil-related purposes."[476]

Although Israel promised the U.S. government that safeguards would ensure networked supercomputers would not be used for designing nuclear weapons, in practice the U.S. was incapable of enforcing simple and mandatory end-use safeguards on even conventional U.S. weapons sales to Israel. Results of a classified audit published in March of 1992[cxxviii] revealed that outright refusals by Israeli government officials and infighting at the U.S. State Department were enough to render such safeguards — as a practical matter — impossible.

U.S. State Department Inspector General Sherman Funk issued his *Report of Audit* about efforts to stem illegal Israeli transfers of U.S. military technology. The report found Israel "is systematically violating U.S. arms control laws." America's "Blue Lantern" system conducted in-country inspections by deploying Customs officials and other qualified U.S. embassy personnel to verify that "sensitive U.S. Munitions List items and technology are used only for authorized purposes." The audit uncovered a breakdown in U.S. inspection regimes within Israel. The State Department relied on "government to government" assurances that items were not "retransferred" or "used for unauthorized purposes." Shipments to non-government entities could only be checked if Israeli government officials granted permission.

Sherman Funk found this trust-based approach to be wholly inadequate. "After reviewing the end use procedures, we stated to post officials that relying entirely on government-to-government assurances is an inadequate verification procedure. This is especially true for a country which, according to numerous intelligence reports, is systematically violating U.S. arms control laws."

But the U.S. State Department's own Bureau of Politico-Military Affairs (PM) reined in the Blue Lantern investigators by demanding that "investigations were generally not to be conducted unless authorized." One of the Blue Lantern audits profiled in the report found U.S. telemetry equipment being put to prohibited uses. When notified, the State Department's Directorate of Defense Trade Control refused to take action and promptly closed the case. Other sensitive equipment included materials useful in chemical and biological weapons programs.[477] A careful reading of history reveals that the U.S. has never had credible conventional or nuclear weapons safeguards systems when it comes to Israel.

When CIA Tel Aviv Station Chief John Hadden called the Nuclear Materials and Equipment Corporation an "Israeli operation from the beginning" he pushed the period of concern back to the late 1950s. To evaluate Hadden's assertion, it has been necessary to reconstruct the inception of NUMEC from a spider's web curious mergers of bankrupt companies in NUMEC's parent corporation, the conglomerate called Apollo Industries, Inc. The group of people who came together as Apollo and NUMEC's directors had only one common denominator: their utter devotion to the advancement of Israel.

---

cxxviii Declassified in part and released to the author in August of 2011.

On paper, Apollo Industries sprang from the 1958 merger of three completely separate corporations, the San Toy Mining Company of Maine; American Nut and Bolt Fastener Company of Pennsylvania, and Apollo Steel Company, also of Pennsylvania which became the site of NUMEC reprocessing activities.

Some researchers, including "Gideon's Spies: The Secret History of the Mossad" author Gordon Thomas, exonerate Zalman Shapiro as Israel's "inside man" for a nuclear materials heist due to his alleged patriotism. In a chapter devoted to NUMEC based on Gordon Thomas interviews with Rafael Eitan, Israel's top spy targeting U.S. technology and military secrets, it was Eitan who nearly single-handedly swooped in to pillage the naïve and trusting Shapiro's plant. Arnold & Porter's later moves to seek U.S. government exoneration of only Zalman Shapiro is well-aligned with this evolving accountability framework.

Yet any cursory review of the evidence and chronology calls into question both Shapiro's commitment to the U.S. and the Gordon Thomas theft timeline. While Eitan may have been involved in material movements before 1968, his only documented visit to NUMEC occurred in 1968, a peak year for MUF. Unacceptable levels of MUF were already detected by the mid-1960s.

And just how patriotic was Shapiro, despite his many contributions to the nuclear submarine and other U.S. military programs? An FBI informant recalled Shapiro telling Jack Goldman, a VP of Ford Motor Company, on October 16, 1968 that he "prioritized" Israel over the U.S., at times even comparing America to Nazi Germany. According to the transcript "he had given some consideration to resettling in Israel on a permanent basis. Goldman, who left Ford to become Group Vice President of Research and Development for the Xerox Corporation, Rochester NY, also indicated he had given this matter some thought in the past. According to source, Goldman and Shapiro agreed that with the situation as it is in the United States, what with George C. Wallace running for president, and the general unrest in the country, conditions here were similar to those in Germany in the 1930's. Both also expressed the belief that if fighting became necessary, they would prefer to fight in Israel where a just cause existed...."[478] Shapiro's insistence that U.S. and Israeli interests were identical, that helping one by definition meant advancing the other, seems to be more of a coping mechanism to overcome cognitive dissonance over the crimes he was committing. Shapiro is far from the only Israel lobbyist publicly advancing such a unified purpose ideology.

So why did Shapiro remain in the U.S.? Another informant recounted the reason on November 8, 1968. "Shapiro, while discussing the possibility of living in Israel permanently, was advised...that he is of more value to Israel if he continues to reside in the United States, where Israel's problems can be more readily resolved."[479] Another Israeli government agent told him not to leave NUMEC even while chafing under Atlantic Richfield's management because of his value to Israel. Shapiro followed such guidance, logically pursuing a position with access to advanced U.S. weapons designs when he finally left NUMEC.

Shapiro's parents, Abraham and Minnie were born in Poland, but became naturalized citizens of the United States. Shapiro was trusted enough to pass a background check of eighteen neighbors and a professor at Johns Hopkins

University in 1949 to work on a "confidential research project."[480] But despite the security clearance he held, once investigation got underway the FBI never believed they could get straight information from Shapiro. One special agent's June 15, 1966 opinion was that interviewing Shapiro about nuclear diversion "does not appear to be expedient at this time. Pittsburgh [FBI field office] has received no information to indicate subject would be cooperative if interviewed in this matter. Pittsburgh is continuing to develop valuable information concerning the activities of subject and associates in gathering technical and scientific data for Israeli nationalists."[481]

In transcripts Shapiro comes across as disgruntled and deeply conflicted. "He believed the American news media were very biased in reporting Israeli events."[482] He was highly dissatisfied with the state of Middle East news reportage in the United States, and maneuvered himself into a position allowing him to tip the scales to Israel's advantage in a fundamental way. In hindsight, it would have been far better for Shapiro—and the U.S.—if he had actually followed his dream by emigrating to Israel in the late 1950s. But like another Eitan associate, Jonathan Pollard—Shapiro betrayed his country of birth for another. Unlike Pollard—who is the exception that proves the rule that most crimes committed in the name of Israel are never punished—Shapiro has never paid for his crimes against the U.S. except through backlash generated entirely by his own suspicious actions and serial fabrications.

David Martin's quirky *Newsweek* article on NUMEC emphasized that "In any case, Israel, in the view of most experts, is now capable of producing plutonium bombs from its own low-grade uranium ore. 'Even if Shapiro never existed, they would have the bomb,' says one official." Yet this truncated analysis is not entirely true. Absent the parastatal international network, of which Shapiro and NUMEC were but small cogs, it is very probable that Israel would not have been able to develop nuclear weapons, and that a regional peace deal of enormous benefit to the U.S. could have been crafted long ago. Propaganda and political pressure have been as important to the Israeli nuclear program as the actual material and technology transfers.

Absent that pressure, detected by Henry Kissinger and intensely felt by Richard Nixon, the U.S. might have rejected the "ambiguity" pact with Golda Meier. Absent the Israel lobby's campaign contribution funding network boosted by foreign inflows, Israel's weapons funding coordinator Abraham Feinberg might not have been able to reach the heart of the LBJ administration. Absent the American Israel Public Affairs Committee, whose leader sprang from the Israeli Ministry of Foreign Affairs, Feinberg would have been unable to find a funding vehicle with a suitably powerful transmission belt to the news media in order to deny Israel had any capability to make weapons in the early 1960s.

Regulators, politicians and law-enforcement officials who advanced very far investigating Israeli nuclear diversion soon found themselves stepping onto difficult terrain. According to the public relations framework relentlessly promoted by Israel and its global lobby, any who question the state's means for acquiring advanced weapons is standing on an absolute plane between Israel and a potential second holocaust. This view is as extreme as it is insulting, but has become a territory few now willingly tread. NUMEC lies along a continuum of corruption in America that began in earnest long before 1948. That is the year Israel's future nuclear weapons funding coordinator Abraham Feinberg began making his first large cash infusions into the office of the U.S. president. This under-the-table presidential payola has bought many things for Israel, most at significant cost to America.

It is alarming to see how for almost half a century the establishment news media and investigative reporters, while occasionally revealing some important aspects of the NUMEC diversion, have generally suppressed key facts or avoided the most troubling questions. What impact are large, well-financed parastatal organizations and operatives dedicated to the advancement of a foreign state having on U.S. rule of law, governance and proliferation? NUMEC provides many answers. But the consensus of establishment newspapers such as the *New York Times* is identical to the Israel lobby's. Criminal activity in the name of Israel's perceived security needs always trumps U.S. rule of law and governance. So while a major *New York Times* article such as *"Nixon Papers Recall Concerns on Israel's Weapons"* devoted many words emphasizing and justify those security needs, they rarely dwell on the harmful fallout of Israel's organized criminal activity in the U.S.

There has been a verifiable explosion of Israel lobby crime on U.S. soil[cxxix] ranging from election fraud, charitable and for-profit criminal fronts, money laundering, 1938 Foreign Agent Registration Act and Logan Act violations, industrial and economic espionage, health and environmental outrages and even a few unexplained assassinations[cxxx] — the key facts around which are only slowly emerging. In the pantheon of Israel lobby crimes NUMEC is only unique because it dealt in nuclear material and is fairly well-documented by declassified primary sources. The audacity of the theft, lack of criminal or diplomatic consequences, and the unrepentant stance of Shapiro are unfortunately the norm. That the residents of Park Township and Apollo and taxpayers are denied the government information connecting the $170 million toxic waste cleanup and ongoing health crisis in any way to Israel is also quite typical.

Unfortunately for America, the policies designed to hide and fill in this yawning void in rule of law have increasingly dire consequences. U.S. presidents and major news media are so wedded to maintaining the "big lie" implicit in Israeli "nuclear ambiguity" that they can't speak honestly about it in the largely fabricated showdown with Iran over its nuclear program. By not seriously discussing Israel's vast, deployed arsenal of advanced weapons when warranted, political leaders and news media leave Americans with few informed insights into relevant regional dynamics. The coercive power of a nuclear-tipped foreign state and its ever-active parastatal organizations remains largely opaque as U.S. taxpayers dig deep to clean up the nuclear mess while shipping huge aid packages overseas to Israel. Understanding how Israeli criminal enterprises function in tandem with business, charitable and lobbying organizations engaged in or supportive of illicit activities is the first step toward their overdue

---

[cxxix] For elaboration, see the author's previous books listed among the introductory pages.
[cxxx] See the case of Alex Odeh.

dismantlement. Israel's parastatal organizations and nationalist operatives have invested vast amounts of time, effort and resources into establishing the political, media and propaganda infrastructure necessary to maintain such privileges. But that infrastructure is beginning to crack.

The advent of the Internet has spawned alternative media, peer-to-peer communications, and independent research teams developing broad new audiences. Disgruntled youth not yet saturated by ideology have begun to detect and challenge the legitimacy of Israel's illegal activities and those individuals and organizations that unquestioningly fund and promote them. Skepticism and disbelief now stand like immovable bulkheads against waves of Israeli parastatal propaganda—especially those washing around establishment U.S. media. A new age of truth and reconciliation—distant during the time NUMEC was stealing America's most precious strategic material—is fast approaching

Epilogue

The following was written by a Walter Patrick "Pat" Lang Jr., who after retiring as a U.S. Army colonel held high-ranking civilian posts as a member of the Senior Executive Service providing military intelligence analysis of the Middle East and South Asia for DOD. He also led world-wide human intelligence activities at a rank equivalent to a lieutenant general. Lang analyzed a review copy of Divert and submitted the following observations.

It seems clear that Shapiro was a committed Zionist from early in his life and someone for whom his American citizenship was a mere convenience. He is remarkably like [Jonathan] Pollard in that way. It also seems clear that it was the brainstorm of people in the Jewish Agency to create a company like this one for the specific purpose of establishing a "front" for the illegal export of refined nuclear materials in support of the weapons program. To that end they found Shapiro (or he found them) and the Israelis and he recruited other Jewish technologists, venture capitalists, and U.S. government officials—probably in the AEC and Justice Department—to make the whole thing hold together.

It is also quite obvious that this front company could not have functioned for as long as it did without continuing protection from within the U.S. government. One of the telltale signs of the company's "front" status is how thinly it was capitalized and its great difficulty in performing the tasks explicit in the contracts that it was mysteriously awarded by probable "implants" in the U.S. government. The indifference of the company to safety considerations for its workers and the townspeople is yet another indication of the fact that it was never taken seriously by its ultimate "owners" in Israel. I can say from experience that the temptation to not fully develop a "cover" entity is severe. Such development eats up available funds and there is only so much money around, even if some of it comes from domestic Jewish donors in the USA.

CIA knows very well how thoroughly penetrated the various ministries of government and political entities are in the USA. It knows that to disclose very much to the rest of the U.S. government is to ensure the destruction through disclosure overseas of its hard won HUMINT assets. I noticed that at several points in the history of this company, the AEC commissioner and others suggested to CIA that they make a maximum effort against Dimona to see if the material there had started its voyage to Israel in Pennsylvania. If I had been the DO [Director of Operations at CIA] my antennae would immediately have gone up as I sensed the possibility of a trap intended to enable a roll up in Israel. In short, CIA does not trust the rest of the government. They do not accept anyone else's investigations of personnel. They are right in doing that.

The continuing foot-dragging by elected and appointed officials in this matter is mute proof of the deep, deep, penetration of the U.S. government by the Zionist apparat. When I was in DIA, it was understood that to raise this issue was a career killer. I once was present at a luncheon in SECDEF's [Secretary of

Defense] dining room in which the Israeli charge d'affaires openly mocked the ease with which the embassy manipulated U.S. congressional elections. That is just one of many examples that I could give. The FBI's investigative findings and the way they were thwarted in all their efforts is so reminiscent of many such affairs that I don't find it odd at all. It is a long time since loyal Americans have been well-served by their federal government in the matter of Israel. When still working for the DIA, I went to the FBI a number of times with evidence of illegal activities involving politically appointed officials. They told me that they had a large number of cases of this kind ready for trial but that the Department of Justice invariably blocked prosecution.

— *Pat Lang*

List of Acronyms/Abbreviations

AEC – The Atomic Energy Commission
AIPAC – American Israel Public Affairs Committee
AZC – American Zionist Council
CIA – Central Intelligence Agency
DIA – Defense Intelligence Agency
DIVERT – FBI code name for reopened NUMEC investigation
DOD – Department of Defense
DOE – Department of Energy
DOJ – U.S. Department of Justice
ERDA – Energy Research and Development Agency
FARA – 1938 Foreign Agents Registration Act
FBI – Federal Bureau of Investigation
FO – Field Office
FOIA – Freedom of Information Act
FUSRAP – Formerly Utilized Sites Remedial Action Program
GAO – Government Accounting Office, later Accountability Office
HEU – Highly Enriched Uranium
HEX – Uranium Hexaflouride
IAEC – Israel Atomic Energy Commission
ISORAD - Isotopes and Radiation Enterprises, Ltd.
JCAE – Joint Committee on Atomic Energy[cxxxi]
KBI – Kawecki Berylco Industries
LAKAM – Israeli economic and technology espionage unit
MDR – Mandatory Declassification Review
MUF – Materials Unaccounted For
NERVA – Nuclear Engine for Rocket Vehicle Application
NMTI - National Medal for Technology and Innovation
NPT and NNPT – Nuclear Non-Proliferation Treaty
NRC – Nuclear Regulatory Commission
NSA –National Security Agency
NUMEC – The Nuclear Materials and Equipment Corporation
OGA – Other Government Agency
SAC – FBI Special Agent in Charge (of a Field Office)
SEC — Special Exposure Cohort
SLDA –"Shallow Land Disposal Area" located in Parks Township, Pennsylvania
SNM – Special Nuclear Materials
STATE – U.S. Department of State
SNPO- Space Nuclear Propulsion Office
SNPO-C - Space Nuclear Propulsion Office, Cleveland
U – Uranium
U-235 - Uranium 235
USACE – U.S. Army Corps of Engineers
WANL – Westinghouse Astronuclear Laboratory
WFO – FBI Washington Field Office

[cxxxi] Succeeded by the House Interior Committee in 1977

Photo and Image Credits

Appendices

NUMEC HEU exports reported to the AEC[483]

OFFICIAL USE ONLY

APPENDIX B

per DOE letter 6-6-85
ALL INFORMATION CONTAINED
HEREIN IS UNCLASSIFIED
DATE 6-18-85 BY

TRANSFERS TO FOREIGN ENTITIES
License No. SNM-145 – Uranium Enriched in the Isotope 235
Nuclear Materials and Equipment Corporation, Apollo, Pennsylvania
for the Period December 1, 1957 to October 31, 1965

Unit: Gram

Date Shipped	Destination	Material Description	Uranium	Percent Isotope	U-235
8/7/58	U.S. Exhibit, Switzerland	UO₂	7,521	19.94	1,500
10/30/56	France	UO₂	4,407	1.50	66
12/30/58	"	UO₂	487,969	1.50	7,359
12/26/58	"	UO₂	489,855	1.50	7,387
12/12/58	"	UO₂	487,422	1.51	7,350
12/17/58	"	UO₂	488,567	1.51	7,368
12/19/58	"	UO₂	486,600	1.51	7,336
1/30/59	"	UO₂	321,461	1.50	4,848
1/9/59	"	UO₂	485,360	1.50	7,319
1/14/59	"	UO₂	324,227	1.50	4,853
2/25/59	"	UO₂	74,330	1.50	1,171
2/25/59	"	UO₂	43,923	3.49	1,533
4/22/59	"	UO₂	170,119	3.49	5,935
5/14/59	Canada	UO₂ Powder	39,989*	6.99	2,794*
5/29/59	France	UO₂	70,241	3.49	2,651
5/4/59	"	UO₂	200,451	3.49	6,994
7/3/59	Italy	Uranyl Sulfate	7,523	19.94	1,500
9/4/59	France	UO₂	70,006	3.49	2,443
9/18/59	"	UO₂	72,059	3.49	2,515
10/16/59	"	UO₂	16,966	3.49	592
11/10/59	Australia	Metal Powder	500*	93.40	467*
4/20/60	France	U Dioxide Powder	127*	19.83	25*
4/20/60	"	Metal Blend	{21, {80	93.00 Normal	20. 0
11/9/60	Japan	Uranium Dioxide	54,067	20.00	10,732
11/10/60	"	Uranium Dioxide	22,231	20.00	4,413
7/11/61	France	UO₃ Powder	107,384	2.984	3,204
7/11/61	"	UO₃ Powder	1,475	90.00	1,328
8/24/61	"	U Dioxide Powder	15,000	20.0568	3,009
4/19/62	"	U Dioxide Pellets	9,130	4.025	367
4/19/62	"	U Dioxide Pellets	9,110	4.52	412
4/19/62	"	U Dioxide Powder	5,265	4.50	238
6/15/62	Italy	U Dioxide Pellets	47,976	19.96	9,576
7/31/62	Japan	UO₂(NO3)₂ and U₃O₈ Powder	21*	93.16	20*
8/24/62	Netherlands	U Dioxide Pellets	350,187	3.136	11,23d
9/7/62	"	U Dioxide "	370,669	3.136	11,62d
10/11/62	"	UO₂ Pellets	315,139	3.613	12,05d

OFFICIAL USE ONLY
- 2 -

Date Shipped	Destination	Material Description	Unit Gram Uranium	Percent Isotope	U-235
10/12/62	Netherlands	UO₂ Pellets	313,986	3.813	11,972
11/2/62	"	UO₂ Pellets	58,385	3.813	2,227
11/2/62	"	UO₂ Pellets	32,553	3.136	1,021
11/2/62	"	UO₂ Pellets	104,754	3.813	3,994
11/23/62	France	UO₃ Powder	4,000	89.82	3,593
11/23/62	"	ADU Powder	10,027	19.86	1,991*
11/30/62	Netherlands	UO₂ Pellets	19,423	3.136	609
11/30/62	"	UO₂ Pellets	1,664	3.813	63
1/27/63	Italy	Al clad U₃O₈ Fuel Plates	12,360	19.83	2,451
5/9/63	France	UO₂ Powder	300,227	4.027	12,090
4/23/63	"	ADU Powder	20,998	60.03	12,605
4/26/63	"	ADU Powder	20,998	60.03	12,605
9/26/63	United Kingdom	Fused UO₂	88,125*	2.90	2,555*
3/27/64	Canada	UO₂ Pellets	131,008	6.00	7,860
3/30/64	Germany	UO₂ Pellets	286	1.00	3
3/30/64	"	UO₂ Pellets	282	1.50	4
3/30/64	"	UO₂ Pellets	283	2.00	6
3/30/64	"	UO₂ Pellets	286	2.50	7
3/30/64	"	UO₂ Pellets	285	3.00	8
3/30/64	"	UO₂ Pellets	286	3.50	10
3/30/64	"	UO₂ Pellets	281	4.00	11
3/30/64	"	UO₂ Pellets	282	4.50	13
3/30/64	"	UO₂ Pellets	446,266	3.03	22,603
4/20/64	France	ADU Powder	66,009	6.00	3,089
4/24/64	Japan	UF₆	48,230	5.704	2,806
4/24/64	"	UF₆	5,297	4.983	264
5/18/64	France	UO₂ Powder	300,000	4.00	11,970
7/13/64	"	ADU Powder	100,000	59.98	59,980
9/2/64	"	UO₂ Pellets	150,313	3.99	5,207
9/15/64	Japan	UF₆	166,723	2.598	4,280
10/13/64	Sweden	UO₂ Powder	52,575*	5.00	2,629*
12/14/64	France	UO₂ Powder	48,916	3.99	1,952
1/13/65	Italy	Al clad U₃O₈ Fuel Plates	5,034	19.83	998
3/13/65	France	UO₂ Powder	481,690	3.977	19,157
4/5/65	"	ADU Powder	100,000	59.93	59,930
10/4/65	Japan	Foils and UO₂ Powder	4	93.00	4

Total NUMEC Foreign Transfers 12/1/57 to 10/31/65 8,765,216   425,396

---

*Indicate sales transactions which equal or total 191 kgs uranium and 11 kgs U-235. All other transactions represent material which is leased.

Excerpt from AEC's Hanford contract with Atlantic Richfield – 1967

CONFORMED COPY

OPERATING CONTRACT NO. AT(45-1)-2130

BETWEEN

UNITED STATES OF AMERICA

REPRESENTED BY

UNITED STATES ATOMIC ENERGY COMMISSION

AND

ATLANTIC RICHFIELD HANFORD COMPANY

- - - - - - - - - -

CHEMICAL PROCESSING OPERATION

· OFFICIAL USE ONLY

ATLANTIC RICHFIELD HANFORD COMPANY NEGOTIATIONS

Don Williams went to Seattle on the afternoon of July 5, 1967, to meet
with Commissioner Johnson and the representatives from Atlantic Richfield
to plan the agenda for the next two days. Atlantic Richfield representatives were:

            R. D. Bent, Senior Vice President
            L. M. Richards, Vice President
            J. M. Schultz, Controller
            R. P. Corlew, Manager of Coordination, Richfield Division
            H. B. Off, Senior Counsel, Labor Relations
            D. R. Gerner, Manager, Employee Relations
            L. W. Cook, Labor Relations
            N. E. Birch, Patent Counsel
            D. W. McPhail, Manager, Evaluation

The first contract discussion took place on the afternoon of July 6, 1967,
in the Federal Building with the Atlantic Richfield representatives,
Commissioner W. E. Johnson, D. G. Williams, B. P. Helgeson, H. E. Parker,
A. M. Waggoner and O. J. Elgert. RL representatives stressed the need for
Atlantic Richfield to take over existing procedures, personnel, salaries
and fringe benefits. Atlantic Richfield agreed to do this in the interest
of a speedy, orderly takeover of the operations of the 200 Area.

Commissioner Johnson stressed the need for proper emphasis on nuclear
materials management. He pointed out the recent interest that the Joint
Committee had shown, the Headquarters recent reorganization of nuclear
materials and the troubles NUMEC had encountered. Mr. Schultz stated that
prior to takeover they would hire a consultant specialist in the nuclear
material management field to review the inventory procedures for transfer
of material. Waggoner pointed out that the RL situation was different
from NUMEC's as under the terms of our operating contract Atlantic Richfield
had no financial liability unless due to willful misconduct. It was
Atlantic Richfield's position that recognizing the lack of liability
their public image could be tarnished if provision was not made for a
prudent review of procedures and testing of physical inventories of plant,
equipment, stores material and nuclear materials. In addition to the
nuclear materials consultant, Atlantic Richfield proposed to have Lybrand,
their outside public accountants, furnish one or two men to participate
in the inventory procedures review and physical inventory. It was pointed
out that to assure an early takeover only a minimum inventory of plant,
equipment and stores material would be made. Reliance will be placed on
individual custodians, not the corporation.

It was agreed that Purex would be shut down, cleaned out, and the inventory
taken. The plant would not start back into operation until after takeover
by Atlantic Richfield.

FBI File: NUMEC invites Israeli operatives in for a visit - 1968[484]

UNITED STATES GOVERNMENT

Memorandum

TO :   DIRECTOR, FBI (117-2564)   DATE: 9/11/68

SAC, WFO ▮▮▮▮▮ (P) b7c/4

SUBJECT: ▮▮▮▮▮▮▮▮▮▮▮
          ATOMIC ENERGY ACT

APPROPRIATE AGENCIES
AND FIELD OFFICES
ADVISED BY ROUTING
SLIP (S) OF ___
DATE ___

ReWFOlet 8/5/68.   Classified by ▮▮▮▮▮
Declassify on: OADR

On 9/6/68, ▮▮▮▮▮▮▮ Security Office, AEC, b7c/2
Germantown, Md., advised that the AEC Security Office in
NYC had contacted him that date to advise of the requested
visit on 9/10/68, to NUMEC, Apollo, Pa., by the following:

AVRAHAM HERMONI, Scientific Counselor,
Israeli Embassy, Washington, D. C.;

Dr. EPHRAIM BEIGON, Dept. of Electronics,
Israel, born 7/15/32, in London;

ABRAHAM BENDOR, Dept. of Electronics,
Israel, born 7/7/28, in Israel;

RAPHAEL EITAN, Chemist, Ministry of
Defense, Israel, born 11/23/26, in Israel

▮▮▮▮▮▮▮▮▮▮▮▮▮▮▮▮▮▮▮▮▮▮▮▮▮

② - Bureau
2 - Pittsburgh ▮▮▮▮▮ (RM)   REC ▮▮ 7-2564-72
1 - WFO   b7c/4
(5)   b7c/1   2 SEP 13 1968

SECRET CONFIDENTIAL

Buy U.S. Savings Bonds Regularly on the Payroll Savings Plan

CONFIDENTIAL NUME

Nuclear Materials and Equipment Corporation   Apollo, Pennsylvania   Telephone   Cable NUMEC

September 12, 1968

Mr. Barry R. Walsh, Director
Security & Property Management Division
United States Atomic Energy Commission
New York Operations Office
376 Hudson Street
New York, New York 10014

Dear Mr. Walsh:

Your permission is hereby requested for the visit of a non-citizen of
the United States to Nuclear Materials & Equipment Corporation's facilities.
The information relative to this visit is as follows:

a. Full name of the visitor: 1. Dr. Abraham Hermoni

                                2. Mr. Ephram Belgon

b. Date and place of birth: 1. May 10, 1926; Tel-Aviv, Israel

                                2. July 15, 1932; London, England

c. Citizenship: Israeli

d. Visitor's affiliation and position (company or government organization
   name):   1. Israeli Embassy - Scientific Councilor

                                2. Israeli Ministry of Defense - Group Leader,
                                    Dept. of Electronics

e. Proposed dates of visit: September 10, 1968

f. Purpose of visit: Discuss thermoelectric devices (unclassified)

g. Areas to be visited: Energy Conversion Laboratory

REG 11

CONFIDENTIAL

Mr. Harry H. Walsh   CONFIDENTIAL   September 12, 1968

h. Names of NUMEC personnel to be contacted: Z. M. Shapiro

i. NUMEC official recommending visit approval: Z. M. Shapiro

Zalman Mordecai

Please advise me of your approval action. Thank you very much for your cooperation.

Yours very truly,

Bruce D. Rice
Manager, Security

ch

CONFIDENTIAL

117-2524-81

UNITED STATES
ATOMIC ENERGY COMMISSION
NEW YORK OPERATIONS OFFICE
376 HUDSON STREET
NEW YORK, N.Y. 10014

20 SEP 1968

Mr. Bruce D. Rice, Manager
Security Division
Nuclear Materials & Equipment Corporation
Apollo, Pennsylvania 15613

Dear Mr. Rice:

This is to confirm the telephonic approval furnished by
of my staff, regarding the unclassified visit of four (4) Israeli
citizens to your facility on September 10, 1968. These visitors are
identified in your two letters to me, dated September 12, 1968.

Your continuing cooperation in this matter is appreciated.

Very truly yours,

Harry R. Walsh, Director
Security Division

brb:HJK

bcc: RO, W/Encl. - Two separate ltrs.,
        dtd. 9/12/68

ALL INFORMATION CONTAINED
HEREIN IS UNCLASSIFIED
DATE 6-17-85 BY

NOT RECORDED
6 OCT 2 1968

117-2564

59 OCT 11 1968

Document 134

Nuclear Materials and Equipment Corporation   Apollo, Pennsylvania 15613   Telephone 412-842-0111   Cable NUMEC

September 27, 1968

Mr. Harry R. Walsh, Director
Security & Property Management Division
New York Operations Office
U. S. Atomic Energy Commission
376 Hudson Street
New York, New York   10014

Dear Mr. Walsh:

Reference your telephone call concerning the September 10 visit of Messrs.
Hermoni, Bendor, Eitan and Biegun, Israeli citizens. Please be advised of
the following.

The above mentioned gentlemen met with Dr. Shapiro, D. Purdy, T. Hurson,
J. Williams, and S. Kolenik. With the exception of Dr. Shapiro, all of
the NUMEC personnel are in our Energy Conversion Department and are thermoelectric generator specialists.

Discussion with the Israeli nationals concerned the possibility of developing
plutonium fueled thermo-electric generator systems in the 5 and 50 milliwatt
power level. Specifically, they were interested in 10 generators in the
5 milliwatt range. Each of which would be fueled with about 2 grams of
plutonium. The 50 milliwatt generator is considered a remote possibility,
but would use approximately 20 grams of plutonium. The generators are of
the terrestrial type.

We are proceeding to make a proposal to these gentlemen for this work using,
of course, only unclassified information which is already in the public
domain. It is also our understanding that these same gentlemen have visited
several of the major nuclear organizations in the United States to develop
proposals from them on these items.

I trust this satisfies your needs.

Very truly yours,

Bruce D. ...
Manager, Security

A Subsidiary of Atlantic Richfield Company   DOCUMENT 149

FBI Security Index file on Zalman Shapiro - 1969[485]

SECRET

ZALMAN MORDECAI SHAPIRO   AEC #CH-2923-NY-SC-OR-   ALBUQUERQUE
                                 HA-IN-HA-AB

[SECURITY STATEMENT](S) (U)   Classified by
                                  Declassify on: OADR

I. Reliable information indicates that you have knowingly established an
    association with individuals reliability reported as suspected of espionage,
    within the meaning of Paragraph (b) (2), Section 10.11, 10 CFR-Part 10,
(U) Criteria and Procedures for Determining Eligibility for Access to Restricted
    Data or Defense Information. The information which provides the basis
    for such belief is that: (S)

    (a) It was reported that on November 3, 1968, you attended and
         served as chairman of a meeting held at your home at 5452 Bartlett
         Street, Pittsburgh, Pennsylvania, in connection with the collection
         of scientific and chemical data for Israel; and that the meeting
         was attended by several individuals, including Dr. Avraham Hermoni,
         Scientific Counselor of the Israeli Embassy, Washington, D.C.
         It was reported that among other topics discussed at the meeting,
         it was brought out that Israel needed information regarding
(U)   image intensifier tubes; information concerning the over-all
         field of reliability and failure rates of certain manufactured
         products including any studies completed in this field; information
         concerning the training of personnel in the operation of
         pressurized water reactors, including the placement of such
         personnel in a training program in the Westinghouse Electric
         Corporation at Pittsburgh, Pennsylvania. It was further reported (S)

~~SECRET~~

-2-

that each person at the meeting was given a specific assignment
to obtain information needed by Israel and was to furnish the
information to Dr. Hermoni either orally or in writing. It was
also reported that you expressed the hope that the persons at
the meeting who were given assignments by Dr. Hermoni would be
productive as Dr. Hermoni was probing in many fields and Israel needed
the information on an urgent basis. On August 14, 1969, you
advised AEC representatives that the meeting referred to above was
held in your home, that Dr. Hermoni was in attendance, but could not
recall the names of all the individuals who attended. You further
advised that the general tone of the meeting concerned ways and
means the group could be of assistance to Israel in solving some
of its technical problems. It was reported in 1961 that
Dr. Abraham Hermoni was a member of the Israeli Intelligence
Service; that in 1963 Dr. Hermoni was known to be engaged in
the establishment of a technical intelligence network in the
United States; and that prior to his assignment as Scientific
Counselor of the Israeli Embassy, Washington, D. C., he was
intimately connected with Israeli efforts to develop nuclear
weapons in his capacity as Technical Director, Armament Development

(U)
Authority, Ministry of Defense. (S)

(U)
(FBI Report, Pittsburgh, Pa., 2/18/69, pp. 30, 31.
    Statement of Confidential Informant PG T-3.)
(FBI Memo, Washington, D.C., 12/23/68, p. 1.
    Statement of Confidential Informant WF T-1.)   (S)

SECRET

-3-

(b) It was reported that you associated with and are considered to be
     a long-time personal friend of David Luzer Lowenthal of Pittsburgh,
     Pennsylvania, President of the Raychord Corporation, Apollo,
     Pennsylvania, and an officer of Apollo Industries, which latter
     company was instrumental in establishing NUMEC through the
     investment of a substantial amount of money when NUMEC was formed.

(S)

(c) During an interview with FBI agents in 1966 you advised that you
     have had a business association in connection with Isorad (joint
     company by NUMEC and Israeli Government) during 1965-67 with
     _____ who was constantly here as at the Israeli Embassy,
     Washington, D. C. from October 1963 to May 1967.

(S)

ISRAELI

~~SECRET~~

-4-

(d) During 1968 and 1969 you were reportedly in contact with
Colonel Avraham Bylonie concerning materials of interest to
Israel among them Plutonium 238.

(S)

II. Reliable information indicates that you have engaged in conduct which
tends to show that you may be subject to influence or pressure which may
cause you to act contrary to the best interests of the national security,
within the meaning of Paragraph (b) (15), Section 10.11, 10 CFR-Part 10,
Criteria and Procedures for Determining Eligibility for Access to

(U) Restricted Data or Defense Information. The information which provides
the basis for such belief is that (S)

(a) It was reported that on October 16, 1968 you expressed a belief that
if fighting became necessary, you would prefer to fight in

(U) Israel where a just cause existed. It was further reported that
you also stated that you had given some consideration to resettling
in Israel on a permanent basis. (S)

(U)   [(FBI Rpt, Pittsburgh, Pa., 2/18/69, p. 46, (PG T-3,)] (S)

(b) It was reported that on December 15, 1968 you advised an

(U) individual that you were seriously considering the possibility of
settling in Israel within the next five years. (S)

(U)   [(FBI Rpt, Pittsburgh, Pa., 2/18/69, p. 47, (PG T-3.)] (S)

SECRET

-5-

(U)   (c) It was reported that you have financial ties with the State of Israel and that you feel a natural kinship toward the State of Israel. (S)

(U)   [(FBI Rpt, New York City, 12/23/68, p. 2,   ] (S)

(U)   (d) It was reported that you are an active supporter of Israel; have traveled there frequently; and have directed that all profits from NUMEC vending machines be used for CARE packages for Israel. (S)

(U)   [(FBI Rpt, Pittsburgh, Pa., 2/18/69, p. 27, PG T-1.)] (S)

(U)   (e) It was reported that a meeting was held in your home on November 3, 1968 and the purpose of that meeting was to assist Avraham Hermoni, Scientific Counselor, Israeli Embassy, in gathering necessary data in scientific fields. (S)

(U)   [(FBI Rpt, Pittsburgh, Pa., 2/18/69, p. 30, PG T-9.)] (S)

(U)   (f) It was reported that subsequent to the November 3, 1968 meeting you stated that the meeting was productive and that Avraham Hermoni was also provided answers to most of his questions or was put in contact with people who could get the answers for him. (S)

(U)   [(FBI Rpt, Pittsburgh, Pa., 2/18/69, p. 2,] (S)

(U)   (g) It was reported that in discussing the November 3, 1968 meeting you pointed out that assignments given out by Avraham Hermoni to individuals who attended the meeting were important to Hermoni and (S)

SECRET

-6-

that Hermoni was probing in many fields and needs the requested information on an expedite basis, as Israel's needs are many and urgent. (S)

(h) It was reported that you expressed the hope that people who were given assignments at the November 3, 1963 meeting by Hermoni would be patient and productive. (S)

(FBI Rpt, Pittsburgh, Pa, p. 9, dtd 11/20/63.) (S)

(i) It was reported that on November 8, 1963 you related to an individual that Israel is in need of all available information concerning the poisoning of municipal water supplies. The individual to whom you related this information reportedly has done research work in this field. (S)

(FBI Rpt, Pittsburgh, Pa. 11/29/63, p. 4.) (S)

(j) It was reported that in connection with your expressed interest in migrating to Israel, your son indicated he felt you would be of more value to Israel by remaining in the United States and helping Israel's problems from here (U.S.A.). (S)

(FBI Rpt, Pittsburgh, Pa. 11/29/63, p. 5.) (S)

(k) It was reported that you planned to head up a new corporation to operate a chemical plant for the production of high grade boric acid, itrium oxide, hydrobromic acid and uranium and to operate this plant between Dimona and Arad near the Dead Sea. (S)

SECRET

-7-

(U) (1) It was reported that you have a very strong alliance to Israel and have contributed heavily to Israeli organizations and fund drives. It was further reported that during the 1967 Israeli-Arab six-day war you organized an Israeli fund drive in connection with a social affair in Pittsburgh and raised over one million dollars for Israel. (S)

(U) [(FBI Rpt, Pittsburgh, Pa., 6/21/68, p. 2.)] (S)

(U) (m) It was reported that you have a very strong allegiance to the Government of Israel and would do anything in your power to aid that country. It was also reported that you have been very active in fund-raising drives and bond sales to obtain money for Israel and are a heavy contributor to Jewish activities in the United States. (S)

(U) [(FBI Rpt, Pittsburgh, Pa., 7/25/68, p. 4.)] (S)

(U) (n) It has been reported that you have had substantial interests involving the commercial application of atomic energy with the Government of Israel. These interests include an equal partnership between NUMEC and the Government of Israel in the Israel NUMEC Isotopes and Radiation Enterprises and 30 thousand curies of Cobalt have been shipped to Israel for use in experimental facilities. (S)

(U) [(FBI Rpt, Pittsburgh, Pa., 8/21/68, p. 5.)] (S)

SECRET

~~SECRET~~

-8-

(o) It has been reported that you are a long-time close personal

     friend of David L. Lowenthal, who is reportedly closely associated

     with high ranking officials of the Israeli Government, fought for

     Israel as a Freedom Fighter in 1956, and who currently travels to

     Israel on the average of approximately once each month.

(S)

(p)

APPROVED:_____   Date:_____

~~SECRET~~

FBI transcript of a 1969 NUMEC spill caused by improper storage[486]

Time	Initial	IC OG	Activity Recorded

FD-297 (1-28-57)   SECRET   DECLASSIFIED BY 60322 uc lrp/plj/lsc ON 02-09-2011

b6
b7C

8:50p rjc — No activity until;

7:52p rjc ic ▢ is leaving his office now. He used EV'S car today and had a flat tire on the way, which he had to change.

7:53p rjc ic ▢ Nothing pertinent in their conversation.

9:05p rjc ic ▢ but ▢ said that he just came in and is sitting down to supper. She said ▢ will call him back.

9:18p rjc ic ▢ reported on a spillage at the plant. They have the area roped off and it will take some pick and shovel work to dig up the contaminated areas. ▢ said they are getting 100,000 counts. ▢ said, "Oh, God". They are dampening it down to avoid dust and will cover it if it looks like it may rain. ▢ asked if there is anything on AAI? ▢ said that very few are being rejected- about 16. ▢ heard they are having trouble at National Lead.

9:20p rjc og ▢ Only last four digits of number were available- 946: ▢ said that he heard that American Instrument Company up for sale. He said he heard this from the former sales manager. ▢ said that he ought to get his agent on that right away. ▢ said it is an old established company, and that maybe i tax trouble. ▢ said he will not see B tomorrow since he has to be in Washington tomorrow morning, but that he is not leaving real early. ▢ said, If we could get them (American Instrument Co.) and buy up Autoclave Engineering. ▢ said they had a bad spill at the plant. ▢ called him on it about 11am. ▢ said its not only a bad spill but "actually they are operating outside compliance". They had the drums all together. They have about 200 drums and estimate that about 6 a day will corrode through. The trouble lay with a flouride which was put in to help the decay and this was not checked.   (Cont.)

PG-213-S*(a) 1   Employee's Name   Date Stamp
Log   Page   SEARCHED _____ INDEXED_____
Mon.   5/5/69   SE ▢ RBG   SERIALIZED_____ FILED_____
Class. & Ext. By _____   MAY ─ 6 1969   FCI ─ PITTSBURGH
Reason FORM II, 1-_____
Date of Review _____   CONFIDENTIAL   SECRET   Classified by _____ Declassify on _____

FD-297 (1-28-57)

Time	Initial	IC OG	Activity Recorded
	rjc		☐ to ☐ said they are also about $230,000 over on their construction costs for the scrap plant. Z said if they could get other people, there would be a lot of firing.
9:26p	rjc	og	☐ to a ☐ and then ☐ (-- 2 3547: ☐ said he understood that ☐ has been trying to get him the last couple of days. ☐ said he is not going to be in to work tomrrow. ☐ wants to go over the plans for the Sal Gel thing. ☐ went into some detail on this- ☐ said he will be back on Wednesday and will see ☐ then. ☐ said they had a bad spill from the scrap collection drums. The materials ate through the stainless steel drums. It got out into the field and it is a real problem. ☐ said they have a very serious problem and are running up a bill of $100,000 on use charges and asked ☐ to review the whole container program and give him a report.
10:08p	rjc	ic	☐ , collect to ☐ Conversation wholly concerned school work with nothing heard of a pertinent nature. (It is believed that ☐ listened to this conversation although he did not say a word.)
12:00N	rjc		No other activity.

PG-213-S*(n)   2
Log .... Page ....
Non.   5/5/69
Day .... Date ....

Employee's Name

SE☐

Date Stamp

Henry Kissinger: NUMEC diversion and Israeli nukes – July 19, 1969[487]

Tab A

TOP SECRET/NODIS
SENSITIVE   July 19, 1969

SUMMARY OF THE SITUATION AND ISSUES

      This paper is designed (1) to summarize the situation that we now face and (2) to brief the issues which two discussions in the Ad Hoc Review Group have raised. A paper on the operational decisions required is at a following tab.

I.   Summary: Elements in the Present Situation

    1. Our general intelligence judgment is that:

[redacted]

    --Israel has 12 surface-to-surface missiles delivered from France. Israel has set up a production line and plans by the end of 1970 to have a total force of 24-30, ten of which are programmed for nuclear warheads. The first domestically produced missile is expected to be completed this summer. Preparation of launch facilities is under way.

    --There is circumstantial evidence that some fissionable material available for Israel's weapons development was illegally obtained from the United States by about 1965.

    2. The intelligence community agrees on the general judgment above. The issue dividing it is the more specific question of whether Israel has already produced completed nuclear weapons. [redacted]

[redacted] Although views in State differ, the institutional position emphasizes that concrete proof is lacking and that Israel is concerned enough about its relations with us -- and aware enough of our opposition to nuclear proliferation -- to think twice about putting nuclear weapons openly in its arsenal.

    3. This difference of assessment raises the choice between recording a judgment that Israel may have nuclear weapons and recording only a general judgment as to Israel's capability.

TOP SECRET/NODIS
SENSITIVE

a. The advantage of recording only the general judgment
is that it permits us the freedom of acting as if we believe Israel is
still short of assembling a weapon and of leaving to Israel the choice
of whether to hide what it has or dismantle it. It also retains our freedom
to press Israel to sign the NPT and prevent the USSR from reacting.

b. The disadvantage of not recording the more precise
estimate is that only this underscores the immediacy of the problem if
we are called on in the Congress, for instance, to justify our position.

4. In signing the contract for sale of the Phantom F-4 aircraft
last December, Israel, in a letter, committed itself not to be "the first
to introduce nuclear weapons into the area." The US stated in reply that
circumstances requiring cancellation of the agreement would exist in the
event of "action inconsistent with your policy and agreement as set forth...."

5. We and Israel differ on what "introducing" nuclear weapons
means. Ambassador Rabin believes only testing and making public the fact
of possession constitute "introduction." We stated in the exchange of
letters confirming the Phantom sale that we consider "physical possession
and control of nuclear arms" to constitute "introduction."

6. Before negotiation of the sale, President Johnson and Secretary
Rusk told Foreign Minister Eban we felt strongly about Israel's signature
on the NPT and stated that political discussions on this issue would precede
negotiation. Later, after strong pressure from the Israeli government and
approaches from American Jewish leaders, the President instructed
Secretary Clifford to sell the planes without conditions. Since the Israelis
had already given us the commitment not to be the first to introduce
nuclear weapons in connection with the 1966 sale of the Skyhawk A-4
aircraft, Secretary Clifford permitted its repetition in the 1968 sale.
What was new in the 1968 talks was the inconclusive attempt to define
the word "introduction."

7. No one in Congress is yet officially aware of the exchange of
letters on Israel's promise not to be the first to introduce nuclear weapons
or our reply. Nevertheless, the Administration might have to defend
someday the delivery of a nuclear weapons carrier despite our intelligence
and the exchange of letters at the time of the sale.

8. Delivery of the Phantoms is scheduled to begin in September,
1969. The planes are almost ready, and the Israelis have asked to begin
taking delivery in August.

9. We do not know exactly how much the Soviets know about Israel's nuclear development. However, the Director of Central Intelligence believes that, while Moscow may not have quite as much detail as we do, the Soviets must be aware of the general state of Israel's nuclear weapons and missile development, though they may not want it publicly known.

10. We do not know exactly how much the Arabs know, but they are aware that Israel's capability in the nuclear field is well-advanced. Both Soviets and Arabs have been surprisingly quiet about this subject.

II.   A Central Issue

    A. As our response to the above situation is considered, the basic question to keep in mind is: Exactly what development do we most want to prevent? There are two aspects to the question:

        1. Israel's secret possession of nuclear weapons would increase the danger in the Near East and, ideally, should be prevented.

        2. But the significant international act is public acknowledgement that Israel possesses nuclear weapons. This might spark Soviet nuclear guarantees to the Arabs, tighten the Soviet hold on the Arabs and increase the danger of US-Soviet nuclear confrontation.

III.   The Major Issues

                BASIC U.S. INTEREST

    A. How detrimental to US interests would Israeli possession of nuclear weapons be?

        1. Danger of US-Soviet confrontation.

            a. Israeli possession of nuclear weapons could substantially increase the danger of a Soviet-American confrontation in the Middle East.

[NLN02-04/4A p30/18]

-4-

--If the Israelis are known to have nuclear weapons
the Russians might feel obliged either before or during
a crisis to indicate that they would retaliate if the
Israelis use nuclear weapons. We might feel obliged
to indicate that we would respond to Soviet use of
nuclear weapons.

--The Israelis, who are one of the few peoples whose
survival is genuinely threatened, are probably more
likely than almost any other country to actually use
their nuclear weapons.

--Because of these dangers, both we and the Russians
might find it harder to stay aloof from conflicts in the
Middle East.

b. On the other hand, it can be argued that we and the
Russians managed in June, 1967 to agree to remain
aloof from the conflict and we might do so again, albeit
with some greater difficulty, even if the Israelis are
known to have nuclear weapons.

2. Effect on chances for an Arab-Israeli political settlement.

a. If Israeli possession of nuclear weapons became
known, it would sharply reduce the chances for any
peace settlement in the near future.

--At the least, diplomatic efforts to achieve a
settlement would be delayed until the Arabs and the
Soviets assessed this development.

--Negotiations would be put off for the foreseeable
future. The Arabs believe they cannot negotiate
from a position of conventional military inferiority,
much less nuclear inferiority.

--Moscow would probably be in a position of resisting
Arab pressures for nuclear weapons or nuclear
guarantees and would find it more difficult to press
the Arabs for diplomatic concessions.

TOP SECRET/NODIS
SENSITIVE   - 5 -

b. While accepting these judgments, some would argue
    that it will also harm chances for a political settlement
    if we tackle this issue head-on. They would argue that
    we can persuade the Israelis to give up their nuclear
    option only in the context of peace and that trying to
    deny Israel that option will only make the Israelis less
    willing to make the concessions on territory that will
    be necessary in a settlement.

3. Charge of US complicity.

    a. If Israel's possession of nuclear weapons became known,
        the US would be highly vulnerable to charges of complicity
        in helping Israel become a nuclear power:

        --Regardless of what we say, the Arabs will assume that
        we could have stopped Israel.

        --The Administration would have delivered to Israel a
        nuclear weapons delivery system (Phantoms) ███████
        █████████ despite a
        contract stating that it would be cancelled if Israel
        violated its pledge not to be the first to introduce nuclear
        weapons into the Middle East.

    b. On the other hand, there is the danger that we will become
        accomplices by talking to the Israelis, pressing them and
        failing to get what we want. Then we might look as if we
        acquiesced, especially if we talked and then went ahead
        and delivered the Phantoms -- a nuclear weapons carrier --
        anyway. Even if we get what we want and the Israelis
        violate their pledge, we might look like accomplices.
        There could be an argument for acting in pretended
        ignorance, ████████████████████

4. Effects on nuclear proliferation.

    a. World-wide knowledge that the Israelis had nuclear
        weapons would almost certainly wreck the Non-
        Proliferation Treaty.

TOP SECRET/NODIS
SENSITIVE

[NLN 02-04/4A p 5 of 16]

--The Arab states would refuse to ratify the treaty.

--Other powers who might be prepared to sign
and ratify the treaty if only the five great powers
have nuclear weapons might find it more difficult
to accept non-nuclear status if a small power
such as Israel is known to have nuclear weapons.

b. Others would argue that adherence by other
potential nuclear powers such as the FRG and
Japan would be little affected by Israeli behavior.

5. Conclusions: Israeli acquisition of nuclear weapons
would: Impose a substantial cost on US relations with
Arabs and Soviets. Setback NPT efforts. Substantially
increase the probability that someone will use nuclear
weapons in anger. Increase the risk of Soviet-US
confrontation. Make a political settlement all but
impossible.

WHAT SHOULD WE WANT?

B. Can we prevent Israel's acquisition of nuclear weapons?
Or to put it more precisely since Israel may already have
some nuclear weapons: Could we persuade Israel to freeze
its nuclear program where it is?

1. We assume that it is impossible to deprive Israel of
option to put together an operational nuclear capability.

a. Regardless of what we think of the military or
deterrent value of nuclear weapons in Israel's
hands, Israelis feel that in conventional war
numbers will eventually tell and that over the
long term this makes nuclear weapons necessary.

[NLN 02-04/4A p6 of 19]

- 7 -

    b.   The Israeli program is very near fruition, and--
given strong Israeli feeling that Israel's very
survival is at stake--it would seem all but impossible
politically for an Israeli Prime Minister to give up
completely an advantage deemed vital and achieved
at considerable cost.

    c.   We have no way of forcing Israel to destroy any
nuclear devices or components it may now have--
much less the design data or the technical
knowledge in people's minds.

2.   If it is impossible to persuade Israel to give up its
nuclear option completely, <u>could we persuade Israel
to stop its nuclear program where it is</u>?

    a.   On the face of it, this seems a difficult but not
unattainable objective. It would satisfy Israel's
principal aim of being able to put together an
operational nuclear capability on short notice--
while avoiding a harsh collision with the US,
possible nuclear threats from the USSR and a
fatal blow to near-term chances for peace with
the Arabs. It could even be consistent with signing
the NPT, which has its own escape clause.

    b.   <u>The argument against setting this as our sole aim
is that this by itself is not a practical objective:</u>

        --Its attainment is unverifiable. We might con-
ceivably persuade Israel to agree to freeze its
nuclear program, but it is unrealistic to think
that such an agreement would mean that Israel
had actually stopped. We would have no way of
assuring compliance. Inspection would not work
because we could never cover all conceivable
Israeli hiding places. This is one program on
which the Israelis have persistently deceived us--
and may even have stolen from us.

--It is not in our interest to verify failure to attain
it. We do not want to prove to the world that Israel
has nuclear weapons, and we would put ourselves in
an even more difficult situation than we are in now
if we proved it to ourselves.

--It is unreal. Israel may already have nuclear
weapons. We may very well want to keep Israel's
nuclear program from going further, but that by
itself would be small gain if Israel agreed and then
made public weapons it may already have.

--We may be better off not talking to the Israelis
about where their program stands. We may be in
a much better position telling them that we do not
want them to possess nuclear weapons and then
letting them figure out how to meet our request.

--Putting this in the record as our objective leaves
us vulnerable to the charge of complicity in
Israel's nuclear program.

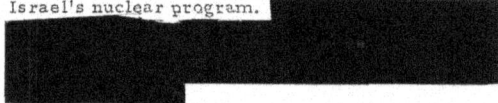

3. Conclusions: Talking about preventing Israel's acquisition of nuclear weapons may be a reasonable way to state
our purpose to the Israelis or for the record, because
keeping nuclear   weapons out of the Near East
would be safer. Neither of these formulations is precise
enough for describing to ourselves what we really want.
We cannot prevent acquisition of weapons that may
already be there, and it is impossible by inspection to
learn what is there.   We do not
simply want to ask for a freeze because that makes
accomplices of us. Therefore, for the sake of our own
understanding at least we may want to try describing our
objective another way. They might be willing to freeze
their program about where it is today, but it is impractical
for us to state our objective this way.

C.   If there are too many pitfalls in saying to ourselves that we
     want to stop the Israeli nuclear program where it is, could we
     state our objective as trying to persuade Israel not to announce
     its possession of nuclear weapons?

     1.   It can be argued that the real impact of Israel's nuclear
          weapons, if any, would be felt only when ███████████
          ████████████████████████████

SANITIZED
3.3(b)(6)

          a.   As long as Israel keeps them secret, both the
               Arabs and the Soviets can act as if they did not
               exist. The moment Israel's program becomes
               an established international fact, the Arab
               governments will have to cope with another major
               demonstration of Israeli superiority, and the
               Soviets will have to cope with substantial Arab
               pressures for a guarantee against nuclear attack.

          b.   Many Israelis would also argue that the first
               purpose of having nuclear weapons is achieved
               only when the Arabs know they exist. As Ambassador
               Rabin said to Assistant Secretary Warnke last
               fall: No one who has nuclear weapons expects to
               use them; their first purpose is as a deterrent.
               And there is no deterrent unless the enemy is
               aware of it.

     2.   It can also be argued that Israelis might be persuaded
          to promise us not to announce their possession of nuclear
          weapons:

          a.   In fact, by Israeli definition they have already made
               this promise. When Warnke asked Rabin what would
               constitute "introduction" of nuclear weapons into
               the Middle East, Rabin replied that "introduction"
               would not occur until a weapon had been tested and
               its existence become publicly known. With that
               definition in the record, the Israeli government
               reaffirmed in writing its commitment not to be the
               first to introduce nuclear weapons into the Mid-East.

- 10 -

b. Israel's conventional superiority will be
sufficient to meet any Arab attack in the foreseeable future.

3. The arguments against stating this as our objective--
at least to the Israelis--are that:

a. It would establish an indefensible record for us.
We would accept complicity in Israel's possession
of nuclear weapons by saying in effect: We know
what Israel has, but we will close our eyes to it--
and deliver the Phantoms--provided the Israelis
promise not to announce what they have. That
would not make an easy record to defend before
the world against a background of our professed
desire to limit nuclear proliferation.

b. It puts the Israelis in a position--with our
acquiescence--to let the world know indirectly
but unmistakably what it has without violating
any pledge to us.

4. Conclusions:

a. Saying that we want to keep Israel's possession
of nuclear weapons from becoming an established
international fact may come very close to describing
what we really want in this case. Our interest is in
preventing Israel's possession of nuclear weapons.
But since we cannot--and may not want to try to--
control the state of Israel's nuclear program and
since Israel may already have nuclear weapons,
the one objective we might achieve is to persuade
them to keep what they have secret. This would
meet our objective because the international implications of an Israeli program are not triggered
until it becomes public knowledge.

- 11 -

b. While this may be a reasonable description of
our real objective to ourselves, it makes an
indefensible public record. It leaves us highly
vulnerable to the charge of acquiescing in the
proliferation of nuclear weapons--and even of
abetting it by delivering the Phantom, a nuclear
weapons carrier.

c. Even though keeping Israeli weapons secret may be
a fair statement of what we most want, we should
not lose sight of the fact that it would also be
desirable to stop the Israeli nuclear program where
it is, or even roll it back a little. Even though
that alone may not be a practical objective, keeping
it in our sights does help us keep in mind that our
public purpose is preventing proliferation.

d. We may, therefore, want to differentiate between
our private understanding of what we want and what
we ask the Israelis for:

--We may want to consider saying to ourselves that our
aim is to keep Israel's possession of nuclear weapons
from becoming public knowledge and to do what we
can to stop further development.

--But in talking to the Israelis and for the record--
as well as because it is not in our interest for them
to have nuclear weapons--we may want to state our
position as opposing Israel's "possession" of nuclear
weapons, leaving it to the Israelis to figure out how
to comply. If they committed themselves not to
"possess" nuclear weapons, they would at the same
time be promising not to test, deploy or announce.

COURSE OF ACTION

D. If we decide that Israel's known possession of nuclear weapons
would be highly detrimental to our interests and that we might
persuade the Israelis to say they do not "possess" such weapons,
what is the best tactic to follow?

1. Should we raise the issue and seek specific Israeli
   assurances or content ourselves with a general statement of our opposition to proliferation?

   a. The arguments for raising the issue directly are:

      --This is the only approach that stands any chance
      of persuading the Israelis to take our interests
      seriously. Their practice is to read silence as
      consent.

      --If it becomes known that Israel has nuclear
      weapons, it will be to our advantage to have built
      a record of attempting to prevent introduction of
      nuclear weapons into the Mid-East.

   b. The arguments against raising the issue in a
      specific way are:

      --While this is debatable, it can be argued that
      the Israelis are unlikely in the near future to
      detonate a nuclear device or to publicly announce
      that they have a nuclear capability. Thus, the
      distinction between where they themselves will
      stop and where we might try to get them to stop
      is too small to risk a confrontation.

      --We cannot hold a detailed dialogue with the
      Israelis and sustain our position publicly without
      risking making Israel's nuclear capability public
      knowledge. That could bring on the crisis and
      the sharp Soviet reaction we are trying to avoid.

      --The only hope of getting the Israelis to agree
      with us to maintain secrecy and sign the NPT is
      to get an Arab-Israeli political settlement. We
      should save our leverage with them for this issue.

- 13 -

2. If we raise the issue, should we hold up delivery of
    the Phantoms (and even shipment of other conventional
    weapons) until we get what we want?

    a. Con.

        --It is important to the US for Israel to be able to
        defend itself. Halting delivery of the Skyhawks
        and suspending plans for delivery of the Phantoms
        would leave Israel with a highly disadvantageous
        ratio in supersonic aircraft vis-a-vis the UAR
        next year. While Israel could probably still hold
        its own on the Suez Canal, its vulnerability would
        increase.

        --A conventional arms embargo might make Israel's
        recourse to nuclear weapons more--not less--likely.

        --The American body politic would generate intolerable
        political trouble for the Administration--damaging
        Congressional attacks on Administration programs.
        Yet, we could not defend our position without making
        the nuclear issue public.

        --If Israel's going nuclear may force us to dissociate
        ourselves from Israel eventually, we want to set it
        up to defend itself first so we will not later face the
        excruciating choice of going to its aid if it gets in
        trouble.

    b. Pro.

        --If we believe stopping the Israelis is important
        enough, this is the only prospect serious enough
        to have a chance of success.

        --They may not want a confrontation with us on this
        issue. If we make a reasonable request that gives
        them some flexibility of interpretation and not make
        a direct threat they might agree to our limited
        requests and we might not have to carry through our
        threat. If they are at a good stopping place, they
        might be able to agree to freeze their program and

- 14 -

keep it secret with little cost. The only loss
to them would be giving up holding the threat of
potential nuclear weapons over the Arabs.

--If Israel openly became a nuclear power, we
would have little choice anyway but to take our
distance. Once Israel's possession of nuclear
weapons was known, it would be difficult for
Israel to confront us publicly on the nuclear
proliferation issue. Our position could be
presented as acting in the US interest without
jeopardizing Israel's security in the near term
as long as we were willing to deliver conventional
weapons to a non-nuclear Israel.

c. Conclusion: There is a serious issue whether we
should make this threat now and risk undercutting
whatever chance we may have via our diplomatic
effort to achieve a peace settlement. The dilemma
on that front is that if we don't stop Israel's nuclear
development, that will jeopardize the peace effort
and increase the danger to us besides.

The real dilemma is how to get Israel to take us
seriously without making the nuclear issue public
and bringing on a crisis. The only way out of
this dilemma seems to be to make the firmest but
gentlest approach possible on the assumption that
Israel does not want a showdown with us on this
issue. There seems little question, however,
that we shall make no dent on the Israelis unless
we put something they very much want into the
balance--at least by implication.

3. Should we try for Israeli assurance that it will stop its
strategic missile as well as its nuclear weapons program?

a. Con.

--Getting the Israelis to abandon their surface-to-
surface missile program seems impossible. Their

- 15 -

assembly line is turning the missiles out now.

--We are on very weak ground provoking a show-down over another sovereign nation's decision to deploy a delivery system that it believes makes sense.

--Nuclear weapons and not missiles are our main objective. We should not overload the circuit.

b. Pro.

The main military justification for these missiles is the nuclear warhead (though the Israelis have also talked of chemical warheads).

--Therefore the deployment of the missiles may provoke the same reaction as the actual deployment of the warheads. Everyone will assume they have nuclear warheads whether they do or not.

--It is a lot easier for us to police Israeli assurances if the missiles are not deployed. We can see missile deployment, and it can be an indicator for us. If missiles are on the launching pads, it is difficult for us to determine whether they have nuclear warheads or not.

c. Conclusion: Our main objective is to keep secret Israeli nuclear weapons. But because the public impact of missile deployment might be almost the same as nuclear weapons deployment, we might start by trying to persuade the Israelis not to deploy SSM's. We probably cannot persuade them to stop the production line.

RELATION TO THE PEACE EFFORT

4. Might anything be done to have this effort complement rather than undercut our efforts to achieve a political settlement?

[NLN 02-04/4A p 15 of 16]

a. Con.

    --The Israelis already doubt our support, as a
result of our talks with the Russians on the terms
of a settlement. Threatening them on the nuclear
issue now would confirm their worst fears.

    --If we threatened to cut off Israel's conventional
arms supply, it would harden its demand for
expanded borders. It would want the added security
of strategic borders if it lost what it considers to be
the security of advanced weapons.

    --Carrying out our threat to cut military supply
would make the nuclear issue public and it would
be harder for the Arabs to make the concessions
necessary for a settlement.

    --It is better to play out the present diplomatic
effort first and then tackle the nuclear problem.

    --Any US effort to encourage the Israelis to get
something from the USSR in return for their
signature on the NPT would, in effect, involve
us in nuclear blackmail.

b. Pro.

    --If we don't settle the nuclear problem soon, it
could itself wreck the diplomatic effort to achieve
a settlement. In fact, the Israelis could well use
it at some point to sabotage the peace talks if they
did not like the way the talks were going.

    --If we want to press the Israelis on the terms of
peace, we would be in a more defensible position
applying pressure ostensibly for the sake of non-
proliferation. If we come to a showdown on either
issue--withdrawal or non-proliferation--the main
leverage will not be jet aircraft but the total US-
Israeli relationship. If we were going to have

- 17 -

that kind of confrontation, it would be easier for
us to manage on the issue of proliferation than of
borders, though it is doubtful that Israel would
give on both.

c. Conclusion: There is probably little constructive
relationship between this nuclear problem and our
diplomatic effort to achieve peace. The main issue
is to structure our dialogue on the nuclear issue,
if any, so as to leave Israel enough flexibility to
minimize the damage on the peace effort.

IV. Conclusions

A. We must reach some sort of understanding with Israel about
its plans for its nuclear weapons program before we can
deliver the Phantom aircraft.

B. The logical bilateral Israeli commitment to press for is:

1. Israeli ratification of the NPT within a stated period.

2. Reaffirmation in writing that Israel will not be the
first to introduce nuclear weapons into the Middle East--
this time with a precise definition of what "introduce"
means. /We may want to agree to ourselves that it will
be sufficient if the Israelis live up to their own definition--
not test and not make public--but in talking to them and
for the record we should stick to our own definition--
"introduce" means "possess." It is not in our interest
that they possess nuclear weapons, but we do have to
take into account the practical limits of what we can
achieve and enforce./

3. Agreement at least not to deploy strategic missiles,
though we may want to consider at the outset asking
them to halt production.

C. If we are to approach the Israelis, they will not take us
seriously unless they believe we are prepared to withhold
something they very much want. The problem is to couch

- 18 -

our request in such a way that they can accede without
paying too high a price. These factors must be taken
into account:

1. Israel has already--in buying the Phantoms--
    committed itself in writing not to be the first
    to introduce nuclear weapons into the Mid-East.
    Ambassador Rabin has defined "introduction"
    as testing and publicizing.

2. The proposal which represents the consensus
    of our special group--ask Israel to define
    "introduction" as "possession"--might just
    allow Israel enough flexibility of interpretation
    to permit acceptance without a showdown.

3. The positive side of implying a threat to withhold
    aircraft could be to promise to meet new Israeli
    needs if we can reach an understanding on this
    issue. They have already said they want more
    Skyhawks and more Phantoms. The hope of a
    positive response on those could be held out as
    an incentive.

Seaborg/DOE: Portsmouth U-235 picked up in Israel – June 21, 1978[488]

June 21, 1978
2:15 p.m.

INTERVIEW: Bill Knauf
            Jim Anderson
            Department of Energy
            Division of Inspection

I met from 2:15 to 3:15 p.m. with Bill Knauf and Jim Anderson of the Division of Inspection of the Department of Energy. Their purpose was to interview me on the allegation that Zalman Shapiro of the Nuclear Materials and Equipment Corporation of Apollo, Pennsylvania diverted large amounts of highly enriched Uranium-235 to Israel in the 1960's.

They questioned me about the degree of surveillance of the Atomic Energy Commission commissioners on the NUMEC and the actions of the Commission when the loss of material was reported. I described the manner in which the commission operated and the responsibility of the staff in this connection.

They focussed a good deal on the dispute which the commissioners had with John Mitchell in 1970 when he wanted to deny the upgrading of Shapiro's clearance without granting him due process.

In response to this questioning I said that the commissioners were motivated by the desire to give Shapiro a proper hearing as well as by their concern that the scientific and legal community would disapprove of any denial of due process.

They were interested in how the matter was finally settled. They told me that they had already discussed this with Ramey and I agreed with them that Ramey served as the means by which a position was found for Shapiro with the Westinghouse Corporation, hence, rendering the question of clearance upgrading as moot. They told me that as late as 1971 the CIA wanted to pursue this further but Mitchell declined to do so.

They asked about any discussions I have had with Helms about this matter and I described the luncheon meeting I had with him in 1967 or 1968 during which I asked Helms if he had any evidence beyond that which I had and Helms replied that he did not. They are going to interview Helms. They are probably going to interview Hardin but not John Mitchell.

They have interviewed Howard Brown and the BBC has also interviewed Howard Brown, giving him a hard time. They indicated that BBC may try to interview me. They said that Shapiro has now engaged the law firm of Arnold and Porter and this law firm may get in touch with me.

I asked them if any responsible persons feel that Shapiro actually diverted material to Israel. They replied that nobody with a scientific background believes this but that it is difficult to convince some members

Interview: Bill Knauf   -2-   June 21, 1978
          Jim Anderson

of Congress. They said that some enriched Uranium-235 which can be
identified as coming from the Portsmouth, Ohio plant has been picked
up in Israel which, of course, has excited some members of Congress.
However, such enriched material has been sold on an official basis
to Israel and this could be the source of the clandestine sample.

          They indicated that they would let me read the draft of
their summary of our conversation today in order that I might make
any necessary corrections.

                                        Glenn T. Seaborg

GTS/scd

I met from 2:15 to 3:15 p.m. with Bill Knauf and Jim Anderson of the Division of Inspection of the Department of Energy. Their purpose was to interview me on the allegation that Zalman Shapiro of the Nuclear Materials and Equipment Corporation of Apollo, Pennsylvania diverted large amounts of highly enriched Uranium-235 to Israel in the 1960's.

They questioned me about the degree of surveillance [surveillance] of the Atomic Energy Commission commissioners on the NUMEC and the actions of the Commission when the loss of material was reported. I described the manner in which the commission operated and the responsibility of the staff in this connection.

They focused a good deal on the dispute which the commissioners had with John Mitchell in 1970 when he wanted to deny the upgrading of Shapiro's clearance without granting him due process.

In response to this questioning I said that the commissioners were motivated by the desire to give Shapiro a proper hearing as well as by their concern that the scientific and legal community would disapprove of any denial of due process.

They were interested in how the matter was finally settled. They told me that they had already discussed this with Ramey and I agreed with them that Ramey served as the means by which a position was found for Shapiro with the Westinghouse Corporation, hence rendering the question of clearance upgrading as moot. They told me that as late as 1971 the CIA wanted to pursue this further but Mitchell declined to do so.

They asked about any discussions I have had with Helms about this matter and I described the luncheon meeting I had with him in 1967 or 1968 during which I asked Helms if he had any evidence beyond that which I had and Helms replied that he did not. They are going to interview Helms. They are probably going to interview Mardian but not John Mitchell.

They have interviewed Howard Brown and the BBC has also interviewed Howard Brown, giving him a hard time. They indicated that BBC may try to interview me. They said that Shapiro has now engaged the law firm of Arnold and Porter and this law firm may get in touch with me.

I asked them if any responsible persons feel that Shapiro actually diverted material to Israel. They replied that nobody with a scientific background believes this but that it is difficult to convince some members of Congress. They said that some enriched Uranium-235 which can be identified as coming from the Portsmouth, Ohio plant has been picked up in Israel which, of course, has exited some members of Congress. However, such enriched material has been sold on an official basis to Israel and this could be the source of the clandestine sample.

They indicated that they would let me read the draft of their summary of our conversation today in order that I might make any necessary corrections.

CIA Carl Duckett leaked classified briefing summary page - 1978[489]

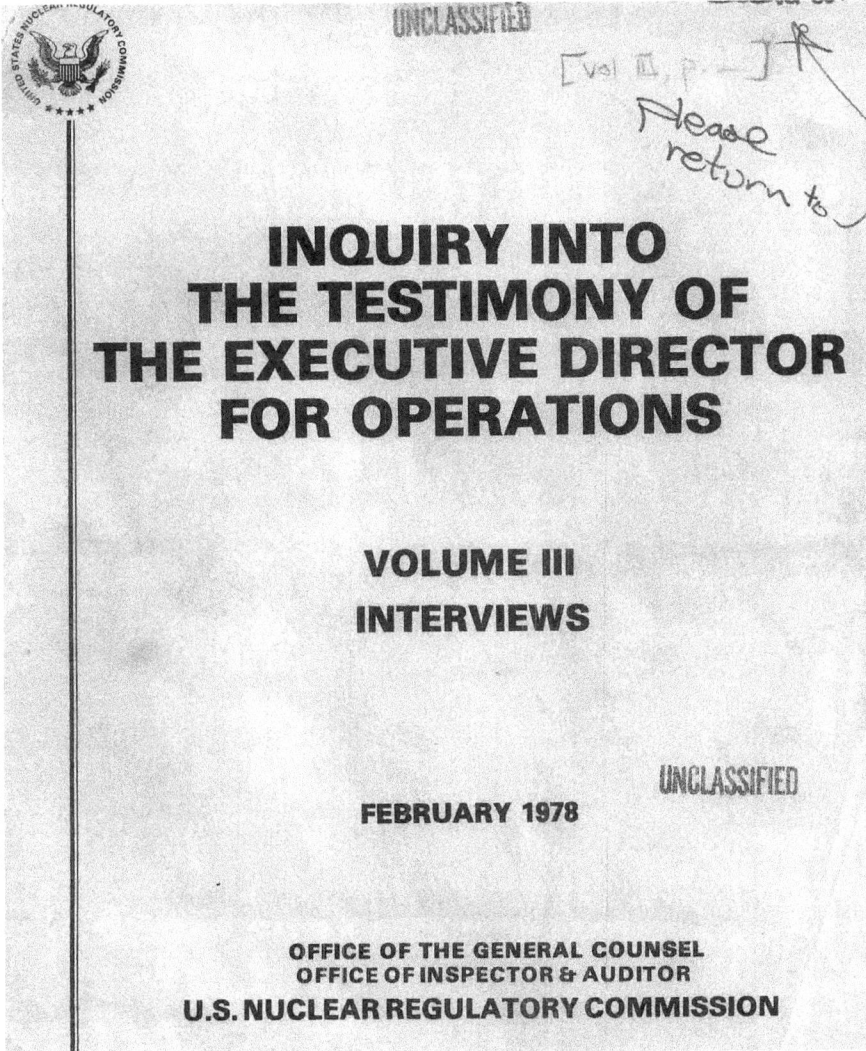

UNCLASSIFIED

[Vol III, p.___]

Please
return to

INQUIRY INTO
THE TESTIMONY OF
THE EXECUTIVE DIRECTOR
FOR OPERATIONS

VOLUME III

INTERVIEWS

UNCLASSIFIED

FEBRUARY 1978

OFFICE OF THE GENERAL COUNSEL
OFFICE OF INSPECTOR & AUDITOR
U.S. NUCLEAR REGULATORY COMMISSION

3

information. They had other information such as a type of bombing
practice done with A-4 aircraft that would not have made sense unless it
was to deliver a nuclear bomb.

By the time of the NRC briefing the question of whether U-235 had been
diverted from NUMEC was academic for the CIA because plutonium from the
Dimona reactor was believed to be available. Therefore, from the CIA's
intelligence point of view the diversion did not matter. The last
inspection of Dimona was in 1969. In his view it was less than an
adequate investigation to determine whether plutonium was there. After-
wards Israel refused to permit inspections. Furthermore, a shipment of
200 tons of non-enriched uranium from Argentina had been diverted to
Israel through a West German cut out.

Mr. Duckett raised the question of whether the U.S. had intentionally
allowed material to go to Israel. He said that if any such scheme was
under consideration, he would have known about it and he never heard so
much as a rumor about this. He, therefore, does not believe there is
any substance to this allegation. In support of this view, he related
that CIA had drafted a National Intelligence Estimate on Israel's
nuclear capability in 1968. In it was the conclusion that the Israelis
had nuclear weapons. He showed it to Mr. Helms. Helms told him not to
publish it and he would take it up with President Johnson. Mr. Helms
later related that he had spoken to the President, that the President
was concerned, and that he had said "Don't tell anyone else, even
Dean Rusk and Robert McNamara."

Mr. Duckett was asked about the reactions of NRC officials who were
present at his briefing. He said that Mr. Anders was very concerned and
felt that already too many people had been exposed to the information.
After the briefing Mr. Duckett went to Mr. Kennedy's office. Mr. Kennedy
wanted to talk about more frequent interchange of information between
the NRC and the CIA. Mr. Anders came in and wanted to apologize for
having so many people present. He said he did not realize how sensitive
the information was and if he had he would have restricted the attendance
even more. Mr. Anders said that, in the future, he should deal only
with Mr. Kennedy and him, and that in light of the sensitive nature of
the information he was going to go to the White House. During this
session, Mr. Duckett recalls that one Commissioner, probably Mr. Mason,
commented with mock jocularity "My God, I almost went to work for
Zal Shapiro. I came close to taking a job with him." By the end of the
meeting it was a pretty somber group. Mr. Duckett does not recall that
the staff actively participated in the briefing. He pointed out that it
was not a formal briefing. It was more of a discussion for the whole
session.

NUMEC diversion employee eyewitness FBI testimony – 1980[490]

In Reply, Please Refer to
File No.

UNITED STATES DEPARTMENT OF JUSTICE

FEDERAL BUREAU OF INVESTIGATION

Pittsburgh, Pennsylvania

March 25, 1980

FORMER EMPLOYEE OF NUCLEAR
MATERIALS AND EQUIPMENT CORPORATION,
APOLLO, PENNSYLVANIA;
ATOMIC ENERGY ACT

On March 21, 1980, _____ was interviewed
at his residence and provided the following information:

ALL INFORMATION CONTAINED
HEREIN IS UNCLASSIFIED
EXCEPT WHERE SHOWN
OTHERWISE

9-9-03   # R030758
           60220 BCE/MJ/#76
         4 0
           6 AS AMENDED

This document contains neither recommendations nor conclusions of the FBI. It is
the property of the FBI and is loaned to your
agency; it and its contents are not to be
distributed outside your agency.

Classified by XR-6/3JHAKB6
Declassify   DADR 4/18/84

FD-302 (REV. 3-8-77)

FEDERAL BUREAU OF INVESTIGATION

~~CONFIDENTIAL~~   3/24/80

Date of transcription _____

[   ] telephone [   ]
was interviewed at his residence. After being advised of
the identities of the interviewing Agents and the nature
of the interview, [   ] provided the following information:

[   ] advised that he was employed by Nuclear
Materials and Equipment Corp. (NUMEC) in February, 1965,
(exact date unknown) and was continuously employed
at the Apollo, Pa., facility through two ownership changes
until October, 1978. [   ] advised that he was fired
in October, 1978, by the present owner, Babcock and Wilcox,
Inc., for job abandonment following an alleged job related
illness.

[   ] advised that upon being hired at NUMEC,
he was given three days of schooling on the equipment he
was to operate and briefed by the Personnel Manager and
Low Enrichment Facility Foreman concerning the security
measures at the Apollo facility nuclear plant. He then
commenced his production line job upon completion of this
brief schooling. [   ] related that his exact position
was Senior Ammonator Operator in the Low Enriched Operations
area, which was immediately adjacent to the loading dock
area of the Apollo nuclear facility. [   ] further
described the NUMEC Apollo plant as being broken down into
four areas: the Low Enriched area, the High Enriched area,
the Sphere area, and the Peletizer area. He advised that
although his full-time job was on the Low Enriched Area
Ammonator, he worked overtime in the High Enriched area
on several occasions.

[   ] advised in late March or early April,
1965 (exact date unknown) while working on a swing shift
from 3:30 p.m. until 12:00 a.m., his Ammonator was shut
down between approximately 9:00 and 10:00 p.m. in the evening.
He stated that because of the negative air pressure within
the plant area, conditions were usually very warm so he
walked out to the loading dock for a breath of air. The
loading dock was located approximately 20 feet from his
equipment through a single door. [   ] advised that

Investigation on __3/21/80__ at __Apollo, Pa.__   File # __Pittsburgh 117-108__

by __SAs [   ]__   Date dictated __3/24/80__

This document contains neither recommendations nor conclusions of the FBI. It is the property of the FBI and is loaned to your agency;
it and its contents are not to be distributed outside your agency.

PG 117-108   2   CONFIDENTIAL

employees often went to the loading dock to get a breath
of air and further said he thought he remembered an employees'
eating area on the dock.

[   ] related that when he entered the loading
dock area on this particular evening, he noticed a flatbed
truck backed up to the loading dock with some strange equipment on it. He described the equipment as several steel
cabinets with some kind of gauges on the front of them and
other equipment which looked like a lathes. [   ]
opined the equipment may have come from the Peletizer area
of which he was not familiar. [   ] advised he then
noticed the NUMEC owner, Dr. Zalman Shapiro, pacing around
the loading dock while [   ] (Shipping and Receiving
Foreman) and [   ] truckdriver for NUMEC) were
loading "stove pipes" into the steel cabinet type equipment that he observed on the truck. [   ] recalled
that there were four or five of the steel cabinets on the
flatbed truck. [   ] stated that [   ] and [   ]
never loaded trucks themselves, always employing other workers.

[   ] stated that the "stove pipes" are cylindrical
storage containers used to store canisters of high enriched
materials in the vaults located at the Apollo nuclear facility.
[   ] stated that the "stove pipes" contained three
or four canisters which were described as highly polished
aluminum with standard printed square yellow labels, approximately three inches in diameter by six inches tall, that
normally were used to store high enriched uranium products
which [   ] defined as 95 percent uranium.

[   ] stated that he observed two workmen,
whose names he could not recall, bringing the "stove pipes"
from the High Enriched vault area located approximately
150 feet from the docks to the dock area where [   ] and
[   ] opened the "stove pipes" and withdrew the canisters
located in the "stove pipes". He then said [   ] checked
the label on each canister for information and checked it
off on a shipping order he had attached to a clipboard.
[   ] advised that the canisters were then replaced
in the "stove pipe" and then the "stove pipe" itself was
loaded into the cabinet type equipment after being wrapped
with a brown paper type insulation. [   ] advised that
he observed one cabinet being loaded and that the "stove
pipes" were placed one in each back corner of the cabinet
and one in the front center of the cabinet directly behind
the door.

[   ] described the canisters found in the
"stove pipes" as approximately three inches by six inches,
bright polished aluminum canisters with yellow labels containing typewritten information and nuclear "fan" symbols
in the upper corners of the label. [   ] said he had

CONFIDENTIAL

PG 117-108   3
never observed typewritten information on the labels that
he had previously seen on the dock.   ~~CONFIDENTIAL~~
                                                                    b6
                                                                    b7C
[   ] advised he was sure this was High Enriched uranium products due to the size and shape of the
container and the labeling. He stated that the containers
he used in the Low Enriched area were much larger than the
canisters he observed and used a different label

[   ] stated he had never seen "stove pipes"
used as shipping containers but whenever High Enriched
uranium products were shipped, the canisters were unloaded
from the "stove pipes" and loaded into cement lined steel
drums. [   ] further advised that the route the workmen
transporting the "stove pipes" used took them away from
the Low Enriched area and brought them onto the dock through
a different door. The Low Enriched materials vaults were
located approximately 50 feet from the dock area down an
angled corridor. [   ] said the normal route for High
Enriched materials from the High Enriched vaults was down
the same corridor where the Low Enriched vaults were located.

[   ] citing his natural curiosity, stated
he observed [   ] lay his clipboard down on an empty
drum located on the dock, whereupon [   ] proceeded
to read the information contained on the shipping order.
He said he noticed that the destination for the equipment
on the truck was Israel, and that it was to be transported
by ship. He recalled that the ship had a long foreign name
which he believed to be Greek, and its location at the time
(U)   was in New York City. ⊗ ⊗

[   ] advised that he believed the ship's
name was Greek because when he was in the U.S. Navy (1956-60),
he was a radio man third class stationed at the Naval Radio
Facility, Londonderry, Northern Ireland, and had handled
(U)   messages from Greek shipping among others. ⊗ ⊗

[   ] stated that after he had quickly read [   ]
the information contained on the shipping order, [   ]
grabbed the clipboard away from him, telling him in words
to the effect that the material contained in the shipping
order was confidential and not for his eyes. [   ]
advised that shortly thereafter, an armed guard ordered
him off the loading dock. [   ] stated he did not observe anybody call the armed guard nor did he see the guard
on the dock, but that he believed the guard came from one
of the hallways adjoining the dock. [   ] stated that
he was on the loading dock for approximately 15 minutes
and that at no time did Dr. Shapiro, [   ] or
[   ] or anybody else ask him to leave.
                                          4

PG 117-108   4   ~~CONFIDENTIAL~~

[   ] further advised that it was highly unusual to see Dr. Shapiro in the manufacturing section of the Apollo nuclear facility; it was unusual to see Dr. Shapiro there at night; and very unusual to see Dr. Shapiro so nervous as to pace around. [   ] described Dr. Shapiro as a very calm, cool and collected man who never got upset.

[   ] advised that the only records and documentation he had access to were the shift productions records for the Low Enrichment Area and then only during the specific shift on which he was working. He stated that at the completion of each shift, the records were removed from the manufacturing area and taken across the street to the administrative offices.

[   ] advised that most of the shipping was normally done at daytime but did state that occasionally there was some shipping activity in the early evenings. He stated it was highly unusual though that any equipment would be shipped at night.

[   ] advised that he had not seen previously the equipment he noted on the loading dock and flatbed trailer and that he had not seen the equipment subsequent to that incident or any equipment like it in the NUMEC Apollo nuclear facility.

[   ] stated he became aware of the alleged diversion of nuclear materials through newspaper articles which caused him to think. He said that "everyone" at the plant knew there were losses of materials from the High Enriched area but nobody seemed to care during the time the facility was owned by NUMEC. He stated when Atlantic Richfield Company purchased NUMEC, the losses stopped. [   ] further stated that newspaper accounts of the alleged diversion mentioned Doctor Shapiro, and he recalled that just prior to the previously mentioned incident, it was an open plant rumor that Doctor Shapiro had just returned from an extensive vacation in Israel.

[   ] advised he had not come forward before because he had a large family to support and the day following the incident, the plant Personnel Manager (name unrecalled) of NUMEC threatened to fire [   ] if he "did not keep his mouth shut" concerning what he had seen on the loading

5

dock the night before, [   ] further advised he mentioned the threat he received from the Personnel Manager
to his union steward, whereupon [   ] claims he was
visited by "sone union goons" from Kittanning, Pa., and
again told to keep his mouth shut.

[   ] stated the prevailing attitude at the
plant in 1965 by management, union and the employees was
that the Atomic Energy Commission was the enemy looking
for a reason to shut the facility down with the resultant
job losses. In addition, he stated he did not know how
or who to contact in authority who would take action.

[   ] advised that he could recall no other
information concerning this incident which occurred in late
March or early April, 1965.

Shapiro letter to Seaborg claiming all MUF had been located – 1993[491]

I was delighted to receive your revealing new book with your personal inscription. Needless to say, me immediate focus was on the NUMEC chapter. I cannot thank you enough for your determination and perseverance in upholding your principles and the rules of fairness established by the AEC, in spite of the pressures to which you were subjected by Attorney General Mitchell and his henchmen. It is interesting to note, in passing, that apparently those who are themselves dishonest have the attitude that everyone else is , as well.

Your chapter on the NUMEC affair brought back very painful memories. Not only were my career and health adversely affected and my family traumatized, but it cost me a fortune in lawyers' fees to defend myself against charges which had no basis in fact. It was particularly uncomfortable to be the butt of snide remarks made by my peers and superiors, and to be shunted aside from projects to which I know I could have made significant contributions, especially since, as Admiral Rickover reluctantly admitted, I worked so hard and effectively for the security of this country while at Bettis and NUMEC.

ducts and accounting for all of its own losses, reported a large overage in its SNM to the NRC. This overage obviously explained a significant portion of the losses reported and paid for by NUMEC. Thus, if the NRC staff were fair, they could have put an end to all of the concern at that time and saved the government the expense, and me the additional expense and embarrassment.

Since that time, B & W has decommissioned and leveled the plant in which we processed uranium, and found that the concrete floor under the process equipment was laden with SNM caused by spills which our workers unfortunately neglected to clean up. These finds prove, beyond a shadow of a doubt, that there was no diversion, only process losses. What really hurts is the fact that the NRC has not seen fit to make this public and clear my name. They were quick to accuse, but obviously loath to admit error.

Your book has make a great contribution to the history of atomic energy. I am especially grateful for your treatment of l'affair NUMEC.

Again, many, many thanks and warmest best wishes for good health, continued fulfillment, and all that you would wish for yourself.

Sincerely,

Zalman M. Shapiro

While I know that there was a great deal of concern evoked at the AEC by the loss of the SNM at NUMEC, until I read the chapter which you devoted to it, I had no idea of the activity and the magnitude of the difficulties it created for you because of interest in the matter at the Cabinet and Presidential levels.

Your account was fair and factual, as I knew it. What you may have been unaware of was the new information which came to light when the matter was reopened during the Ford Administration. In reporting on one of its inventories, B&W, after cleaning up some of the air ducts and accounting for all of its own losses, reported a large overage on its SNM to the NRC. This overage obviously explained a significant portion of the losses reported and paid for by NUMEC. Thus, if the NRC staff were fair, they could have put an end to all of the concern at that time and saved the government the expense, and me the additional expense and embarrassment.

Since that time, B&W has decommissioned and leveled the plant in which we processed uranium, and found that the concrete floor under the process equipment was laden with SNM caused by spills which our workers unfortunately neglected to clean up. These finds prove, beyond a shadow of a doubt, that there was no diversion, only process losses. What really hurts is the fact that the NRC has not seen fit to make this public and clear my name. They were quick to accuse, but obviously loath to admit error.

Your book has make [made] a great contribution to the history of atomic energy. I am especially grateful for your treatment of l 'affair NUMEC.

Again, many, many thanks and warmest best wishes for good health, continued fulfillment, and all that you would wish for yourself.

Senator Arlen Specter requests NRC exoneration of Shapiro - 2009[492]

ARLEN SPECTER
PENNSYLVANIA

COMMITTEES:
JUDICIARY
APPROPRIATIONS
ENVIRONMENT AND PUBLIC WORKS
VETERANS' AFFAIRS
AGING

☐ 711 HART SENATE OFFICE BUILDING
WASHINGTON, DC 20510–3802
202–224–4254

United States Senate

WASHINGTON, DC 20510–3802
specter.senate.gov
August 27, 2009

STATE OFFICES:
☐ 600 ARCH STREET, SUITE 9400
PHILADELPHIA, PA 19106
215–597–7200
☐ REGIONAL ENTERPRISE TOWER
425 SIXTH AVENUE, SUITE 1450
PITTSBURGH, PA 15219
412–644–3400
☐ STE B–120, FEDERAL BUILDING
17 SOUTH PARK ROW
ERIE, PA 16501
814–453–3010
☐ ROOM 1104, FEDERAL BUILDING
HARRISBURG, PA 17101
717–782–3951
☐ SUITE 3814, FEDERAL BUILDING
504 W. HAMILTON
ALLENTOWN, PA 18101
610–434–1444
☐ 310 SPRUCE STREET, SUITE 201
SCRANTON, PA 18503
570–346–2006
☐ 7 NORTH WILKES-BARRE BLVD.
SUITE 377M
116 S. MAIN STREET
WILKES-BARRE, PA 18702
570–826–6265

Ms. Rebecca Schmidt
Director
Office of Congressional Affairs
Nuclear Regulatory Commission
Washington, D.C. 20555

Dear Ms. Schmidt:

My office has been contacted by Hadrian R. Katz of Arnold & Porter, LLP on behalf of his client and my constituent, Dr. Zalman Shapiro.

According to the information I have received, Dr. Shapiro organized the Nuclear Materials and Equipment Corporation (NUMEC), which was the subject of Atomic Energy Commission, Department of Justice, and Joint Committee on Atomic Energy investigations for alleged diversion of special nuclear material. I have been advised that, following the closure of the NUMEC facility, an amount of uranium equal to the amount believed to have been diverted to Israel was later identified and collected from the decommissioned facility.

It is my understanding that no formal charges were ever brought against Dr. Shapiro; however, numerous articles and books on the subject have referenced the investigations, which Dr. Shapiro reports have lent credibility to the accusations and significantly damaged his professional reputation. Therefore, Dr. Shapiro is requesting that the Nuclear Regulatory Commission issue a formal public statement confirming that he was not involved in any activities related to the diversion of uranium to Israel. I am enclosing a copy of the correspondence that I have received, in which Dr. Shapiro's concerns are explained in greater detail.

I would greatly appreciate your reviewing this matter and affording Dr. Shapiro's request your full and fair consideration. Please direct your reply to my assistant, Mr. Bill Bayer, at the following address:

425 6th Avenue, Suite 1450
Pittsburgh, PA 15219
(412) 644-3400 (T); (412) 644-4871 (F)

Thank you for your assistance with the aforementioned matter.

Sincerely,

Arlen Specter

ARNOLD & PORTER LLP

Hadrian R. Katz
Hadrian_Katz@aporter.com

202.942.5707
202.942.5999 Fax

555 Twelfth Street, NW
Washington, DC 20004-1206

August 7, 2009

BY HAND

The Honorable Arlen Specter
United States Senate
711 Hart Building
Washington, DC 20510

Re: Dr. Zalman M. Shapiro

Dear Senator Specter:

We write at Mr. Bayer's suggestion on behalf of our long-time pro bono client and friend, Dr. Zalman M. Shapiro of Pittsburgh, a distinguished engineer and innovator, who has contributed with distinction to the advancement of science and technology, and the success of the United States nuclear program, over a 60 year career. Despite his many accomplishments, Dr. Shapiro has for many years been the subject of repeated defamatory statements, and we respectfully request your assistance in clearing a great American scientist's good name. In particular, we would ask that this letter be forwarded to the Office of the Chairman of the Nuclear Regulatory Commission with your recommendation that the Commission issue a formal statement confirming once and for all that Dr. Shapiro did not participate in the unlawful diversion of nuclear material to the State of Israel.

From 1950 to 1957, Dr. Shapiro worked at the Naval Reactor Facility at Bettis, Pennsylvania, earning Westinghouse's highest employee award for his work on zirconium. He was cited by Admiral Rickover as one of the four men most responsible for the success of the first nuclear powered submarine. Thereafter, when the government was encouraging the development of a private nuclear industry, Dr. Shapiro resigned from Westinghouse to organize the Nuclear Materials and Equipment Corporation ("NUMEC"), which was engaged primarily in the conversion of enriched UF_6 (uranium hexafluoride) into UO_2 (uranium oxide) powder for fuel fabrication, and in the reprocessing of enriched uranium scrap. NUMEC also fabricated UO_2 powder into pellets for commercial reactors, and developed and fabricated fuel for advanced reactors such as the propulsion system for the proposed NERVA rocket.

As you may be aware, during the late 1960s and 1970s, NUMEC became the subject of investigation by Attorney General Mitchell and the FBI for alleged diversion of special nuclear material to the State of Israel. In the course of processing at NUMEC,

ARNOLD & PORTER LLP

particularly scrap recovery operations, there were inevitably losses of small amounts of uranium. Because of the low product yields for the exotic fuels produced by NUMEC, an exceptional amount of scrap was generated that had to be reprocessed and recycled. Each time the material was remanufactured, the total process losses increased. Determining the amount of material lost in processing is not an exact science, and it appears that for some period of time NUMEC underestimated the amount of material expended. When it was later determined that actual processing losses exceeded NUMEC estimates, suspicion of possible diversion was raised, and intensive investigations followed.

In 1965, the Atomic Energy Commission sent a team of nuclear material management personnel to NUMEC to conduct an audit and determine, if possible, the reason for the processing losses, a cumulative total of approximately 100 Kg (220 lbs) of enriched uranium. The investigators concluded that there was no indication of any diversion, and that a diversion would have been as a practical matter impossible. In the course of our representation of Dr. Shapiro in the 1970s, we spoke with every significant individual involved in these investigations personally, and all of them repeated their conclusion that there was no diversion.

Following the AEC investigation, the FBI and the Joint Committee on Atomic Energy sent in their own teams to investigate. Various possibilities were investigated, but no evidence indicating diversion was ever found. One CIA analyst concluded that Dr. Shapiro had diverted uranium to Israel, and that view was picked up by journalists. One journalist would rely on another's misinformation, and the diversion suspicions were treated as fact. The distortions snowballed; books and articles magnified and embellished damaging falsehoods. The more these maligning assertions were repeated in print and online media, the greater the perception of credibility.

As described in a book by well-known journalist Seymour Hersh, in a chapter entitled "Injustice," the CIA operative who advanced the diversion theory later recanted: "'With all the grief I've caused,' he said, referring to Shapiro's ruined career, 'I know of nothing at all to indicate that Shapiro was guilty.'" S.M. Hersh, *The Samson Option* 255 (1991). Nevertheless, with the advent of the Internet, these discredited falsehoods are now globally available and are still kept alive and recirculating.

Though no charges were ever brought against Dr. Shapiro, as a result of all of the allegations, investigations, and being tried in the press, his life was made miserable. He was the butt of snide remarks made by peers and superiors and was shunted aside from projects to which he could have made significant contributions. Dr. Shapiro's reputation, career and health were all adversely affected, his family was traumatized, and in order to

ARNOLD & PORTER LLP

The Honorable Arlen Specter
August 7, 2009
Page 3

defend himself against charges that had no basis in fact he was forced to deplete his savings.

Dr. Glenn Seaborg summarized the Shapiro investigations with characteristic eloquence in his autobiography:

> Shapiro continued in a successful career, occupying positions of increasing responsibility with Westinghouse until his retirement in 1983. But his career might have been even more successful if not for this undeserved blemish on his record. Later in the 1970s, the story came to light after enterprising journalists filed Freedom of Information Act requests. Unfortunately, however, some of their articles left the impression that Shapiro had in fact diverted the uranium.

> Lest I be considered a biased source, with an interest in claiming that no uranium diversion happened on my watch, let me quote from Seymour Hersh's intensively researched book on Israel's quest for nuclear weapons, *The Samson Option*: "Despite more than ten years of intensive investigation involving active FBI surveillance, however, no significant evidence proving that Shapiro had diverted any uranium from his plant was ever found. Nonetheless, he remained guilty in the minds of many in the government and the press. . . . Zalman Shapiro did not divert uranium from the processing plant to Israel." Hersh relates that the "missing" uranium was found during the cleanup of Shapiro's plant: "More than one hundred kilograms of enriched uranium – the amount allegedly diverted to Israel by Zalman Shapiro – was recovered from the decommissioned plant by 1982, with still more being recovered each year."

G.T. Seaborg, *Adventures in the Atomic Age* 221-22 (2001).

Dr. Shapiro recently received his fifteenth patent, this one covering an innovative process for manufacturing both jewel-grade and industrial diamonds cheaply and efficiently. This accomplishment by an 89-year-old scientist has received favorable

ARNOLD & PORTER LLP

The Honorable Arlen Specter
August 7, 2009
Page 4

coverage in the press, but some of the stories have seen fit to repeat the old allegations of diversion.

In addition, Dr. Shapiro has recently been nominated to be one of the recipients of the 2009 National Medal of Technology and Innovation. Numerous letters of recommendation, strongly supporting that nomination, attest to Dr. Shapiro's profound contributions to the defense and well-being of the United States over his professional lifetime, and demonstrate that he richly deserves this award. But Dr. Shapiro's nomination will lead to another FBI background check, and we are concerned that repetition of the diversion innuendo could adversely affect a distinguished scientist's opportunity to receive a well-deserved honor.

Dr. Shapiro has never had an opportunity to obtain a formal statement from any government agency clearing him of the false accusations made long ago. We respectfully suggest that the time has come for the NRC once and for all to confirm that Dr. Shapiro committed no diversion. An attack on Dr. Shapiro is necessarily an attack on the Atomic Energy Commission as well, and the NRC would itself benefit from putting the stories of diversion to final rest. The NRC's unequivocal statement that Dr. Shapiro did not divert nuclear material to Israel, that the material has been accounted for, and that he is, and has always been, a loyal citizen of the United States who has contributed significantly to its defense should be conveyed to the FBI, with a recommendation that this statement be given a prominent position in the files on Dr. Shapiro.

We appreciate your attention to this letter, and respectfully ask that the NRC assign this matter to an appropriate member of the Commission staff to assist us in bringing the defamation of Dr. Shapiro to a close. We would welcome the opportunity to discuss the situation at the Commission's convenience with whomever is designated.

Respectfully yours,

Hadrian R. Katz

cc: Mr. William J. Bayer

NRC denies Zalman Shapiro exoneration request – 2009[493]

November 2, 2009

The Honorable Arlen Specter
United States Senator
425 6th Avenue, Suite 1450
Pittsburgh, PA. 15219

Dear Senator Specter:

On behalf of the U.S. Nuclear Regulatory Commission (NRC), I am responding to your letter, dated August 7, 2009, regarding Dr. Zalman M. Shapiro and the previous Federal investigation of his activities as head of the Nuclear Material and Equipment Corporation (NUMEC). You requested that NRC give full and fair consideration to Dr. Shapiro's request for a formal public statement confirming that Dr. Shapiro was not involved in any activities related to the diversion of uranium. This request was based on a letter that you received from Dr. Shapiro's attorney.

The Atomic Energy Commission (AEC) and other Federal agencies investigated whether Dr. Shapiro played a role in the possible diversion of nuclear material. The AEC concluded that it had no evidence that diversion had occurred. However, the AEC also determined that it did not have sufficient evidence to conclude, unequivocally, that uranium had not been diverted. In the time period that these statements were made, the conclusions reflected the minimal nature of the material control and accounting features that were in use to detect the loss or diversion of special nuclear material.

Your request is based on information that during the decommissioning of the facility, an amount of uranium equal to the amount alleged to have been diverted had recovered at the facility. NRC staff has reviewed agency documents related to the "material unaccounted for" (MUF) discovered at the site and investigated by the AEC in 1965, including those pertaining to additional inspections and MUF evaluations during subsequent operations, and the decommissioning activities for the facility.

Accordingly, after a thorough review of its records, NRC found no documents that provided specific evidence that the diversion of nuclear materials occurred. However, consistent with previous Commission statements, NRC does not have information that would allow it to unequivocally conclude that nuclear material was not diverted from the site, nor that all previously unaccounted for material was accounted for during the decommissioning of the site.

Sincerely,

/RA/

R. W. Borchardt
Executive Director
for Operations

CIA denies release of Carl Duckett's NRC briefing file – 2011[494]

Central Intelligence Agency

Washington, D.C. 20505

15 July 2011

Mr. Grant F. Smith
Director of Research
Institute for Research: Middle Eastern Policy
Calvert Station
P.O. Box 32041
Washington, D.C. 20007

Reference: F-2011-00873

Dear Mr. Smith:

On 28 February 2011, the office of the Information and Privacy Coordinator received your 18 February 2011 Freedom of Information Act request, made on behalf of the Institute for Research: Middle Eastern Policy, for:

1. all contents of a CIA folder NUMEC classified as Top Secret.

2. any related notes made by CIA Deputy Director Carl Duckett's in relation to his February, 1976 briefing to the Nuclear Regulatory Commission.

We have assigned your request the reference number above. Please use this number when corresponding so that we can identify it easily.

In regards to Item 1, we cannot accept your FOIA request in its current form because it would require the Agency to perform an unreasonably burdensome search. The FOIA requires requesters to "reasonably describe" the information they seek so that professional employees familiar with the subject matter can locate responsive information with a reasonable amount of effort. Because of the breadth and lack of specificity of your request, and the way in which our records systems are configured, the Agency cannot conduct a reasonable search for information responsive to your request.

In regards to Item 2, we have previously conducted searches on behalf of earlier requesters for records concerning the subject of your request. No responsive records were located. Although our searches were through and diligent, and it is highly unlikely that repeating those searches would change the result you nevertheless have the legal right to appeal the finding of no records responsive to your request. Should you choose to do so, you may address your appeal to the Agency Release Panel, in my care, within 45 days from the date of this letter. Please include the basis of your appeal.

Sincerely,

Susan Viscuso
Information and Privacy Coordinator

Sources

1 U.S. House of Representatives, Transcript of Proceedings, "Informal Meeting Between Interior Committee Representatives and Dr. Zalman M. Shapiro." – December 28, 1978 – Morris K. Udall papers, University of Arizona.

2 Abraham Feinberg FBI File, released under the Freedom of Information Act 1147992-000 on November 5, 2010.
   Israel Lobby Archive, http://www.IRmep.org/ila/feinberg/

3 Nahum A. Bernstein, FBI File, released under the Freedom of Information Act 1161210-000 on July 12, 2011
   Israel Lobby Archive http://www.IRmep.org/ila/Bernstein/

4 Nahum A. Bernstein, FBI File, counterintelligence 65-59184-1, 2, 3, released under FOIA 1161210-000 on July 12, 2011
   http://IRmep.org/ILA/Bernstein/default.asp

5 Slater, Leonard, *The Pledge*, New York: Simon and Schuster, 1971 pp. 137-140

6 Calhoun, Ricky-Dale, "Arming David: The Haganah's Illegal Arms Procurement Network in the United States, 1945-1949," *Journal of Palestine Studies*, Vol. 36, No. 4 (Summer 2007)

7 Calhoun, Ricky-Dale, "Arming David: The Haganah's Illegal Arms Procurement Network in the United States, 1945-1949," *Journal of Palestine Studies*, Vol. 36, No. 4 (Summer 2007)

8 Brackman, Harold, "Hawaii Residents Aided Underdog Israel's Struggle," *Honolulu Star Bulletin*, October 15, 2006

9 Calhoun, Ricky-Dale, "Arming David: The Haganah's Illegal Arms Procurement Network in the United States, 1945-1949," *Journal of Palestine Studies*, Vol. 36, No. 4 (Summer 2007)

10 Central Intelligence Agency, "Clandestine Air Transport Operations: Memorandum for the Secretary of Defense," May 28, 1948; declassified and released on September 27, 2001, CIA Freedom of Information Act electronic reading room

11 Clark, Alfred E., "Henry Montor is Dead at 76; U.J.A. and Israel Bond Leader," *New York Times*, April 16, 1982

12 Slater, Leonard, *The Pledge*, New York: Simon and Schuster, 1971 p. 301

13 Schiesel, Seth, "Robert R. Nathan, 92, Dies; Set Factory Goals in War," *New York Times*, September 10, 2001

14 Slater, Leonard, *The Pledge*, New York: Simon and Schuster, 1971 p. 76

15 Brackman, Harold, "Hawaii Residents Aided Underdog Israel's Struggle," *Honolulu Star Bulletin*, October 15, 2006

16 Burston, Bradley, "Al Schwimmer," *The Jewish Daily Forward*, March 2, 2001

17 NUMEC/Zalman Shapiro FBI file, released under the Freedom of Information Act 1146454-000 – 105-188123   p   20   Israel   Lobby   Archive http://www.IRmep.org/ila/numec

18 NUMEC/Zalman Shapiro FBI file, 1091168-000-117-2564 Section 4 (805090) p 87 Israel Lobby Archive http://www.IRmep.org/ila/numec

19 NUMEC/Zalman Shapiro FBI file FBI 1091168-000 – 117-2564 – Section 1 (805474) p 18 Israel Lobby Archive http://www.IRmep.org/ila/numec

20 Cohen, Avner "The Worst Kept Secret: Israel's Bargain with the Bomb" Columbia University Press, New York, 2010 p. 57

21 NUMEC/Zalman Shapiro FBI file, 1091168-000-117-2564 Section 4 (805090) p 87 Israel Lobby Archive http://www.IRmep.org/ila/numec

22 Cohen, Avner "The Worst Kept Secret: Israel's Bargain with the Bomb" Columbia University Press, New York, 2010 pp. 178-179

23 NUMEC/Zalman Shapiro FBI file FBI 1146454-000 – 105-188123 p 9 Israel Lobby Archive http://www.IRmep.org/ila/numec

24 NUMEC/Zalman Shapiro FBI file 1091168-000 – 117-2564 – Section 1 (805474) p 18 Israel Lobby Archive http://www.IRmep.org/ila/numec

25 "Nuclear Diversion in the U.S.? 13 Years of Contradiction and Confusion." General Accounting Office, December 18, 1978. Partially Declassified on May 6, 2010, p. 5 http://IRmep.org/ILA/nukes/NUMEC/co1162251.pdf

26 Martin, David C. "Mysteries of Israel's Bomb" *Newsweek* , January 9, 1979

27 Shapiro, Zalman, Application for National Medal for Technology and Innovation, 2009 obtained under FOIA, Israel Lobby Archive http://IRmep.org/ILA/nukes/specter/shapiro_nomination.pdf

28 Cooke, Stephanie (2009). *In Mortal Hands: A Cautionary History of the Nuclear Age*, Black Inc., p. 252.

29 NUMEC Articles of Incorporation, Department of State, Commonwealth of Pennsylvania, notarized December 28, 1956, and Certificate of Incorporation issued December 31, 1956, which became the official date of incorporation.

30 *Numec Spark*, company newsletter, February, 1969

31 "NUMEC's Founders Say it Takes Only Common Sense" *The New York Times*, September 25, 1960

32" NUMEC's Founders Say it Takes Only Common Sense," *The New York Times*, September 25, 1960

33 *Numec Spark*, a company newsletter, December, 1968

34 Maclean, John "Making the Bomb: 3 Theories," *Chicago Tribune*, November 20, 1970, pg A2

35 Burnham, David, special to *The New York Times*, "U.S. Documents Support Belief Israel Got Missing Uranium for Arms," November 6, 1977

36 "Nuclear Diversion in the U.S.? 13 Years of Contradiction and Confusion." General Accounting Office, December 18, 1978. Partially Declassified on May 6, 2010, p. 5 http://IRmep.org/ILA/nukes/NUMEC/co1162251.pdf

37 "Nuclear Nightmare" *Pittsburgh Tribune*, April 12, 2008

38 Company promotional booklet "NUMEC 1966" The booklet contains a calendar, reactive property charts and useful references such as a periodic and conversion tables. It promotes 66 distinct NUMEC goods and services within the following categories: Fuel materials, moderating materials, control materials, crystal bar, powder, preparation and fabrication; power generators, quality control equipment, chemical and laboratory equipment, coatings, scrap recovery, laboratory services, decontamination, consulting services and architectural and engineering services. The booklet consulted is available from the National Agricultural Library.

39 Acton, Robin "Nuclear settlement money little solace for survivors in Armstrong County. *Pittsburgh Tribune*, May 4, 2008
    http://www.pittsburghlive.com/x/pittsburghtrib/business/s_565716.html

40 "City Water Fails; 45 Dead, 350 Hurt: Receding Rivers Reveal Vast Desolation" *Pittsburgh Sun-Telegraph*, March 20, 1936
    http://www.clpgh.org/exhibit/neighborhoods/downtown/down_n41.html

41 Ameno, Patricia J. "Testimony on Radiation Levels at Landfills Before the Senate Environmental Resources and Energy Committee, Harrisburg, PA" June 28, 2009

42 Ameno, Patricia J. "Testimony on Radiation Levels at Landfills Before the Senate Environmental Resources and Energy Committee, Harrisburg, PA" June 28,2009

43 House of Representatives, Transcript of Proceedings, "Informal Meeting Between Interior Committee Representatives and Dr. Zalman M. Shapiro." –December 21, 1978

[44] David Lowenthal FBI File, released under the Freedom of Information Act 1146454-000 on August 29, 2011 FBI FD-297 (log for technical surveillance), May 5, 1969

[45] "Nuclear Nightmare" *Pittsburgh Tribune*, April 12, 2008

[46] "Nuclear Nightmare" *Pittsburgh Tribune*, April 12, 2008

[47] NUMEC/Zalman Shapiro FBI file 1091168-000 --- 117-2564 --- Section 1 (830179) p 25. Israel Lobby Archive http://www.IRmep.org/ila/numec

[48] Acton, Robin "Nuclear settlement money little solace for survivors in Armstrong County. *Pittsburgh Tribune*, May 4, 2008
http://www.pittsburghlive.com/x/pittsburghtrib/business/s_565716.html

[49] Yerace, Tom "Service Ministers to former nuke workers," *McClatchy – Tribune Business News*, April 28, 2008

[50] "Nuclear Nightmare" *Pittsburgh Tribune*, April 12, 2008

[51] Keller, Charles A, Assistant manager for manufacturing and support, U.S. Department of Energy, Oak Ridge. Summary of diary entries on NUMEC, p. 1 Benjamin S. Loeb Papers, Manuscript Division, Library of Congress

[52] Keller, Charles A, Assistant manager for manufacturing and support, U.S. Department of Energy, Oak Ridge. Summary of diary entries on NUMEC, p. 2 Benjamin S. Loeb Papers, Manuscript Division, Library of Congress

[53] Keller, Charles A, Assistant manager for manufacturing and support, U.S. Department of Energy, Oak Ridge. Summary of diary entries on NUMEC, p. 1 Benjamin S. Loeb Papers, Manuscript Division, Library of Congress

[54] Keller, Charles A, Assistant manager for manufacturing and support, U.S. Department of Energy, Oak Ridge. Summary of diary entries on NUMEC, p. 2 Benjamin S. Loeb Papers, Manuscript Division, Library of Congress

[55] Keller, Charles A, Assistant manager for manufacturing and support, U.S. Department of Energy, Oak Ridge. Summary of diary entries on NUMEC, pp. 2-3 Benjamin S. Loeb Papers, Manuscript Division, Library of Congress

[56] Keller, Charles A, Assistant manager for manufacturing and support, U.S. Department of Energy, Oak Ridge. Summary of diary entries on NUMEC, pp. 2-3 Benjamin S. Loeb Papers, Manuscript Division, Library of Congress

[57] Keller, Charles A, Assistant manager for manufacturing and support, U.S. Department of Energy, Oak Ridge. Summary of diary entries on NUMEC, p. 3 Benjamin S. Loeb Papers, Manuscript Division, Library of Congress

[58] Keller, Charles A, Assistant manager for manufacturing and support, U.S. Department of Energy, Oak Ridge. Summary of diary entries on NUMEC, p. 3 Benjamin S. Loeb Papers, Manuscript Division, Library of Congress

[59] Keller, Charles A, Assistant manager for manufacturing and support, U.S. Department of Energy, Oak Ridge. Summary of diary entries on NUMEC, p. 3 Benjamin S. Loeb Papers, Manuscript Division, Library of Congress

[60] Keller, Charles A, Assistant manager for manufacturing and support, U.S. Department of Energy, Oak Ridge. Summary of diary entries on NUMEC, p. 4 Benjamin S. Loeb Papers, Manuscript Division, Library of Congress

[61] Keller, Charles A, Assistant manager for manufacturing and support, U.S. Department of Energy, Oak Ridge. Summary of diary entries on NUMEC, p. 4 Benjamin S. Loeb Papers, Manuscript Division, Library of Congress

[62] Keller, Charles A, Assistant manager for manufacturing and support, U.S. Department of Energy, Oak Ridge. Summary of diary entries on NUMEC, p. 5 Benjamin S. Loeb Papers, Manuscript Division, Library of Congress

[63] Keller, Charles A, Assistant manager for manufacturing and support, U.S. Department of Energy, Oak Ridge. Summary of diary entries on NUMEC, p. 5 Benjamin S. Loeb Papers, Manuscript Division, Library of Congress

[64] NUMEC/Zalman Shapiro FBI file, 1091168-000 --- 117-2564 --- Section 1 (830179)) pp 23-24. Israel Lobby Archive http://www.IRmep.org/ila/numec

[65] Keller, Charles A, Assistant manager for manufacturing and support, U.S. Department of Energy, Oak Ridge. Summary of diary entries on NUMEC, Benjamin S. Loeb Papers, Manuscript Division, Library of Congress

[66] Keller, Charles A, Assistant manager for manufacturing and support, U.S. Department of Energy, Oak Ridge. Summary of diary entries on NUMEC, Benjamin S. Loeb Papers, Manuscript Division, Library of Congress

[67] Keller, Charles A, Assistant manager for manufacturing and support, U.S. Department of Energy, Oak Ridge. Summary of diary entries on NUMEC, p. 9 Benjamin S. Loeb Papers, Manuscript Division, Library of Congress

[68] Keller, Charles A, Assistant manager for manufacturing and support, U.S. Department of Energy, Oak Ridge. Summary of diary entries on NUMEC, p. 10 Benjamin S. Loeb Papers, Manuscript Division, Library of Congress

[69] Memo to Henry Myers, House Interior Committee, Charles A. Keller, Assistant Manager for Manufacturing and Support, December 14, 1979, Benjamin S. Loeb Papers, Office Diary, Glenn T. Seaborg, Chair of the AEC 1961-1972, Box 7, Folder 9

[70] Memo to Henry Myers, House Interior Committee, Charles A. Keller, Assistant Manager for Manufacturing and Support, December 14, 1979, Benjamin S. Loeb Papers, Office Diary, Glenn T. Seaborg, Chair of the AEC 1961-1972, Box 7, Folder 9

[71] Keller, Charles A, Assistant manager for manufacturing and support, U.S. Department of Energy, Oak Ridge. Summary of diary entries on NUMEC, p. 10 Benjamin S. Loeb Papers, Manuscript Division, Library of Congress

[72] Keller, Charles A, Assistant manager for manufacturing and support, U.S. Department of Energy, Oak Ridge. Summary of diary entries on NUMEC, p. 11 Benjamin S. Loeb Papers, Manuscript Division, Library of Congress

[73] Keller, Charles A, Assistant manager for manufacturing and support, U.S. Department of Energy, Oak Ridge. Summary of diary entries on NUMEC, Benjamin S. Loeb Papers, Manuscript Division, Library of Congress

[74] Keller, Charles A, Assistant manager for manufacturing and support, U.S. Department of Energy, Oak Ridge. Summary of diary entries on NUMEC, p. 12 Benjamin S. Loeb Papers, Manuscript Division, Library of Congress

[75] Keller, Charles A, Assistant manager for manufacturing and support, U.S. Department of Energy, Oak Ridge. Summary of diary entries on NUMEC, pp 12-13 Benjamin S. Loeb Papers, Manuscript Division, Library of Congress

[76] Keller, Charles A, Assistant manager for manufacturing and support, U.S. Department of Energy, Oak Ridge. Summary of diary entries on NUMEC, p. 12 Benjamin S. Loeb Papers, Manuscript Division, Library of Congress

[77] Keller, Charles A, Assistant manager for manufacturing and support, U.S. Department of Energy, Oak Ridge. Summary of diary entries on NUMEC, p. 13 Benjamin S. Loeb Papers, Manuscript Division, Library of Congress

[78] Keller, Charles A, Assistant manager for manufacturing and support, U.S. Department of Energy, Oak Ridge. Summary of diary entries on NUMEC p. 13 Benjamin S. Loeb Papers, Manuscript Division, Library of Congress

[79] Glenn T. Seaborg papers, Library of Congress, Manuscript Division, box 59, folder 3

[80] Keller, Charles A, Assistant manager for manufacturing and support, U.S. Department of Energy, Oak Ridge. Summary of diary entries on NUMEC p. 13 Benjamin S. Loeb Papers, Manuscript Division, Library of Congress

[81] Keller, Charles A, Assistant manager for manufacturing and support, U.S. Department of Energy, Oak Ridge. Summary of diary entries on NUMEC pp 12-13 Benjamin S. Loeb Papers, Manuscript Division, Library of Congress

[82] Keller, Charles A, Assistant manager for manufacturing and support, U.S. Department of Energy, Oak Ridge. Summary of diary entries on NUMEC p. 14 Benjamin S. Loeb Papers, Manuscript Division, Library of Congress

[83] Keller, Charles A, Assistant manager for manufacturing and support, U.S. Department of Energy, Oak Ridge. Summary of diary entries on NUMEC p. `4 Benjamin S. Loeb Papers, Manuscript Division, Library of Congress

[84] Keller, Charles A, Assistant manager for manufacturing and support, U.S. Department of Energy, Oak Ridge. Summary of diary entries on NUMEC p. 15 Benjamin S. Loeb Papers, Manuscript Division, Library of Congress

[85] Keller, Charles A, Assistant manager for manufacturing and support, U.S. Department of Energy, Oak Ridge. Summary of diary entries on NUMEC Benjamin S. Loeb Papers, Manuscript Division, Library of Congress

[86] Keller, Charles A, Assistant manager for manufacturing and support, U.S. Department of Energy, Oak Ridge. Summary of diary entries on NUMEC pp 15-16 Benjamin S. Loeb Papers, Manuscript Division, Library of Congress

[87] Keller, Charles A, Assistant manager for manufacturing and support, U.S. Department of Energy, Oak Ridge. Summary of diary entries on NUMEC p. 16 Benjamin S. Loeb Papers, Manuscript Division, Library of Congress

[88] Keller, Charles A, Assistant manager for manufacturing and support, U.S. Department of Energy, Oak Ridge. Summary of diary entries on NUMEC p. 17 Benjamin S. Loeb Papers, Manuscript Division, Library of Congress

[89] Keller, Charles A, Assistant manager for manufacturing and support, U.S. Department of Energy, Oak Ridge. Summary of diary entries on NUMEC p. 18 Benjamin S. Loeb Papers, Manuscript Division, Library of Congress

[90] Keller, Charles A, Assistant manager for manufacturing and support, U.S. Department of Energy, Oak Ridge. Summary of diary entries on NUMEC p 18 Benjamin S. Loeb Papers, Manuscript Division, Library of Congress

[91] Keller, Charles A, Assistant manager for manufacturing and support, U.S. Department of Energy, Oak Ridge. Summary of diary entries on NUMEC p. 18 Benjamin S. Loeb Papers, Manuscript Division, Library of Congress

[92] Keller, Charles A, Assistant manager for manufacturing and support, U.S. Department of Energy, Oak Ridge. Summary of diary entries on NUMEC Benjamin S. Loeb Papers, Manuscript Division, Library of Congress

[93] Keller, Charles A, Assistant manager for manufacturing and support, U.S. Department of Energy, Oak Ridge. Summary of diary entries on NUMEC pp. 20-21 Benjamin S. Loeb Papers, Manuscript Division, Library of Congress

[94] Keller, Charles A, Assistant manager for manufacturing and support, U.S. Department of Energy, Oak Ridge. Summary of diary entries on NUMEC p.21 Benjamin S. Loeb Papers, Manuscript Division, Library of Congress

[95] Keller, Charles A, Assistant manager for manufacturing and support, U.S. Department of Energy, Oak Ridge. Summary of diary entries on NUMEC p. 21 Benjamin S. Loeb Papers, Manuscript Division, Library of Congress

[96] Memo to FBI Director from NYC SAC, BJL, Appendix B "Transfers to Foreign Entities, License No. SNM-145 – Uranium Enriched in the Isotope 235, Nuclear Materials and Equipment Corporation, Apollo, Pennsylvania for the Period December 1, 1957 to October 31, 1965, Benjamin Loeb Papers, Box 7, Folder 3

[97] Burnham, David, special to *The New York Times*, "U.S. Documents Support Belief Israel Got Missing Uranium for Arms" November 6, 1977.

[98] Transfers to Foreign Entities, License No. SNM-145 – Uranium Enriched in the Isotope 235 – Nuclear Materials and Equipment Corporation, Apollo, PA for the Period December 1, 1957 to October 31, 1965, Atomic Energy Commission, Benjamin Loeb Papers

[99] "Nuclear Nonproliferation: U.S. Agencies have Limited Ability to Account for, Monitor, and Evaluate the Security of U.S. Nuclear Material Overseas" Government Accountability Office, September, 2011

[100] Keller, Charles A, Assistant manager for manufacturing and support, U.S. Department of Energy, Oak Ridge. Summary of diary entries on NUMEC, p. 21 Benjamin S. Loeb Collection, Manuscript Division, Library of Congress

[101] Federal Bureau of Investigation memo, "Atomic Energy Act: Obstruction of Justice," July 28, 1976, file 117-2564-405, Benjamin S. Loeb Papers box 7, folder 2.

[102] Keller, Charles A, Assistant manager for manufacturing and support, U.S. Department of Energy, Oak Ridge. Summary of diary entries on NUMEC, p. 22 Benjamin S. Loeb Papers, Manuscript Division, Library of Congress

[103] Seaborg, Glenn T. – Letter to Benjamin Loeb, Benjamin S. Loeb Papers box 7, folder 12.

[104] Atomic Energy Commission, "Summary Notes of Briefing on Safeguards and Domestic Material Accountability" February 14, 1966 Benjamin S. Loeb Papers box 6, folder 8.

[105] FBI Memo – D.J. Brennan to W.C. Sullivan "Processing Loss of 61 Kilograms of Uranium-235 Nuclear Materials and Electronics Corp – NUMEC Apollo, Pennsylvania, Atomic Energy Act" February 18, 1966 – Benjamin S. Loeb Papers box 6, folder 8.

[106] FBI Memo – John Edgar Hoover "Processing Loss of 61 Kilograms of Uranium-235 Nuclear Materials and Electronics Corp – NUMEC Apollo, Pennsylvania, Atomic Energy Act" March 1, 1966 – Benjamin S. Loeb Papers box 6, folder 8.

[107] Memorandum from Bartlett, Earl F. Lane, and Samuel C.T. McDowell to Howard C. Brown, Assistant General Manager for Administration "NUMEC—Interviews of Former and Present Employees" April 6, 1966

[108] Exhibit B – Instructions of Office of General Counsel RE NUMEC Interviews – Certain Considerations to bear in Mind from Legal Viewpoint", AEC, Benjamin S. Loeb Papers box 6, folder 8.

[109] Exhibit A – "Interviews—General Objectives and Specific Questions", AEC, Benjamin S. Loeb Papers box 6, folder 8.

[110] Keller, Charles A, Assistant manager for manufacturing and support, U.S. Department of Energy, Oak Ridge. Summary of diary entries on NUMEC, p. 21 Benjamin S. Loeb Papers, Manuscript Division, Library of Congress

[111] FBI Report – Zalman Shapiro, page 14, Benjamin S. Loeb Papers Box 8, Folder 5

[112] FBI Report – Zalman Shapiro, page 16, Benjamin S. Loeb Papers Box 8, Folder 5

[113] FBI FD-302 "Dr. Zalman Mordecai Shapiro" June 12, 1966, Benjamin S. Loeb Papers Box 8, Folder 5

[114] NUMEC/Zalman Shapiro FBI file, 1091168-000—117-2564—Section 7 (805473) p 9 Israel Lobby Archive http://www.IRmep.org/ila/numec

[115] Cohen, Avner "The Worst Kept Secret: Israel's Bargain with the Bomb" Columbia University Press, New York, 2010 pp.91-92

[116] FBI memo, Benjamin S. Loeb Papers, box 8, folder 5

[117] The declassified Justice Department files on these key foreign agent battles may be viewed online at the Israel Lobby Archive http://www.IRmep.org/ila/AZCDOJ

[118] NUMEC/Zalman Shapiro FBI file, 1091168-000—117-2564—Section 4 (805648) p 7 Israel Lobby Archive http://www.IRmep.org/ila/numec

[119] Glenn T. Seaborg papers, Library of Congress, Manuscript Division, box 61, folder 7

[120] Keller, Charles A, Assistant manager for manufacturing and support, U.S. Department of Energy, Oak Ridge. Summary of diary entries on NUMEC, p. 26 Benjamin S. Loeb Papers, Manuscript Division, Library of Congress

[121] Keller, Charles A, Assistant manager for manufacturing and support, U.S. Department of Energy, Oak Ridge. Summary of diary entries on NUMEC, p. 26 Benjamin S. Loeb Papers, Manuscript Division, Library of Congress

[122] Keller, Charles A, Assistant manager for manufacturing and support, U.S. Department of Energy, Oak Ridge. Summary of diary entries on NUMEC, p. 28 Benjamin S. Loeb Papers, Manuscript Division, Library of Congress

[123] NUMEC/Zalman Shapiro FBI file, 1091168-000 − 117-2564 − Section 7 (805473) p 369 Israel Lobby Archive http://www.IRmep.org/ila/numec

[124] Gilinsky, Victor and Mattson, Roger J "Revisiting the NUMEC Affair" *Bulletin of the Atomic Scientists*, March/April 2010 citing John J. Fialka, "CIA Found Israel Could Make Bomb: Soil, Air Samples Disclosed Atomic Capability," Washington Star, December 8, 1977; John J. Fialka, "The American Connection: How Israel Got the Bomb," Washington Monthly, January 1979; Seymour Hersh, The Samson Option: Israel's Nuclear Arsenal and American Foreign Policy (New York: Random House, 1991), p. 255; *The Atomic Energy Commission under Nixon*, p. 197; Energy Department memorandum, deputy inspector general to undersecretary, "August 8, 1977 NUMEC-Related Congressional Hearing," April 27, 1979, Udall papers.

[125] McTiernan, Tom "Inquiry into the Testimony of the Executive Director for Operations" Volume III, Interviews, February 1978. The CIA's Carl Duckett briefed NRC commissioners in 1976. In 1978, Tom McTiernan of NRC investigated the 1977 Congressional testimony of NRC's Executive Director for Operations Lee Gossick to see if Gossick lied to Congress about whether officials thought there was evidence of a diversion. The 1978 report of McTiernan's investigation contains recollections by NRC people who attended the Duckett briefing in 1976. There is also a four page summary of an interview with Duckett. Nearly all of what Duckett said or what others recalled he said was redacted from the public version of McTiernan's report that was eventually released to the public. However, one page (number 3) of the four pages summarizing Duckett's interview summary was inadvertently released to the Natural Resources Defense Council when the report was first made public. See the Appendices for the page.

[126] NUMEC/Zalman Shapiro FBI file FBI 1091168-000 − 117-2564 − Section 2 (805648) p 1 Israel Lobby Archive http://www.IRmep.org/ila/numec

[127] FBI Airtel, June 14,, Benjamin S. Loeb Papers, box 8, folder 5

[128] FBI Airtel, May 28, 1968, Benjamin S. Loeb Papers, box 8, folder 5

[129] FBI memo to director, June 13, 1968, Benjamin S. Loeb Papers, box 8, folder 5

[130] Atlantic Richfield Annual Report, 1967, page 35

[131] FBI Pittsburgh, PA report August 21, 1968, Benjamin S. Loeb Papers, box 8, folder 5

[132] FBI SAC New York AIRTEL to Director October 28, 1968, BENJAMIN S. LOEB PAPERS, box 8, folder 5

[133] FBI SAC New York AIRTEL to Director October 16, 1968, Benjamin S. Loeb Papers, box 8, folder 5

[134] NUMEC/Zalman Shapiro FBI file, 1091168-000 − 117-2564 − Section 2 (805648) pp 7 Israel Lobby Archive http://www.IRmep.org/ila/numec

[135] NUMEC/Zalman Shapiro FBI file, FBI 1091168-000 − 117-2564 − Section 2 (805648) pp 5-7 Israel Lobby Archive http://www.IRmep.org/ila/numec

[136] NUMEC *Daily Dispatch* August 18, 1968

[137] NUMEC/Zalman Shapiro FBI file, 1091168-000 − 117-2564 − Section 2 (805648) pp 7 Israel Lobby Archive http://www.IRmep.org/ila/numec

[138] "Israeli Spy Visited A-Plant Where Uranium Vanished," *The Los Angeles Times*, June 16, 1986.

[139] Cohen, Avner "The Worst Kept Secret: Israel's Bargain with the Bomb" Columbia University Press, New York, 2010 pp. 178-179

[140] Cohen, Avner "The Worst Kept Secret: Israel's Bargain with the Bomb" Columbia University Press, New York, 2010 pp. 179

141 FBI SAC WFO New York AIRTEL to Director November 9, 1968, Benjamin S. Loeb Papers, box 8, folder 5

142 Zipporah Schefrin, *Palm Beach Daily News*, October 10, 2011 http://www.palmbeachdailynews.com/news/zipporah-schefrin-1905905.html

143 NUMEC/Zalman Shapiro FBI file, FBI 1091168-000 – 117-2564 – Section 2 (805648) p 8 Israel Lobby Archive http://www.IRmep.org/ila/numec

144 The Battle over the NPT: The Warne-Rabin Dialogue, National Security Archive at George Washington University http://www.gwu.edu/~nsarchiv/israel/documents/battle/index.html

145 NUMEC/Zalman Shapiro FBI file, 1091168-000 – 117-2564 – Section 2 (805648) p 9 Israel Lobby Archive http://www.IRmep.org/ila/numec

146 NUMEC/Zalman Shapiro FBI file, 1091168-000 – 117-2564 – Section 4 (805648) p 1 Israel Lobby Archive http://www.IRmep.org/ila/numec

147 NUMEC/Zalman Shapiro FBI file, 1091168-000 – 117-2564 – Section 4 (805648) p 6-7 Israel Lobby Archive http://www.IRmep.org/ila/numec

148 NUMEC/Zalman Shapiro FBI file, 1091168-000 – 117-2564 – Section 4 (805648) p 7 Israel Lobby Archive http://www.IRmep.org/ila/numec

149 NUMEC/Zalman Shapiro FBI file, 1091168-000 – 117-2564 – Section 4 (805648) p 7 Israel Lobby Archive http://www.IRmep.org/ila/numec

150 NUMEC/Zalman Shapiro FBI file, 1091168-000 – 117-2564 – Section 7 (805473) p 107 Israel Lobby Archive http://www.IRmep.org/ila/numec

151 NUMEC/Zalman Shapiro FBI file, 1091168-000 – 117-2564 – Section 2 (805648) p 11 Israel Lobby Archive http://www.IRmep.org/ila/numec

152 NUMEC/Zalman Shapiro FBI file, 1091168-000 – 117-2564 – Section 4 (805090) p 8 Israel Lobby Archive http://www.IRmep.org/ila/numec

153 NUMEC/Zalman Shapiro FBI file, 1091168-000 – 117-2564 – Section 4 (805090) p 8 Israel Lobby Archive http://www.IRmep.org/ila/numec

154 NUMEC/Zalman Shapiro FBI file, 1091168-000 – 117-2564 – Section 4 (805090) p 10 Israel Lobby Archive http://www.IRmep.org/ila/numec

155 NUMEC/Zalman Shapiro FBI file, 1091168-000 – 117-2564 – Section 4 (805090) p 11 Israel Lobby Archive http://www.IRmep.org/ila/numec

156 NUMEC/Zalman Shapiro FBI file, 1091168-000 – 117-2564 – Section 4 (805090) p 11 Israel Lobby Archive http://www.IRmep.org/ila/numec

157 NUMEC/Zalman Shapiro FBI file, 1091168-000 – 117-2564 – Section 4 (805090) p 12 Israel Lobby Archive http://www.IRmep.org/ila/numec

158 NUMEC/Zalman Shapiro FBI file, 1091168-000 – 117-2564 – Section 4 (805090) p 18 Israel Lobby Archive http://www.IRmep.org/ila/numec

159 NUMEC/Zalman Shapiro FBI file, 1091168-000 – 117-2564 – Section 4 (805090) p 14 Israel Lobby Archive http://www.IRmep.org/ila/numec

160 NUMEC/Zalman Shapiro FBI file, 1091168-000 – 117-2564 – Section 4 (805090) p 1 Israel Lobby Archive http://www.IRmep.org/ila/numec

161 NUMEC/Zalman Shapiro FBI file, 1091168-000 – 117-2564 – Section 4 (805090) pp 34-35 Israel Lobby Archive http://www.IRmep.org/ila/numec

162 Routing Envelope February, 14 1968, Benjamin S. Loeb Papers, box 8, folder 5

163 Recommendation to SAC Pittsburgh for awards to two special agents, February 28, 1969. Benjamin S. Loeb Papers, box 7, folder 6 "NUMEC - FBI Investigation - Progress of Investigation and Findings"

164 FBI SAC New Haven AIRTEL to Director, March 5, 1969, Benjamin S. Loeb Papers, box 8, folder 5

165 Undated FBI - Benjamin S. Loeb Papers, box 8, folder 5 FBI 1091168-000 – 117-2564 – Section 4 (805090) p 34 Israel Lobby Archive http://www.IRmep.org/ila/numec

[166] Seaborg Glenn and Loeb, Benjamin *The Atomic Energy Commission under Nixon*, Palgrave Macmillan, 1993 p. 199.

[167] Gilinsky, Victor and Mattson, Roger J "Revisiting the NUMEC Affair" Bulletin of the Atomic Scientists, March/April 2010, p 65 citing AEC letter, Assistant General Manager Howard C. Brown Jr. to Director of Central Intelligence Richard Helms about transmitting William T. Riley's summary of August 14, 1969 interview of Shapiro, August 28, 1969. CIA response to FOIA request by Natural Resources Defense Council, reference F87-1446, December 15, 1989.

[168] Gilinsky, Victor and Mattson, Roger J "Revisiting the NUMEC Affair" Bulletin of the Atomic Scientists, March/April 2010, p 65 citing AEC letter, general manager to Rep. Chet Holifield, August 27, 1969, Udall papers.

[169] NUMEC/Zalman Shapiro FBI file, 1091168-000 – 117-2564 – Section 5 (805096) p 12 Israel Lobby Archive http://www.IRmep.org/ila/numec

[170] NUMEC/Zalman Shapiro FBI file, 1091168-000 – 117-2564 – Section 5 (805096) p 15 Israel Lobby Archive http://www.IRmep.org/ila/numec

[171] NUMEC/Zalman Shapiro FBI file, 1091168-000 – 117-2564 – Section 5 (805096) p 16 Israel Lobby Archive http://www.IRmep.org/ila/numec

[172] Glenn T. Seaborg papers, Library of Congress, Manuscript Division, box 63, folder 3

[173] NUMEC/Zalman Shapiro FBI file, 1091168-000 – 117-2564 – Section 5 (805096) p 20 Israel Lobby Archive http://www.IRmep.org/ila/numec

[174] NUMEC/Zalman Shapiro FBI file, 1091168-000 – 117-2564 – Section 5 (805096) p 17 Israel Lobby Archive http://www.IRmep.org/ila/numec

[175] NUMEC/Zalman Shapiro FBI file, 1091168-000 – 117-2564 – Section 5 (805096) p 26-27 Israel Lobby Archive http://www.IRmep.org/ila/numec

[176] NUMEC/Zalman Shapiro FBI file, 1091168-000 – 117-2564 – Section 5 (805096) p 26 Israel Lobby Archive http://www.IRmep.org/ila/numec

[177] NUMEC/Zalman Shapiro FBI file, 1091168-000 – 117-2564 – Section 5 (805096) p 26 Israel Lobby Archive http://www.IRmep.org/ila/numec

[178] NUMEC/Zalman Shapiro FBI file, 1091168-000 – 117-2564 – Section 5 (805096) p 28 Israel Lobby Archive http://www.IRmep.org/ila/numec

[179] NUMEC/Zalman Shapiro FBI file, 1091168-000 – 117-2564 – Section 5 (805096) p 29 Israel Lobby Archive http://www.IRmep.org/ila/numec

[180] NUMEC/Zalman Shapiro FBI file, 1091168-000 – 117-2564 – Section 5 (805096) p 36 Israel Lobby Archive http://www.IRmep.org/ila/numec

[181] NUMEC/Zalman Shapiro FBI file, 1091168-000 – 117-2564 – Section 5 (805096) p 35 Israel Lobby Archive http://www.IRmep.org/ila/numec

[182] "Nuclear Diversion in the U.S.? 13 Years of Contradiction and Confusion." General Accounting Office, December 18, 1978. Partially Declassified on May 6, 2010. http://IRmep.org/ILA/nukes/NUMEC/co1162251.pdf

[183] "Highly Enriched Uranium: Striking a Balance" U.S. Department of Energy, 2001 released to the Federation of American Scientists on February 2, 2006 http://www.fas.org/sgp/othergov/doe/heu/striking.pdf

[184] Peabody Energy and NUMEC entry, Glenn T. Seaborg papers, Library of Congress, Manuscript Division, box 59, folder 2

[185] Atlantic Richfield Annual Report, 1967, page 35

[186] Atlantic Richfield Annual Report, 1967, page 29

[187] "Justification of Negotiated Contract" Atomic Energy Commission, Contract Numbers AT(45-1)-2130 Operating Contract, and AT(45-1)-2131 Diversification Agreement, p. 3 1967. Obtained from the Pacific Northwest National Laboratory Public Reading Room.

[188] "Justification of Negotiated Contract" Atomic Energy Commission, Contract Numbers AT(45-1)-2130 Operating Contract, and AT(45-1)-2131 Diversification Agreement, p. 7 1967. Obtained from the Pacific Northwest National Laboratory Public Reading Room.

189 Ritchie, Russell Chemical Processing Division and Fitz, Clyde, Office of the Chief Counsel. – "Summary of Chemical Processing Contract Negotiations" Operating Contract No. AT(45-1)-2130 Between the United States of America Represented by United States Atomic Energy Commission and Atlantic Richfield Hanford Company – Chemical Processing Operation"

190 Ritchie, Russell Chemical Processing Division and Fitz, Clyde, Office of the Chief Counsel. – "Justification of Negotiated Contract", p 21 Operating Contract No. AT(45-1)-2130 Between the United States of America Represented by United States Atomic Energy Commission and Atlantic Richfield Hanford Company – Chemical Processing Operation"

191 Ritchie, Russell Chemical Processing Division and Fitz, Clyde, Office of the Chief Counsel. – "Justification of Negotiated Contract," p 3 Operating Contract No. AT(45-1)-2130 Between the United States of America Represented by United States Atomic Energy Commission and Atlantic Richfield Hanford Company – Chemical Processing Operation"

192 Modification No. 2 – Supplemental Agreement to Contract No. AT (45-1)-2130 Between United States of America Represented by United States Atomic Energy Commission and Atlantic Richfield Hanford Company" May 8, 1972 p. 6, Obtained from the Pacific Northwest National Laboratory Public Reading Room.

193 Oscar J. Bennett, Director of ERDA Contracts and Procurement Division, "Letter to G.T. Stocking, President Atlantic Richfield Hanford Company." November 5, 1975, Obtained from the Pacific Northwest National Laboratory Public Reading Room.

194 Modification No. 6, Supplemental Agreement to Contract No. EY-76-C-06-2130, December 7, 1976. Obtained from the Pacific Northwest National Laboratory Public Reading Room.

195 FBI FD-302, AEC official, November 14, 1979, Benjamin S. Loeb Papers Box 7, Folder 13

196 American Nuclear Insurers And Mutual Atomic Energy Liability Underwriters Versus The Babcock And Wilcox Co. And Atlantic Richfield Co. Civil Action No: 01-2751 United States District Court For The Eastern District Of Louisiana June 14, 2002, Decided - retrieved from Lexis

197 Babcock & Wilcox website, retrieved November 22, 2011 http://www.babcock.com/about/history.html

198 Babcock & Wilcox Annual Report, 1955, page 10-11

199 Babcock & Wilcox Annual Report, 1963, page 27

200 Babcock & Wilcox Annual Report, 1971

201 American Nuclear Insurers And Mutual Atomic Energy Liability Underwriters Versus The Babcock And Wilcox Co. And Atlantic Richfield Co. Civil Action No: 01-2751 United States District Court For The Eastern District Of Louisiana June 14, 2002, Decided - retrieved from Lexis

202 Babcock & Wilcox Co. v. American Nuclear Insurers, nos. GD99-11498 and GD99-16227, Common Pleas Court Of Allegheny County, Pennsylvania, 2001 Pa. Dist. & Cnty. Dec. LEXIS 295; 51 Pa. D. & C.4th 353, April 25, 2001 Decided - Retrieved from Lexis.

203 Babcock & Wilcox website, retrieved November 11, 2011 http://www.babcock.com/about/history.html

204 Mary Ann Thomas and Ramesh Santanam, "Murtha lends hand to local activist for federal investigation" Valley News Dispatch, August 25, 2002 http://www.pittsburghlive.com/x/valleynewsdispatch/s_87939.html

205 NUMEC/Zalman Shapiro FBI file, 1146454-000–105-188123 p 2 Israel Lobby Archive http://www.IRmep.org/ila/numec

206 NUMEC/Zalman Shapiro FBI file, 1146454-000–105-188123 p 3 Israel Lobby Archive http://www.IRmep.org/ila/numec

207 NUMEC/Zalman Shapiro FBI file, 1146454-000—105-188123 p 2 Israel Lobby Archive http://www.IRmep.org/ila/numec

208 NUMEC/Zalman Shapiro FBI file, 1146454-000—105-188123 p 7 Israel Lobby Archive http://www.IRmep.org/ila/numec

209 Thomas, Gordon "Operation Exodus: From the Nazi Death Camps to the Promised Land" Thomas Dunne Books, 2010

210 Thomas, Gordon "Operation Exodus: From the Nazi Death Camps to the Promised Land" Thomas Dunne Books, 2010, pp 8-9

211 Thomas, Gordon "Operation Exodus: From the Nazi Death Camps to the Promised Land" Thomas Dunne Books, 2010 pp 10-12

212 Thomas, Gordon "Operation Exodus: From the Nazi Death Camps to the Promised Land" Thomas Dunne Books, 2010 p 120

213 Thomas, Gordon "Operation Exodus: From the Nazi Death Camps to the Promised Land" Thomas Dunne Books, 2010 p 128

214 Thomas, Gordon "Operation Exodus: From the Nazi Death Camps to the Promised Land" Thomas Dunne Books, 2010 p 165

215 Thomas, Gordon "Operation Exodus: From the Nazi Death Camps to the Promised Land" Thomas Dunne Books, 2010 p 165

216 Thomas, Gordon "Operation Exodus: From the Nazi Death Camps to the Promised Land" Thomas Dunne Books, 2010 p 179

217 "Sale of Apollo Steel Company is Announced," *The Christian Science Monitor*, October 5, 1946

218 "Citizens may buy Apollo Steel Plant, firm gives Chamber of Commerce time to make bid" *The Pittsburgh Press*, June 1, 1949.

219 "Apollo Steel Mill Reopens On Monday: Ghost Town Averted As Industry Returns; 600 to be employed," *Pittsburgh Post-Gazette*, November 1, 1949.

220 Apollo Steel Company, Articles of Incorporation filed on August 8, 1955, Department of State, Commonwealth of Pennsylvania

221 Levin, Steve "David Lowenthal: Innovative industrialist who helped Jews settle in Israel" *Pittsburgh Post-Gazette*, September 14, 2006

222 San Toy Certificate of Organization of a Corporation under the General Law

223 San Toy Mining's Showing, *The Christian Science Monitor*, February 12, 1916

224 Application for a Certificate of Authority by a Foreign Business Corporation, San Toy Mining Company, December 22, 1949 Department of State, Commonwealth of Pennsylvania

225 San Toy Mining, Filing with the State of Maine, Sanford I. Fogg, clerk, February 19, 1951

226 "Ivan Novick, Led ZOA, 81" *Washington Jewish Week*, March 25, 2009 http://washingtonjewishweek.com/print.asp?ArticleID=10471&SectionID=4&SubSectionID=17

227 U.S. Jewish Groups Back Israel on Golan Annexation, *The Washington Post*, December 23, 1981

228 Novick, Ivan J., letter to the editor, "No Words of Hostility", *The Washington Post*, September 7, 1981

229 Novick, Victor "When Westerners Attack the Israeli National Consensus." *The New York Times*, May 19, 1982

230 Novick, Victor, "The Scenario of American Policy is Troubling", *The Sun*, September 1, 1979

231 Ostrovsky, Victor, "By Way of Deception: The Making and Unmaking of a Mossad Officer" 1990, St. Martin's Press, pp 278-283

232 "Rivets and Bolts" *The Wall Street Journal*, August 3, 1943

233 Ford, Payton, Memo to FBI Director J. Edgar Hoover, April, 20 1950, released under FOIA 1169595-000

[234] Ford, Payton, Memo to FBI Director J. Edgar Hoover, June 21, 1950, released under FOIA 1169595-000

[235] Levin, Steve, "Elliot W. Finkel/ Lawyer and dedicated community leader" The *Pittsburgh Post-Gazette*, September 13, 2006 http://www.post-gazette.com/pg/06256/721277-122.stm

[236] Merger of Apollo Steel Company (Pennsylvania) American Nut and Bolt Fastener Company (Pennsylvania) San Toy Mining Company (Maine) into Apollo Industries – Documents filed and approved by the Commonwealth of Pennsylvania, May 31, 1958

[237] FBI-DOJ Correspondence about the Zionist Organization of America, August 22, 1951, FBI Records: The Vault, ZOA Part 5 of 10, pp 7-12 http://vault.fbi.gov/Zionist%20Organization%20of%20America/Zionist%20Organizat ion%20of%20America%20Part%205%20of%2010/at_download/file

[238] David Lowenthal FBI file, 1146454-000−105-188123 February 26, 1969, Memo to the Director, FBI from SAC Pittsburgh – David Luzer Lowenthal

[239] David Lowenthal FBI file, 1146454-000−105-188123 May 22, 1969, Memo to the Director, FBI from SAC Pittsburgh – David Luzer Lowenthal

[240] David Lowenthal FBI file, 1146454-000−105-188123 May 22, 1969, Memo to the Director, FBI from SAC Pittsburgh – David Luzer Lowenthal

[241] Roberts, Sam "For First time, Figure in Rosenberg Case Admits Spying for Soviets" The *New York Times*, September 12, 2008

[242] Glenn T. Seaborg papers, Library of Congress, Manuscript Division, box 18, journal, page 289

[243] Glenn T. Seaborg papers, Library of Congress, Manuscript Division, box 21, office journal, page 003

[244] Glenn T. Seaborg papers, Library of Congress, Manuscript Division, box 21, office journal, page 352

[245] Hersh, Seymour (1991). *The Samson Option: Israel's Nuclear Arsenal and America's Foreign Policy* Random House. Chapter 14. ISBN 0-394-57006-5.

[246] Glenn T. Seaborg papers, Library of Congress, Manuscript Division, box 23, office journal, page 681, December 10, 1965

[247] Glenn T. Seaborg papers, Library of Congress, Manuscript Division, box 23, office journal, page 325, December 10, 1965

[248] Glenn T. Seaborg papers, Library of Congress, Manuscript Division, box 23, office journal, page 264

[249] Glenn T. Seaborg papers, Library of Congress, Manuscript Division, box 59, folder 6

[250] Glenn T. Seaborg papers, Library of Congress, Manuscript Division, box 24, office journal, pp 317-323

[251] Glenn T. Seaborg papers, Library of Congress, Manuscript Division, box 24, office journal, page 695

[252] Glenn T. Seaborg papers, Library of Congress, Manuscript Division, box 25, office journal, vol. 18 page 247

[253] Benjamin S. Loeb Papers, Office Diary, Glenn T. Seaborg, Chair of the AEC 1961-1972, Box 8, Folder 3, Page 98096

[254] Benjamin S. Loeb Papers, Office Diary, Glenn T. Seaborg, Chair of the AEC 1961-1972, Box 8, Folder 3, Page 98101

[255] Benjamin S. Loeb Papers, Office Diary, Glenn T. Seaborg, Chair of the AEC 1961-1972, Box 8, Folder 3, Page 98134

[256] Benjamin S. Loeb Papers, Office Diary, Glenn T. Seaborg, Chair of the AEC 1961-1972, Box 8, Folder 3, Page 100035

[257] Glenn T. Seaborg papers, Library of Congress, Manuscript Division, box 63, folder 7

[258] Benjamin S. Loeb Papers, Office Diary, Glenn T. Seaborg, Chair of the AEC 1961-1972, Box 8, Folder 3, Page 100046-48

[259] Benjamin S. Loeb Papers, Office Diary, Glenn T. Seaborg, Chair of the AEC 1961-1972, Box 8, Folder 3, Page 100053

[260] NUMEC/Zalman Shapiro FBI file, 1091168-000 – 117-2564 – Section 7 (805473) p 10

[261] Benjamin S. Loeb Papers, Office Diary, Glenn T. Seaborg, Chair of the AEC 1961-1972, Box 8, Folder 3, Page 100023

[262] Benjamin S. Loeb Papers, Office Diary, Glenn T. Seaborg, Chair of the AEC 1961-1972, Box 8, Folder 3, Page 100053

[263] Benjamin S. Loeb Papers, Office Diary, Glenn T. Seaborg, Chair of the AEC 1961-1972, Box 8, Folder 3, Page 100053

[264] Benjamin S. Loeb Papers, Office Diary, Glenn T. Seaborg, Chair of the AEC 1961-1972, Box 8, Folder 3, Page 103015

[265] NUMEC/Zalman Shapiro FBI file, 1091168-000 --- 117-2564 --- Section 1 (830179)) p 33. Israel Lobby Archive http://www.IRmep.org/ila/numec

[266] NUMEC/Zalman Shapiro FBI file, 1091168-000 --- 117-2564 --- Section 1 (830179)) p 19-22. Israel Lobby Archive http://www.IRmep.org/ila/numec

[267] NUMEC/Zalman Shapiro FBI file, 1091168-000 --- 117-2564 --- Section 1 (830179)) p 25. Israel Lobby Archive http://www.IRmep.org/ila/numec

[268] NUMEC/Zalman Shapiro FBI file, 1091168-000 --- 117-2564 --- Section 1 (830179)) p 26. Israel Lobby Archive http://www.IRmep.org/ila/numec

[269] NUMEC/Zalman Shapiro FBI file, 1091168-000 --- 117-2564 --- Section 1 (830179)) p 26. Israel Lobby Archive http://www.IRmep.org/ila/numec

[270] NUMEC/Zalman Shapiro FBI file, 1091168-000 --- 117-2564 --- Section 1 (830179)) p 27. Israel Lobby Archive http://www.IRmep.org/ila/numec

[271] Cockburn, Andre and Leslie, "Dangerous Liaison", Harper Collins, 1991, p 92

[272] NUMEC/Zalman Shapiro FBI file, 1091168-000 --- 117-2564 --- Section 1 (830179)) p 30. Israel Lobby Archive http://www.IRmep.org/ila/numec

[273] NUMEC/Zalman Shapiro FBI file, 1091168-000 --- 117-2564 --- Section 1 (830179)) p 48. Israel Lobby Archive http://www.IRmep.org/ila/numec

[274] D.J. Brennan memorandum, May,13 1969, Benjamin S. Loeb Papers, Box 7, Folder 5 "NUMEC FBI Investigations 1969-1978"

[275] NUMEC/Zalman Shapiro FBI file, 1091168-000 --- 117-2564 --- Section 1 (830179) p 50. Israel Lobby Archive http://www.IRmep.org/ila/numec

[276] NUMEC/Zalman Shapiro FBI file, 1091168-000 --- 117-2564 --- Section 1 (830179) p 33. Israel Lobby Archive http://www.IRmep.org/ila/numec

[277] NUMEC/Zalman Shapiro FBI file, 1091168-000 --- 117-2564 --- Section 1 (830179) p 36. Israel Lobby Archive http://www.IRmep.org/ila/numec

[278] NUMEC/Zalman Shapiro FBI file, 1091168-000 --- 117-2564 --- Section 1 (830179) p 39. Israel Lobby Archive http://www.IRmep.org/ila/numec

[279] NUMEC/Zalman Shapiro FBI file, 1091168-000 --- 117-2564 --- Section 1 (830179) p 39. Israel Lobby Archive http://www.IRmep.org/ila/numec

[280] NUMEC/Zalman Shapiro FBI file, 1091168-000 --- 117-2564 --- Section 1 (830179) p 40. Israel Lobby Archive http://www.IRmep.org/ila/numec

[281] NUMEC/Zalman Shapiro FBI file, 1091168-000 --- 117-2564 --- Section 1 (830179) p 40. Israel Lobby Archive http://www.IRmep.org/ila/numec

[282] NUMEC/Zalman Shapiro FBI file, 1091168-000 --- 117-2564 --- Section 1 (830179) p 43. Israel Lobby Archive http://www.IRmep.org/ila/numec

[283] Benjamin S. Loeb Papers, Office Diary, Glenn T. Seaborg, Chair of the AEC 1961-1972, Box 8, Folder 3, Page 104188

[284] Benjamin S. Loeb Papers, Office Diary, Glenn T. Seaborg, Chair of the AEC 1961-1972, Box 8, Folder 3, Page 104203

[285] Benjamin S. Loeb Papers, Office Diary, Glenn T. Seaborg, Chair of the AEC 1961-1972, Box 8, Folder 3, Page 105062 1091168-000 --- 117-2564 --- Section 6 (805155) p 10.

286 Stout, David "Nixon Papers Recall Concerns on Israel's Weapons" *New York Times*, November 28, 2007 I

287 Kissinger, Henry "Israeli Nuclear Program" July 19, 1979 – Nixon Presidential Library and Museum
http://nixon.archives.gov/virtuallibrary/documents/mr/071969_israel.pdf

288 Martin, David C. "Mysteries of Israel's Bomb" *Newsweek* , January 9, 1979

289 Fialka, John J. "The American Connection: How Israel Got the Bomb" *The Washington Monthly*, January, 1979

290 NUMEC/Zalman Shapiro FBI file, 1091168-000 --- 117-2564 --- Section 7 (805473) p 10. Israel Lobby Archive http://www.IRmep.org/ila/numec

291 NUMEC/Zalman Shapiro FBI file, 1091168-000 --- 117-2564 --- Section 1 (830179)) p 54. Israel Lobby Archive http://www.IRmep.org/ila/numec

292 SAC, WFO finance report June 29, 1969 BENJAMIN S. LOEB PAPERS Box 7, Folder 3

293 NUMEC/Zalman Shapiro FBI file, 1091168-000 --- 117-2564 --- Section 6 (805155) p 2 Israel Lobby Archive http://www.IRmep.org/ila/numec

294 Benjamin S. Loeb Papers, Office Diary, Glenn T. Seaborg, Chair of the AEC 1961-1972, Box 8, Folder 3, Page 118143

295 NUMEC/Zalman Shapiro FBI file, 1091168-000 --- 117-2564 --- Section 6 (805155) p 13. Israel Lobby Archive http://www.IRmep.org/ila/numec

296 NUMEC/Zalman Shapiro FBI file, 1091168-000 --- 117-2564 --- Section 6 (805155) p 13. Israel Lobby Archive http://www.IRmep.org/ila/numec

297 Benjamin S. Loeb Papers, Office Diary, Glenn T. Seaborg, Chair of the AEC 1961-1972, Box 8, Folder 3, Page 119018

298 Benjamin S. Loeb Papers, Office Diary, Glenn T. Seaborg, Chair of the AEC 1961-1972, Box 8, Folder 3, Page 119212

299 NUMEC/Zalman Shapiro FBI file, 1091168-000 --- 117-2564 --- Section 7 (805155) p 2. Israel Lobby Archive http://www.IRmep.org/ila/numec

300 Benjamin S. Loeb Papers, Office Diary, Glenn T. Seaborg, Chair of the AEC 1961-1972, Box 8, Folder 3, Page 121236

301 Benjamin S. Loeb Papers, Office Diary, Glenn T. Seaborg, Chair of the AEC 1961-1972, Box 8, Folder 3, Page 121242

302 Benjamin S. Loeb Papers, Office Diary, Glenn T. Seaborg, Chair of the AEC 1961-1972, Box 8, Folder 3, Page 121236

303 Benjamin S. Loeb Papers, Office Diary, Glenn T. Seaborg, Chair of the AEC 1961-1972, Box 8, Folder 3, Page 121286

304 Benjamin S. Loeb Papers, Office Diary, Glenn T. Seaborg, Chair of the AEC 1961-1972, Box 8, Folder 3, Page 121306

305 Benjamin S. Loeb Papers, Office Diary, Glenn T. Seaborg, Chair of the AEC 1961-1972, Box 8, Folder 3, Page 122028

306 Benjamin S. Loeb Papers, Office Diary, Glenn T. Seaborg, Chair of the AEC 1961-1972, Box 8, Folder 3, Page 122034

307 Benjamin S. Loeb Papers, Office Diary, Glenn T. Seaborg, Chair of the AEC 1961-1972, Box 8, Folder 3, Page 122044

308 Benjamin S. Loeb Papers, Office Diary, Glenn T. Seaborg, Chair of the AEC 1961-1972, Box 8, Folder 3, Page 122045

309 FBI 1091168-000 – 117-2564 – Section 5 (805096) p 12 Israel Lobby Archive http://www.IRmep.org/ila/numec

310 Benjamin S. Loeb Papers, Office Diary, Glenn T. Seaborg, Chair of the AEC 1961-1972, Box 8, Folder 3, Page 122056-122058

311 Benjamin S. Loeb Papers, Office Diary, Glenn T. Seaborg, Chair of the AEC 1961-1972, Box 8, Folder 3, Page 122068

312 Benjamin S. Loeb Papers, Office Diary, Glenn T. Seaborg, Chair of the AEC 1961-1972, Box 8, Folder 3, Page 122098

313 Benjamin S. Loeb Papers, Office Diary, Glenn T. Seaborg, Chair of the AEC 1961-1972, Box 8, Folder 3, Page 122102

314 Benjamin S. Loeb Papers, Office Diary, Glenn T. Seaborg, Chair of the AEC 1961-1972, Box 8, Folder 3, Page 122111

315 Benjamin S. Loeb Papers, Office Diary, Glenn T. Seaborg, Chair of the AEC 1961-1972, Box 8, Folder 3, Page 122114

316 Benjamin S. Loeb Papers, Office Diary, Glenn T. Seaborg, Chair of the AEC 1961-1972, Box 8, Folder 3, Page 122134

317 Benjamin S. Loeb Papers, Office Diary, Glenn T. Seaborg, Chair of the AEC 1961-1972, Box 8, Folder 3, Page 122183

318 Benjamin S. Loeb Papers, Office Diary, Glenn T. Seaborg, Chair of the AEC 1961-1972, Box 8, Folder 3, Page 122200-122204

319 Benjamin S. Loeb Papers, Office Diary, Glenn T. Seaborg, Chair of the AEC 1961-1972, Box 8, Folder 3, Page 122135

320 Benjamin S. Loeb Papers, Office Diary, Glenn T. Seaborg, Chair of the AEC 1961-1972, Box 8, Folder 3, Page 122160

321 Benjamin S. Loeb Papers, Office Diary, Glenn T. Seaborg, Chair of the AEC 1961-1972, Box 8, Folder 3, Page 122192

322 Benjamin S. Loeb Papers, Office Diary, Glenn T. Seaborg, Chair of the AEC 1961-1972, Box 8, Folder 3, Page 123228

323 Benjamin S. Loeb Papers, Office Diary, Glenn T. Seaborg, Chair of the AEC 1961-1972, Box 8, Folder 3, Page 124004-124005

324 Benjamin S. Loeb Papers, Office Diary, Glenn T. Seaborg, Chair of the AEC 1961-1972, Box 8, Folder 3, Page 123313

325 Benjamin S. Loeb Papers, Office Diary, Glenn T. Seaborg, Chair of the AEC 1961-1972, Box 8, Folder 3, pp. 124025-124026

326 Benjamin S. Loeb Papers, Office Diary, Glenn T. Seaborg, Chair of the AEC 1961-1972, Box 8, Folder 3, Page 124046

327 Benjamin S. Loeb Papers, Office Diary, Glenn T. Seaborg, Chair of the AEC 1961-1972, Box 8, Folder 3, 124244

328 NUMEC/Zalman Shapiro FBI file, 1091168-000 --- 117-2564 --- Section 6 (805473) p 10 Israel Lobby Archive http://www.IRmep.org/ila/numec

329 Benjamin S. Loeb Papers, Office Diary, Glenn T. Seaborg, Chair of the AEC 1961-1972, Folder Page 130026

330 "Israeli Spy Visited A-Plant Where Uranium Vanished," *The Los Angeles Times*, June 16, 1986.

331 Seaborg Glenn and Loeb, Benjamin *The Atomic Energy Commission under Nixon*, Palgrave Macmillan, 1993 p. 194

332 Seaborg Glenn and Loeb, Benjamin *The Atomic Energy Commission under Nixon*, Palgrave Macmillan, 1993 p. 198

333 Seaborg Glenn and Loeb, Benjamin *The Atomic Energy Commission under Nixon*, Palgrave Macmillan, 1993 p. 196

334 Seaborg, Glenn T. manuscript notes to Benjamin Loeb, May 20, 1991 Box 7, Folder 12

335 Glenn T. Seaborg papers, Library of Congress, Manuscript Division, box 441, Folder "Speech to the American Committee for the Weizmann Institute for Science"

336 Glenn T. Seaborg papers, Library of Congress, Manuscript Division, box 441, Folder "Speech to the American Committee for the Weizmann Institute for Science"

337 See "Israel Crosses the Threshold" at the National Security Archive for primary documents and analysis.
http://www.gwu.edu/~nsarchiv/NSAEBB/NSAEBB189/index.htm

[338] Hersh, Seymour (1991). *The Samson Option: Israel's Nuclear Arsenal and America's Foreign Policy* Random House. Chapter 14. ISBN 0-394-57006-5.

[339] "Contacts of Feinberg with the President of the United States." FBI file released under FOIA in 2010, Israel Lobby Archive, http://www.IRmep.org/ila/feinberg/04171952feinberg_potus.pdf

[340] Abraham Feinberg FBI file released under FOIA in 2010, Israel Lobby Archive, http://www.IRmep.org/ila/feinberg/04171952activities_creation_israel.pdf

[341] Abraham Feinberg FBI file released under FOIA in 2010, Israel Lobby Archive, http://www.IRmep.org/ila/feinberg/

[342] McTierman, Tom "Inquiry into the Testimony of the Executive Director for Operations" Volume III, Interviews, February 1978.

[343] Glenn T. Seaborg papers, Library of Congress, Manuscript Division, alphabetized correspondence.

[344] Hersh, Seymour (1991). *The Samson Option: Israel's Nuclear Arsenal and America's Foreign Policy.* Random House. Chapter 14. ISBN 0-394-57006-5.

[345] O'Toole, Thomas, "Possibility of Attempted Nuclear Thefts Causing Deep Concern." *The Washington Post*, November 10, 1974

[346] Fialka, John J. "The American Connection: How Israel Got the Bomb" *The Washington Monthly*, January, 1979

[347] Gilinsky, Victor and Mattson, Roger J "Revisiting the NUMEC Affair" Bulletin of the Atomic Scientists, March/April 2010 citing the report "Inquiry into the Testimony of the Executive Director for Operations," Nuclear Regulatory Commission (NRC) Offices of General Counsel and Inspector and Auditor, February 1978,

[348] Fialka, John J. "The American Connection: How Israel Got the Bomb" *The Washington Monthly*, January, 1979

[349] Fialka, John J. "The American Connection: How Israel Got the Bomb" *The Washington Monthly*, January, 1979 p 51

[350] Glenn T. Seaborg papers, Library of Congress, Manuscript Division, box 556, Folder "Nuclear Materials and Equipment Corporation", June 21, 1978

[351] The Department of Energy has not yet released its report under a Mandatory Declassification Review filed by the author. FOIA specialist Fletcher Whitworth speculates that the DOE inspectors may have been on loan to the GAO for the report "Nuclear Diversion in the U.S.? 13 Years of Contradiction and Confusion" partially declassified in 2010. He further speculates that working papers, including the Seaborg interview and any correction of the survey questions, would have been destroyed.

[352] Glenn T. Seaborg papers, Library of Congress, Manuscript Division, box 556, Folder "Nuclear Materials and Equipment Corporation", June 21, 1978

[353] Despite numerous FOIA and MDR requests, the NRC refuses to release its classified three volume report on the vast implications of Duckett's briefing.

[354] Fialka, John J. "The American Connection: How Israel Got the Bomb" *The Washington Monthly*, January, 1979 p 52

[355] Glenn T. Seaborg papers, Library of Congress, Manuscript Division, box 62, folder 8

[356] Benjamin S. Loeb Papers, Box 7, Folder 5 "NUMEC FBI Investigations 1969-1978" FBI interview May 11, 1976

[357] Gilinsky, Victor and Mattson, Roger J "Revisiting the NUMEC Affair" *Bulletin of the Atomic Scientists*, March/April 2010 p 62

[358] The FBI interview subject who Gilinsky and Mattson believe was James Connor, reports, "The president is particularly concerned with any indication of a prior cover-up. [Redacted] expressed the feeling that he was appalled at the superficiality of the AEC investigation into the matter." FBI Airtel, special agent in charge, Washington Field Office to FBI director, [subject redacted], June 15, 1976, Loeb papers.

359 McDowell, Samuel C. T. ERDA, Memo to files May26, 1976 as quoted by Burnham, David, special to the New York Times, "U.S. Documents Support Belief Israel Got Missing Uranium for Arms," November 6, 1977.

360 McDowell, Samuel C. T. ERDA, Memo to files May 26, 1976 as quoted by Burnham, David, special to *The New York Times*, "U.S. Documents Support Belief Israel Got Missing Uranium for Arms," November 6, 1977.

361 Gilinsky, Victor "Time for More NUMEC Information" *Arms Control Association,* Letters to the Editor, June 2008 http://www.armscontrol.org/print/2951

362 Gilinsky, Victor "Time for More NUMEC Information" *Arms Control Association,* Letters to the Editor, June 2008 http://www.armscontrol.org/print/2951

363 "Nuclear Regulator Legislation" Volume 1, Number 9. Office of the General Counsel of the U.S. Nuclear Regulatory Commission, January 2011 P. 156 http://www.nrc.gov/reading-rm/doc-collections/nuregs/staff/sr0980/v1/

364 Fialka, John J. "The American Connection: How Israel Got the Bomb" *The Washington Monthly,* January, 1979

365 FBI Memo on Atomic Energy Act investigation June 15, 1976 BENJAMIN S. LOEB PAPERS, box 7, folder 2

366 FBI interview form FD-302 Robert E. Tharp, May 3, 1976 BENJAMIN S. LOEB PAPERS, box 7, folder 2

367 FBI Interview, FD-302, Ralph G. Page, Benjamin S. Loeb Papers, Box 7, Folder 5 "NUMEC FBI Investigations 1969-1978"

368 DAAG John C. Keeny memo to FBI Director, October 22, 1976 BENJAMIN S. LOEB PAPERS, box 7, folder 2

369 FBI correspondence, November 4, 1976, November 10, 1976, BENJAMIN S. LOEB PAPERS, box 7, folder 2

370 Gilinsky, Victor "Time for More NUMEC Information" *Arms Control Association,* Letters to the Editor, June 2008 http://www.armscontrol.org/print/2951

371 Letter FBI Director to AAG, Criminal Division May 4, 1969 Benjamin S. Loeb Papers, box 8, folder 5

372 FBI WFO to FBI Director June 17, 1977

373 NUMEC/Zalman Shapiro FBI file, 1091168-000-117-2564 Section 8 (805763) pp 2-6 Israel Lobby Archive http://www.IRmep.org/ila/numec

374 Jimmy Carter Library correspondence with the author confirming a Mandatory Declassification Review request to the National Security Agency. September 12, 2011

375 Burnham, David, special to *The New York Times*, "U.S. Documents Support Belief Israel Got Missing Uranium for Arms," November 6, 1977.

376 Jimmy Carter Library, Global Issues group evening report to Zbigniew Brezinski, December 7, 1977 declassified April 29, 2008

377 Hersh, Seymour (1991). *The Samson Option: Israel's Nuclear Arsenal and America's Foreign Policy* Random House. Chapter 20. ISBN 0-394-57006-5.

378 "NUMEC MUF" NSC memo from Jerry Oplinger to Zbigniew Brezinski, November 27. 1979, declassified September 12, 2008

379 "Nuclear Diversion in the U.S.? 13 Years of Contradiction and Confusion." General Accounting Office, December 18, 1978. Partially Declassified on May 6, 2010. http://IRmep.org/ILA/nukes/NUMEC/co1162251.pdf

380 "Nuclear Diversion in the U.S.? 13 Years of Contradiction and Confusion." General Accounting Office, December 18, 1978. Partially Declassified on May 6, 2010. http://IRmep.org/ILA/nukes/NUMEC/co1162251.pdf

381 "Nuclear Diversion in the U.S.? 13 Years of Contradiction and Confusion." General Accounting Office, December 18, 1978. Partially Declassified on May 6, 2010, p. ii http://IRmep.org/ILA/nukes/NUMEC/co1162251.pdf

[382] Jimmy Carter Library Confidential Deputy Attorney General letter to President Jimmy Carter, February 28, 1979, partially declassified June 23, 2008

[383] Maddox, Bronwen, "Jimmy Carter says Israel had 150 Nuclear weapons," *The Times*, May 26, 2008

[384] House of Representatives, Transcript of Proceedings, "Informal Meeting Between Interior Committee Representatives and Dr. Zalman M. Shapiro." – December 21, 1978 p 5

[385] "Nuclear Diversion in the U.S.? 13 Years of Contradiction and Confusion." General Accounting Office, December 18, 1978. Partially Declassified on May 6, 2010, p. ii http://IRmep.org/ILA/nukes/NUMEC/co1162251.pdf

[386] Thomas, Mary Ann, and Santanam, Ramesh "Government investigations proved fruitless." *Valley News Dispatch*, August 27, 2002

[387] Thomas, Mary Ann, and Santanam, Ramesh "Government investigations proved fruitless." *Valley News Dispatch*, August 27, 2002

[388] Thomas, Mary Ann, and Santanam, Ramesh "Government investigations proved fruitless." *Valley News Dispatch*, August 27, 2002

[389] Undated FBI - Benjamin S. Loeb Papers, box 8, folder 5 FBI 1091168-000 – 117-2564 – Section 4 (805090) p 34 Israel Lobby Archive http://www.IRmep.org/ila/numec

[390] Undated FBI - Benjamin S. Loeb Papers, box 8, folder 5 FBI 1091168-000 – 117-2564 – Section 4 (805090) p 34 Israel Lobby Archive http://www.IRmep.org/ila/numec

[391] NUMEC/Zalman Shapiro FBI file, 1091168-000 – 117-2564 – Section 1 (805474) p 18 Israel Lobby Archive http://www.IRmep.org/ila/numec

[392] "Meeting with Dr. Lahav Concerning Implementation of Israeli Atomic Energy Program," Atomic Energy Commission memo, December 4, 1956. http://www.gwu.edu/~nsarchiv/israel/documents/before/07-01.htm

[393] Office Diary, Glenn T. Seaborg, Chair of the AEC 1961-1972, Benjamin S. Loeb Papers, Box 8, Folder 3, Page 124025-124026

[394] "Panorama on Israel," *BBC*, transcript of program discussing Israel's acquisition of the Bomb, June 26, 1979, Udall papers.

[395] Thomas, Mary Ann, and Santanam, Ramesh "Government investigations proved fruitless." *Valley News Dispatch*, August 27, 2002

[396] Simkin, John, Historian "Spartacus Educational " website, biographies, http://www.spartacus.schoolnet.co.uk/JFKmartinDC.htm

[397] Martin, David C. "Mysteries of Israel's Bomb" *Newsweek* , January 9, 1979

[398] FD-302, likely Fred Forscher based on previous documents, November 11, 1977, Benjamin S. Loeb Papers, box 7, folder 6 "NUMEC – FBI Investigation – Progress of Investigation and Findings"

[399] FBI FD-302 Leonard Pepkowitz, November 21, 1977 Benjamin S. Loeb Papers, Box 7, Folder 5

[400] FD-302, Charles L. Keller, November 17, 1978, "NUMEC - Fact file concerning allegations" Benjamin S. Loeb Papers, box 7, folder 3

[401] NUMEC/Zalman Shapiro FBI file, 1091168-000-117-2564 Section 10 (805762)p 54. Israel Lobby Archive http://www.IRmep.org/ila/numec

[402] FBI Director William Webster memo to CIA Director Stansfield Turner, September 6, 1979. Benjamin S. Loeb Papers, box 7, folder 6 "NUMEC – FBI Investigation – Progress of Investigation and Findings"

[403] FD-302 Hyman Rickover, March 10, 1978 Benjamin S. Loeb Papers, Box 7, Folder 4

[404] FD-302 William T. Riley, October 19, 1978, Benjamin S. Loeb Papers, Box 7, Folder 4

[405] Attempt to interview Glenn Seaborg, FBI memo, March 3, 1979, Benjamin S. Loeb Papers, Box 7, Folder 4

[406] FD-302 Earle Hightower, October 23, 1978, Benjamin S. Loeb Papers, Box 7, Folder 3

[407] NUMEC/Zalman Shapiro FBI file, 1091168-000-117-2564 Section 10 (805762)p 54. Israel Lobby Archive http://www.IRmep.org/ila/numec

[408] NUMEC/Zalman Shapiro FBI file, 1091168-000-117-2564 Section 10 (805762)p 4. Israel Lobby Archive http://www.IRmep.org/ila/numec

[409] NUMEC/Zalman Shapiro FBI file, 1091168-000-117-2564 Section 10 (805762)p 8 Israel Lobby Archive http://www.IRmep.org/ila/numec

[410] NUMEC/Zalman Shapiro FBI file, 1091168-000-117-2564 Section 10 (805762)p 9-10 Israel Lobby Archive http://www.IRmep.org/ila/numec

[411] Patty Ameno, Founder and Chairperson at Citizen's Action for a Safe Environment, Linkedin profile, retrieved December 2, 2011 http://www.linkedin.com/pub/patty-ameno/12/730/a77

[412] Benjamin S. Loeb Papers, Box 7, Folder 10 NUMEC Loading Dock Incidents, March 11, 1980 FBI interview

[413] NUMEC/Zalman Shapiro FBI file, 1091168-000-117-2564 Section 10 (805762)p 24 Israel Lobby Archive http://www.IRmep.org/ila/numec

[414] NUMEC/Zalman Shapiro FBI file, 1091168-000-117-2564 Section 10 (805762)p 24 Israel Lobby Archive http://www.IRmep.org/ila/numec

[415] NUMEC/Zalman Shapiro FBI file, 1091168-000-117-2564 Section 10 (805762)p 24 Israel Lobby Archive http://www.IRmep.org/ila/numec

[416] NUMEC/Zalman Shapiro FBI file, 1091168-000-117-2564 Section 10 (805762)p 25 Israel Lobby Archive http://www.IRmep.org/ila/numec

[417] NUMEC/Zalman Shapiro FBI file, 1091168-000-117-2564 Section 10 (805762)p 27 Israel Lobby Archive http://www.IRmep.org/ila/numec

[418] NUMEC/Zalman Shapiro FBI file, 1091168-000-117-2564 Section 10 (805762)p 27 Israel Lobby Archive http://www.IRmep.org/ila/numec

[419] NUMEC/Zalman Shapiro FBI file, 1091168-000-117-2564 Section 10 (805762)p 28 Israel Lobby Archive http://www.IRmep.org/ila/numec

[420] NUMEC/Zalman Shapiro FBI file, 1091168-000-117-2564 Section 10 (805762)pp 14-15 Israel Lobby Archive http://www.IRmep.org/ila/numec

[421] NUMEC/Zalman Shapiro FBI file, 1091168-000-117-2564 Section 10 (805762)p 18 Israel Lobby Archive http://www.IRmep.org/ila/numec

[422] NUMEC/Zalman Shapiro FBI file, 1091168-000-117-2564 Section 10 (805762)p 20 Israel Lobby Archive http://www.IRmep.org/ila/numec

[423] NUMEC/Zalman Shapiro FBI file, 1091168-000-117-2564 Section 10 (805762)p 20 Israel Lobby Archive http://www.IRmep.org/ila/numec

[424] NUMEC/Zalman Shapiro FBI file, 1091168-000-117-2564 Section 10 (805762)p 22 Israel Lobby Archive http://www.IRmep.org/ila/numec

[425] Justice Department memo to FBI director, 4/22/1980. Benjamin S. Loeb Papers, Box 7, Folder 10 NUMEC Loading Dock Incidents

[426] Benjamin S. Loeb Papers, Box 7, Folder 10 NUMEC Loading Dock Incidents, May 13, 1980 FBI interview

[427] Benjamin S. Loeb Papers, Box 7, Folder 10 NUMEC Loading Dock Incidents, August 12, 1980 WFO memo to FBI Director

[428] Benjamin S. Loeb Papers, Box 7, Folder 10 NUMEC Loading Dock Incidents, December 13, 1980 FD-302 interview

[429] Benjamin S. Loeb Papers, Box 7, Folder 6 January 23, 1981 memo closing investigation

[430] "Highly Enriched Uranium: Striking a Balance" U.S. Department of Energy, 2001 released to the Federation of American Scientists on February 2, 2006 http://www.fas.org/sgp/othergov/doe/heu/striking.pdf

[431] "Gamma Irradiators for Radiation Processing" International Atomic Energy Agency, July 7, 2005 P 12-13 http://www-naweb.iaea.org/napc/iachem/Brochgammairradd.pdf

[432] Hersh, Seymour (1991). *The Samson Option: Israel's Nuclear Arsenal and America's Foreign Policy* Random House. pp. 243, 250, 252, 255. ISBN 0-394-57006-5.

[433] United States Patent and Trademark Office, Patent Full Text Databases, Retrieved August 30, 2011 for inventor Shapiro, Zalman http://patft.uspto.gov/

[434] Jin, Liyn, "89-year-old Oakland inventor receives 15th patent" *Pittsburgh-Post Gazette*, June 26, 2009.

[435] Shapiro, Zalman, Application for National Medal for Technology and Innovation, 2009 p 8, obtained under FOIA, Israel Lobby Archive http://www.irmep.org/ila/nukes/specter/default.asp

[436] Smith, Grant *Spy Trade: How Israel's Lobby Undermines America's Economy* Institute for Research: Middle Eastern Policy, 2009 page 108

[437] "NMTI Nomination Guidelines, U.S. Patent and Trade Office http://www.uspto.gov/about/nmti/Nomination_Guidelines_page.jsp

[438] Shapiro, Zalman, Application for National Medal for Technology and Innovation, U.S. Patent and Trade Office, NMTI applications, 2009 p 4, obtained under FOIA, Israel Lobby Archive http://www.irmep.org/ila/nukes/specter/default.asp

[439] Shapiro, Zalman, Application for National Medal for Technology and Innovation, U.S. Patent and Trade Office, NMTI applications, 2009 obtained under FOIA, Israel Lobby Archive http://IRmep.org/ILA/nukes/specter/Defense_Safety.pdf

[440] Shapiro, Zalman, Application for National Medal for Technology and Innovation, U.S. Patent and Trade Office, NMTI applications, 2009 obtained under FOIA, Israel Lobby Archive http://IRmep.org/ILA/nukes/specter/gillbrand.pdf

[441] Shapiro, Zalman, Application for National Medal for Technology and Innovation, U.S. Patent and Trade Office, NMTI applications, 2009 obtained under FOIA, Israel Lobby Archive http://IRmep.org/ILA/nukes/specter/McRory_McDowell.pdf

[442] Katz, Hadrian R., Letter to Senator Arlen Specter, "RE Zalman M. Shapiro" 8/7/2009. Obtained from the Nuclear Regulatory Commission under the Freedom of Information Act. http://IRmep.org/08272009specter_numec.pdf

[443] Office Journal, June 21, 1978, Glenn T. Seaborg papers, Library of Congress, Manuscript Division, box 556, Folder 6

[444] Arnold & Porter, Supplemental Foreign Agent Registration Act filings, 1750-Supplemental-Statement-20080820-7.pdf, 1750-Supplemental-Statement-20090223-8.pdf, United States Department of Justice, Foreign Agent Registration Act website, http://fara.gov/quick-search.html

[445] Arnold & Porter, Supplemental Foreign Agent Registration Act filings, 1750-Supplemental-Statement-20101209-14.pdf, United States Department of Justice, Foreign Agent Registration Act website, http://fara.gov/quick-search.html

[446] Arnold & Porter, Supplemental Foreign Agent Registration Act filings, 1750-Supplemental-Statement-20100820-11.pdf, 1750-Supplemental-Statement-20110222-12.pdf, United States Department of Justice, Foreign Agent Registration Act website, http://fara.gov/quick-search.html

[447] "Arnold & Porter Experience" as retrieved from the corporate website on November 26, 2011 http://www.arnoldporter.com/experience.cfm?action=case_study_view&id=1241

[448] Shapiro, Zalman, Application for National Medal for Technology and Innovation, U.S. Patent and Trade Office, NMTI applications, 2009 obtained under FOIA, Israel Lobby Archive http://IRmep.org/08272009specter_numec.pdf

[449] Shapiro, Zalman, Application for National Medal for Technology and Innovation, U.S. Patent and Trade Office, NMTI applications, 2009 obtained under FOIA, Israel Lobby Archive http://IRmep.org/ML092720878.pdf

[450] "GAO Report on the 1965 NUMEC Affair Declassified" Federation of American Scientists, May 13, 2010
http://www.fas.org/blog/secrecy/2010/05/gao_numec.html

[451] Gilinsky, Victor and Mattson, Roger J "Revisiting the NUMEC Affair" *Bulletin of the Atomic Scientists*, March/April 2010

[452] Gilinsky, Victor and Mattson, Roger J "Revisiting the NUMEC Affair" *Bulletin of the Atomic Scientists*, March/April 2010 citing "Highly Enriched Uranium: Striking a Balance" U.S. Department of Energy, 2001 released to the Federation of American Scientists   on   February   2,   2006
http://www.fas.org/sgp/othergov/doe/heu/striking.pdf

[453] Gilinsky, Victor and Mattson, Roger J "Revisiting the NUMEC Affair" *Bulletin of the Atomic Scientists*, March/April 2010

[454] FBI FD-297 (log for technical surveillance), May 5, 1969, released under FOIA on August 29, 2011

[455] *Kyodo News*, March 2011, retrieved October 20, 2011
http://english.kyodonews.jp/news/2011/03/80010.html

[456] FBI FD-297 (log for technical surveillance), May 5, 1969 David Lowenthal FBI File, released under the Freedom of Information Act 1146454-000 on August 29, 2011

[457] Hersh, Seymour (1991). *The Samson Option: Israel's Nuclear Arsenal and America's Foreign Policy* Random House. Chapter 18. ISBN 0-394-57006-5.

[458] Acton, Robin, "Nuclear settlement money little solace for survivors in Armstrong County"   Pittsburgh   Tribune-   Review,   May   4,   2008
http://www.pittsburghlive.com/x/pittsburghtrib/business/s_565716.html

[459] "Final Technical Report: Apollo Decommissioning Project, Apollo PA" Department of Energy assistance instrument number DC-FG01-91EW40017. April 30, 1997 – Prepared by B&W NESI

[460] Rittmeyer, Brian C. "Apollo Officials lay out welcome mat for business" *Valley News Dispatch*, December 18, 2011

[461] "Final Feasibility Study for the Shallow Land Disposal Area Site, Parks Township, Armstrong County, Pennsylvania" U.S. Army Corps of Engineers p 31
http://www.lrp.usace.army.mil/fusrap/sldapp.pdf

[462] "Shallow Land Disposal Area Parks Township, Pennsylvania, U.S. Army Corps of Engineers, Pittsburgh District

[463] "Record of Decision for the Shallow Land Disposal Area Site – Parks Township, Armstrong County, PA" U.S. Army Corps of Engineers, August, 2007

[464] Acton, Robin, "Nuclear settlement money little solace for survivors in Armstrong County"   Pittsburgh   Tribune-   Review,   May   4,   2008
http://www.pittsburghlive.com/x/pittsburghtrib/business/s_565716.html

[465] Thomas, Mary Ann, "B&W case one of the very few even to reach the trial stage" April 25, 2009
http://www.pittsburghlive.com/x/valleynewsdispatch/s_622294.html

[466] "Decision to evaluate a petition to designate a class of employees at NUMEC in Apollo, PA to be included in the special exposure cohort" Federal Register, February 9, 2007
http://www.cdc.gov/niosh/ocas/pdfs/sec/numec/fr020907.pdf

[467] Division of Energy Employees Occupation al Illness Compensation, Special Exposure Cohort Employees" U.S. Department of Labor, retrieved November 1, 2011
http://www.dol.gov/owcp/energy/regs/compliance/law/SEC-Employees.htm

[468] EEOICPA Bulletin, Department of Labor, January 16, 2008 http://www.dol.gov/owcp/energy/regs/compliance/PolicyandProcedures/finalbull etinshtml/EEOICPABulletin08-12.htm

[469] McConnell, Jeff and Higgins, Richard "The Israeli Account" *The Boston Globe*, December 14, 1986, p 16

[470] NUMEC/Zalman Shapiro FBI file, 1091168-000 — 117-2564 — Section 2 (805648) p 7 Israel Lobby Archive http://www.IRmep.org/ila/numec

[471] "Panorama," *BBC*, June 26, 1978

[472] Hersh, Seymour (1991). *The Samson Option: Israel's Nuclear Arsenal and America's Foreign Policy* Random House. pp. 243,250,252,255. ISBN 0-394-57006-5.

[473] "Israel Gets High-Speed Computers" *The Risk Report*, Wisconsin Project on Nuclear Arms Control, Volume 1, Number 1, January-February, 1995 http://www.wisconsinproject.org/countries/israel/highspeedcomputers.htm

[474] Cohler, Larry, "Supercomputers Slow in Coming" *The Jerusalem Post*, May 25, 1990

[475] "Israel Gets High-Speed Computers" *The Risk Report*, Wisconsin Project on Nuclear Arms Control, Volume 1, Number 1, January-February, 1995 http://www.wisconsinproject.org/countries/israel/highspeedcomputers.htm

[476] Cohler, Larry, "Supercomputers Slow in Coming" *The Jerusalem Post*, May 25, 1990

[477] Funk, Sherman "Report of Audit - Defense Trade Control Annex," March 1992, U.S. Department of State Office of Inspector General, Israel Lobby Archive http://www.IRmep.org/ila/audit/1992report_of_audit.pdf

[478] NUMEC/Zalman Shapiro FBI file,, 1091168-000-117-2564 Section 4 (805090)p 85. Israel Lobby Archive http://www.IRmep.org/ila/numec

[479] NUMEC/Zalman Shapiro FBI file,, 1091168-000-117-2564 Section 4 (805090) p 86. Israel Lobby Archive http://www.IRmep.org/ila/numec

[480] NUMEC/Zalman Shapiro FBI file,, 1091168-000-117-2564 Section 7 (805473) p 11. Israel Lobby Archive http://www.IRmep.org/ila/numec

[481] NUMEC/Zalman Shapiro FBI file, 1091168-000 — 117-2564 — Section 5 (805096) p 9 Israel Lobby Archive http://www.IRmep.org/ila/numec

[482] NUMEC/Zalman Shapiro FBI file, 1091168-000 — 117-2564 — Section 5 (805096) p 23 Israel Lobby Archive http://www.IRmep.org/ila/numec

[483] Transfers to Foreign Entities - License No. SNM-145 - Uranium Enriched in the Isotope 235 - Nuclear Materials and Equipment Corporation, Apollo, PA for the Period December 1, 1957 to October 31, 1965" Atomic Energy Commission, Benjamin S. Loeb Papers, box 7, folder 3

[484] NUMEC/Zalman Shapiro FBI file, 1091168-000 — 117-2564 — Section 2 (805648) Israel Lobby Archive http://www.IRmep.org/ila/numec

[485] NUMEC/Zalman Shapiro FBI file, 1091168-000 --- 117-2564 --- Section 6 (805155) pp 14-21 Israel Lobby Archive http://www.IRmep.org/ila/numec

[486] David Lowenthal FBI File, released under the Freedom of Information Act 1146454-000 on August 29, 2011 FBI FD-297 (log for technical surveillance), May 5, 1969

[487] Kissinger, Henry, "Summary of the Situation and Issues", Nixon Presidential Library. This and other documents were released by the National Archives and Records Administration on November 28, 2007 as a result of Mandatory Declassification Review filings. Declassification authorization was given by equity holders on June 4, 2007. But the Kissinger memo about theft of nuclear material was never posted online by NARA with other contemporary documents and is only available to researchers who either travel to the Nixon Presidential Library document room in Yorba Linda, California or file a special request.

[488] Office Journal, June 21, 1978, Glenn T. Seaborg papers, Library of Congress, Manuscript Division, box 556, Folder 6

[489] McTierman, Tom "Inquiry into the Testimony of the Executive Director for Operations" Volume III, Interviews, February 1978.

[490] NUMEC/Zalman Shapiro FBI file, 1091168-000-117-2564 Section 10 (805762). Israel Lobby Archive http://www.IRmep.org/ila/numec

[491] Shapiro, Zalman "Letter to Glenn T. Seaborg" April 13, 1993, Benjamin S. Loeb papers.

[492] Specter, Arlen "Letter to Rebecca Schmidt, Director, Office of Congressional Affairs – Nuclear Regulatory Commission" August 27, 2990. Obtained under a Freedom of Information Act filing made June 18, 2010 for correspondence triggering NRC response G20090508/LTR-09-0441/EDATS: SECY-2009-0406 in the ADAMS database.

[493] Borchardt, R.W., Executive Director for Operations "Letter to the Honorable Arlen Specter, United States Senator" November 2, 2009 G20090508/LTR-09-0441/EDATS: SECY-2009-0406 in the ADAMS database.

[494] Viscuso, Susan, Information and Privacy Coordinator, CIA "FOIA F-2011-00873" July 15, 2011